Your guide to Excel in First Lego League: Robot Architec

s.

To the next generation of innovators,

Hopefully we will see colonization of Mars in our lifetime

Foreword

We are glad to offer this book *Your guide to Excel in First Lego League: Robot Architecture, Design, Programming and Game Strategies* to you. We believe that new users as well as experienced participants will find this book very useful. This book evolves from our experience over years of coaching and working with multiple teams.

Using many of the guiding principles listed out in this book, our teams have not only excelled through multiple levels of progression; they have done so with flying colors. The awards won by our teams include prestigious regional championship awards, robot performance awards at qualifiers, semi-finals, two state championships with two representations at the World Championships. Additionally, we have won multiple design awards, project awards and core value awards.

This book summarizes design principles including different kind of drives, elements of robot architecture, design of robot as system. There are detailed explanation of various programing elements including the flow structure, usage of various sensors and design and programming for a consistent and more predictable movement. A section is dedicated to the development of menu system that users will find very helpful in organizing individual programs for various missions.

In addition to the technical information, the book has a section dedicated to apprising teams, participants and coaches of many other issues that will help them be better prepared for the competition.

The book also describes many mechanisms and fixtures used to reduce the overall timing and repeatable performance. The book concludes with a section dedicated to building a robot that encompasses many design features of well-balanced highly reconfigurable robot.

Many programs described in the book are provided on our website: www.FLLguide.com

While every care has been taken to provide accurate information, feel free to write to us if you notice any discrepancies. We are also open to additional suggestion to improve.

Happy building and good luck,

Go Lego

Sanjeev Dwivedi
Rajeev Dwivedi
August 15th, 2017

Table of Contents

Foreword 3

Table of Contents 4

Quick Introduction to FLL 9

- Summary 14

Robot Architecture 15

- Subcomponents of a Mobile Robot 19
- Robot Design as a System 20
- Guidelines for Robot Design 29
- Other important considerations for overall Robot design 34
- Summary 35

Mechanical System 36

- Mechanical Drives 36
- Gear Drives 38
  - Spur gears 39
  - Bevel Gears 43
  - Worm Gear 47
  - Rack and Pinion 49
  - Gear Trains and Combination gear trains 49
  - Chain and Sprocket Drives 51
  - Belt and Pulley 55
- Mechanical Linkages 56
- Composite Drives 61
- Summary 62

Pneumatic Systems 63

- Pneumatic pump 63
- Air Tank 64
- Flow distribution accessories 65
- Manometer 66
- Pneumatic cylinders 66
- Pneumatic Valve (switch) 68

Pneumatic mechanisms 69
Summary 74
Getting familiar with the programming environment....75
Introduction to the programming 76
Green/Action Blocks 81
Orange/Flow Blocks 82
Yellow/Sensor Blocks 83
Red/Data Operation Blocks 85
Royal Blue/Advanced blocks....86
Myblocks 87
Summary 88
Fundamentals of Robot Movement 89
Parts of the robot....89
Large Motors 89
Connection to Wheels....90
Caster 91
Bumper 91
Chassis 92
Moving Straight 94
Turning 99
Exercises to reinforce basic robot movement 105
Summary 107
Basic Robot Navigation....108
Navigation Paradigms....108
Aligning the robot perfectly....111
Taking care of the Gear Slack 114
Starting Jig....115
Using wall and flat surfaces for navigation....118
Exercises with basic alignment....120
Wall Following using wheels 120
Summary 123
MyBlocks 124

Pre-requisite to Myblocks **124**
Variables 124
Data Wires 127
Display Block 129
The Math Block 130
Exercises 131
MyBlocks for Moving Straight and turning **132**
Summary **138**
Basic programming blocks **139**
The Single motor block **139**
Flow Control Blocks **139**
Flow Control: The Wait block 141
Flow Control: The Switch block **145**
Flow Control: Programming mistakes with the Switch block **150**
Flow Control: The Loop block **151**
The Loop Interrupt 155
Flow Control: Programming Mistakes with the loop block **156**
Summary **159**
Touch Sensor and the Brick buttons **160**
Programming with the touch sensor **161**
Mounting the touch sensor 163
Wall aligning with a touch sensor **165**
Wall Following with a touch sensor **166**
Exercises with the Touch sensor **169**
Brick Button as a sensor **170**
Summary **172**
Ultrasonic Sensor **173**
Mounting the Ultrasonic sensor **176**
Examples of the Ultrasonic sensor use **177**
Wall following with the Ultrasonic Sensor 177
Exercises **179**
Summary **181**
Color Sensor **182**

Wait for color 184
Line squaring to color lines 186
Simple Line following with a color sensor 188
Calibrating the color sensor 192
Summary 194
Gyroscope Sensor 195
Basic Functionality 195
Summary 199
Motor Rotation Sensor and its use in conjunction with other Sensors .200
Exercises 205
Summary 207
PID algorithm for wall following using Ultrasonic and line following using Color Sensors 208
PID line following 208
Summary 218
The Master Program 219
Display Program Names MyBlock 222
Summary 226
Some Design suggestions to improve accuracy and timing 227
Self-aligning features 227
Optimize motor usage by eliminating motorized mechanisms and attachment by Robot Movement 229
Use passive non-motorized end effector for collecting objects 231
Designs based on Out of box thinking for objects placed on ground -effector for collecting objects 233
Save time by using single fixture for simultaneous multiple deliveries235
Robot need not go behind the wall 238
Axle based connectors for quickly switching static attachment 242
Touch sensor Engagement 245
Use Level fix vehicle for aligning and make deliveries above the ground level 248
Summary 251
Robot Mission and Path Planning Strategies for FLL 252

One at a time mission planning **252**
Zone based mission planning **253**
Path and Opportunity based Mission planning **254**
Island of Certainty 255
Overshoot your goal 257
Use techniques when you can be certain of them 259
Minimum number of segments on a path 260
Opportunistic Mission completion 262
Taking penalties as a strategy 263
Tying it all together **264**
Summary **264**
Appendix A: Winning Strategies at FLL - beyond Robotics **265**
Appendix B: Things to remember for the competition day **268**
Appendix C: The Essential kit for a successful FLL season **272**
Appendix D: A FLL Robot **277**

Quick Introduction to FLL

Started in 1999 through partnership between FIRST (For Inspiration and Recognition of Science and Technology) and Lego group, First Lego League® or FLL as it is commonly called, is a competition that focuses on STEM with a goal to develop kids with a *play nice* attitude. Many or most people who have heard of FLL associate FLL with robotics; however, it is key to understand the differentiation.

FLL is a STEM (Science, Technology, Engineering and Mathematics) oriented competition with a robotics component to it. It is, NOT a Robotics competition. This differentiation is important to understand for most coaches to avoid frustration and to better navigate the competition for great results and to *win* or move from one level to another in FLL. Before we go deep into explaining how to Win at FLL events, it is important to understand the format as well as the various metrics that FLL judging occurs on. In our experience, even experienced coaches quite often do not fully understand the dynamics of judging or even various parts, so it is crucial to lay it out.

FLL as a competition has 4 parts to it:

Core Values - we like to call this teamwork. That is how team works together, cooperates and exhibits gracious professionalism
Project - This is essentially a scientific research endeavor. The teams need to identify a real-life problem, find a solution to it, or improve an existing solution and share it with others.
Robot Design – Also referred to as technical judging is about the innovative strategies, design and programing.
Robot Games - The robotics component where you get graded objectively based on how well your robot performs in the competition.

In our experience, most participants love the robotics portion of FLL and that is where they spend most of their time. Even though the FLL competition is heavily weighted towards teamwork, project and robot design, most participants that we have interacted with have a very positive experience that they gained out of the robotics portion of FLL. Thus, we have decided to keep the majority of the book dedicated to robotics with Lego and EV3 Mindstorms.

The central theme of this book is the robotics and programming that goes along with FLL although we discuss some of the most important metrics needed to *win* in FLL in the appendices. In this chapter, we setup the terminology and basic concepts of FLL so that you can follow along in the rest of the book.

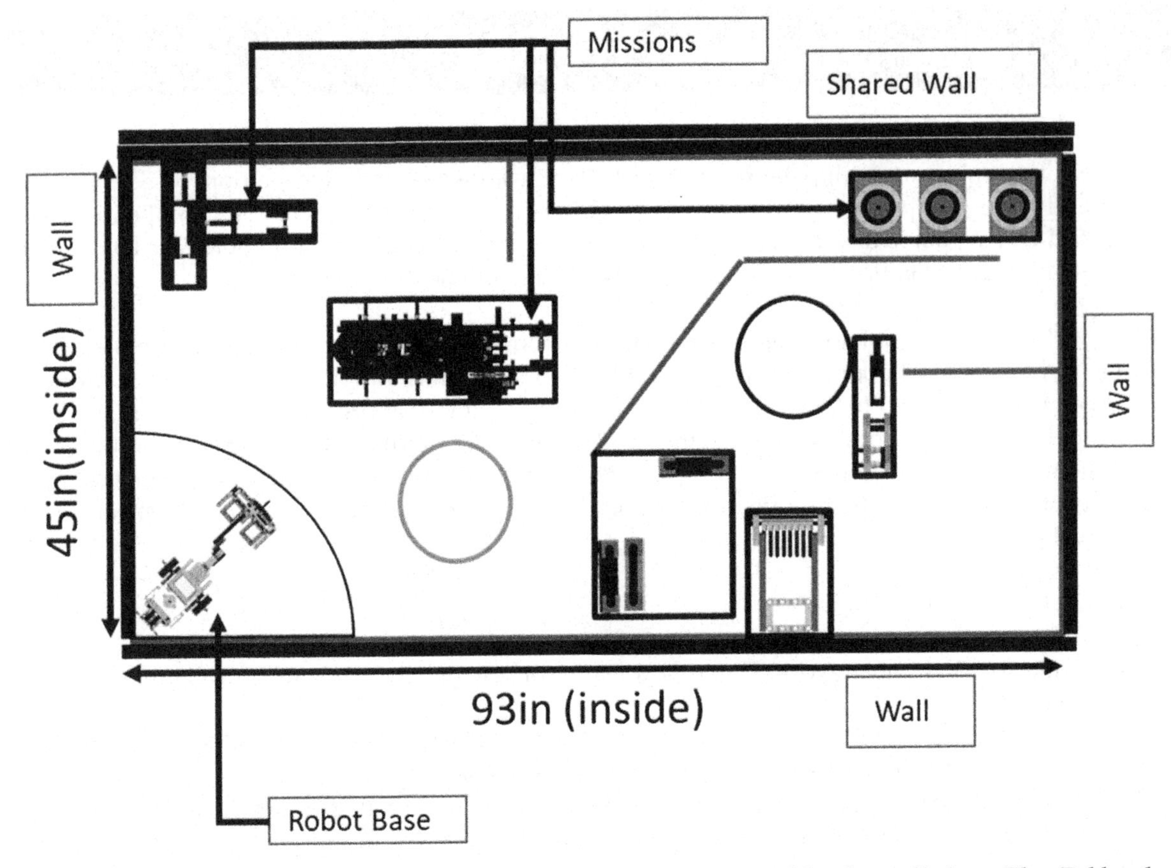

Figure 1: A sample FLL Robot Challenge Field, also called as The Table shown for illustration purposes

For most of the teams, a FLL season runs from roughly the beginning of September to Mid-December when the FLL Qualifiers take place. At the beginning of each FLL season, FIRST, announces a challenge with a specific theme. The challenge contains the description of the research area that the kids should work in and more importantly, it contains the robotics challenge environment in which the team's EV3 robot must work in. The challenge environment is built with the help of the FLL field kit that includes a mat and various models that are built with the help of Lego pieces. As described in Figure 1, the challenge area is 45 in x 93 in with one area reserved as the Robot Base. The challenge area is surrounded by single 2 x 3 stud (actual 1-1/2 in x 2-1/2in or 38mm x 64mm) walls on three sides and double walls one side. During actual challenge two tables are placed next to each other so that the walls are touching. The shared walls described in Figure 1 represent the second table.

Figure 2: A FLL table with the mat and the models shown. The table is from the 2016 Animal Allies season.

Figure 3: An FLL table shown with the base and a few models

On an actual challenge table (Figure 2), multiple Lego models are dispersed at known locations on top of a printed mat and the robot needs to interact with the Lego models and fulfill certain conditions. Each travel of the robot outside the base in order to interact with one or more Lego Models to perform certain objectives is known as a *mission*. The missions may include, push, pull, engage, collect, redistribute or object launching tasks. While attempting a mission, the Robot must be autonomous. Autonomy of the robot means that once the robot exits the Robot Base, no member of the team may touch or aid the robot until the robot returns back to the Robot Base on its own. Thus, as per the rules, the robot can be touched only in the robot base. While in Robot base, the robot may be reconfigured by changing the attachment and suitable program selected to execute a mission. If due to some reason the robot is stranded outside the Robot base, participants may accept a penalty point and pick the robot to bring it back to base.

In the FLL Robot game where the robots are driven using autonomous programs by kids, two tables are placed next to each other (Figure 4) and two teams run their robot on them simultaneously. It is key to note that even though two teams are running their robots simultaneously, there is limited impact they can have on each other. The two tables have one wall touching each other and a shared model is placed on this double wall. In the 2016 Animal Allies season, scoring points on the shared model would grant points to both the teams irrespective of which team was the one that took on that mission. Apart from the shared models, the scoring as well as running of the robot on the two sides is completely independent and does not impact the other team.

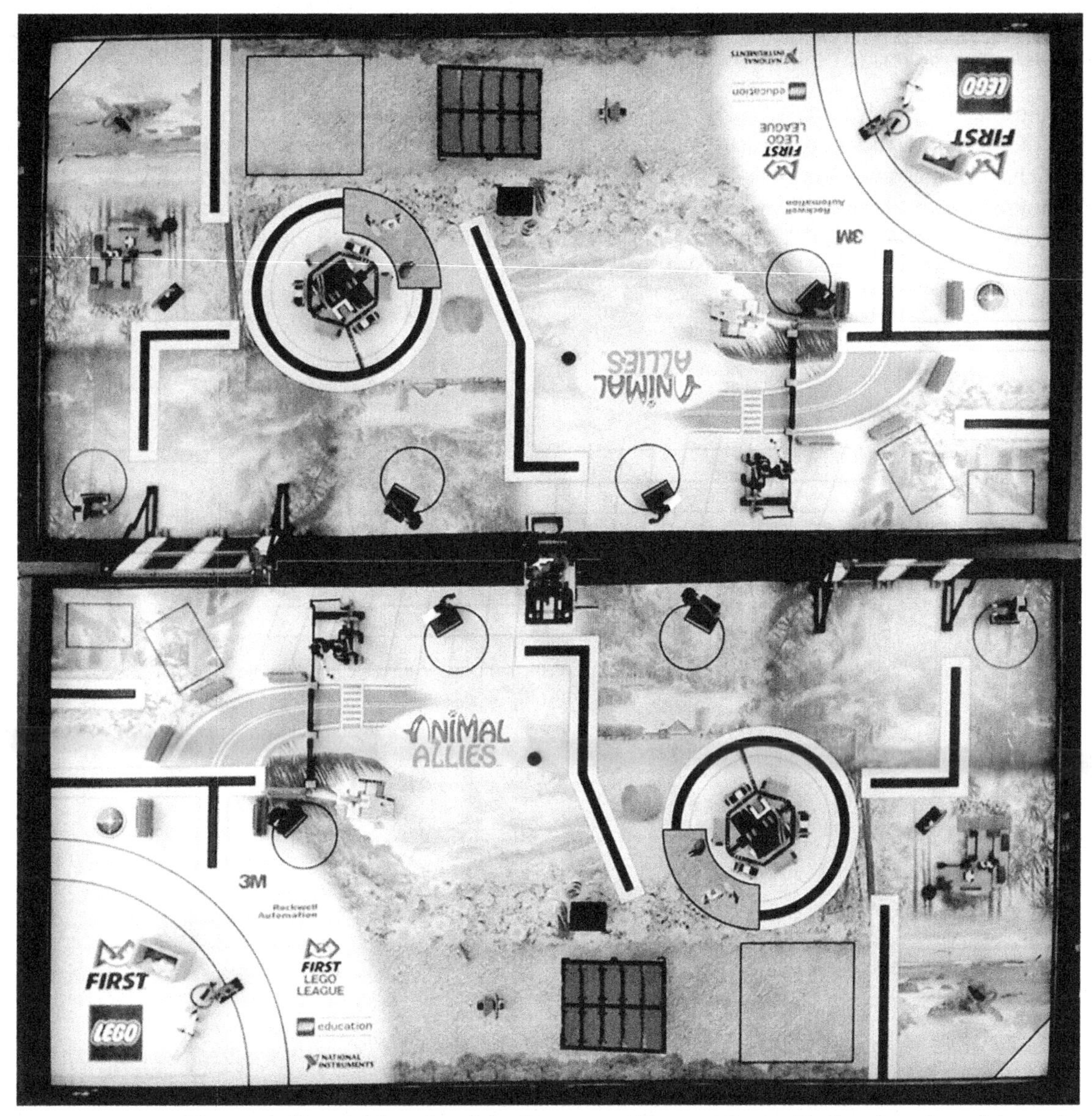

Figure 4: Two tables placed in the competition setting with the shared model overlapping the two walls. Note that there are quite often loose pieces in the base which are required to finish some missions. During the competition, you may move them off the competition table and onto a small stool placed next to the table for keeping the spare pieces. Apart from the one shared model between the two walls, the two sides are completely independent of each other.

Summary

In this chapter, we quickly touched upon some core concepts of FLL. The most exciting thing that participants find in FLL is the robotics and the Robot Games. The Robot Game is performed on top of a 45inchx93inch table which, contains a mat and lots of Lego Models. The participating team builds and programs a robot to go and interact with the models in an autonomous way within a 2 minute and 30 seconds timeframe. Each autonomous run of the robot is known as a mission and usually involves the robot interacting with multiple models using a variety of techniques. With this introduction, we are ready to delve in the world of Lego based robotics.

Robot Architecture

Ever wondered what a Robot is? Fundamentally, a Robot is a machine. We see all manners of machines around us. There are machines that help us with transportation, machines that help make our live more comfortable, machines that help us explore and machine that help us manufacture and produce goods. So, what differentiates the Robots from all the rest of the machines?

Come to think about it, most of the machines that we see around us, and use are dedicated to a single task. They perform the same task over and over. A juicer mixer grinder - makes juice and grinds food. A car transports us from one place to another. A drilling machine drills and similarly a washing machine washes clothes. What differentiates these machines from Robots?

Robots are special kinds of machines distinguished by following three characteristics:

1. Robots are multifunctional machines
2. Robots are Reconfigurable machines and
3. Robots are reprogrammable machines

Contrary to most of the machines, the Robot can perform many different functions. Attach a welding torch to a Robot and it welds to join two pieces of metal. Attach a gripper to the Robot and it can perfectly pick and place and sort items. Attach a surgical implement to a Robot and it can perform surgery just like a skilled surgeon. Attach a laser to the Robot and it can custom cut a piece of cloth to be sewn into a great dress. Let's take the multi-functionality to the next level - attach a baseball bat to the Robot and it can hit perfect homeruns. In fact, some enthusiasts have configured Robots to paint and mock sea-surf. For each different task, the Robots needs a different attachment. The unique attachment for the Robot required to perform a specific task is also referred to as an *end effector*.

This takes us to the next subject, that is, the reconfigurability. Think of your arm, it is a perfect multi-functional implement. Hold a brush to paint, hold a bat to hit home runs and hold a fork to enjoy a perfect meal. It is worth noticing that, in order to perform a task, we need to *configure* our arms in a certain manner. Similarly, for a robot to perform different tasks, robot needs to configure its shape accordingly.

All the robots used for specific functionality, industrial applications, entertainment and education, are classified into two categories:

1. Mobile robots
2. Non-mobile robots

Non-mobile robots are used in manufacturing and assembly industries where the raw material(s) arrive near the Robot and Robots picks and assembles, solders or welds pieces. By contrast, Mobile robots are used for material movement and exploratory purposes; Mars rover is a popular example of a mobile robot.

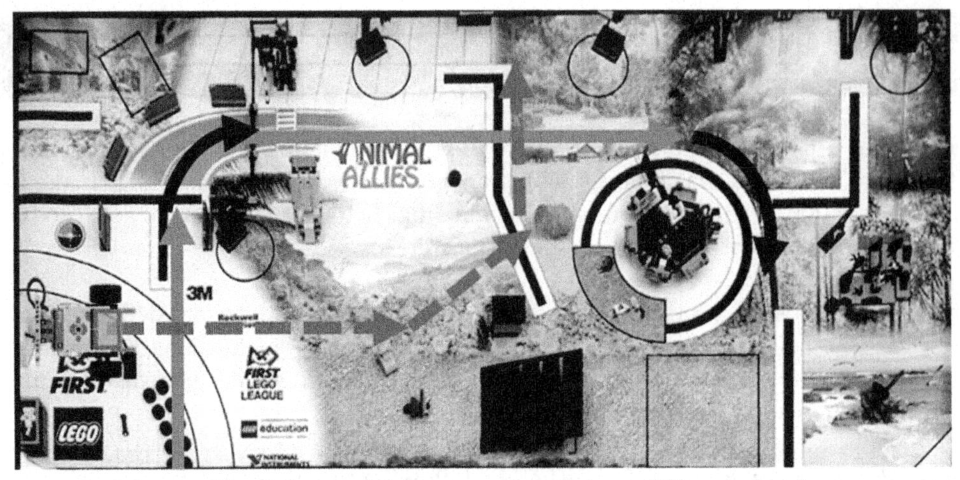

Figure 5: The Robot path may comprise of a set of linear, circular or a task specific profile as indicated by the differently colored (or solid vs dashed) arrows. The robot environment above is from the FLL 2016 Animal Allies Season.

The Robots used in FLL are mobile Robots. That is, they need to move from one location to another. As shown in Figure 5, the movement of a robot may comprise of linear segments, circular arcs; or need based unique path profiles. Different attachments on the Robot help the robot complete mission tasks. In FLL, the maximum number of allowed motors are limited by the availability of ports on EV3/NXT module. Therefore, significant reusability of attachments as well as easy and quick reconfigurability of the Robots is critical to solving maximum number of missions in fixed time window. Typically, the Robots are multifunctional since the variety of problems that are solved may range from push, pull to pick and place, engage and actuate among others. The end effector connected to the robot helps solve various mission problems.

Each mission and reconfigured design of the robot may go hand in hand with a specific program.

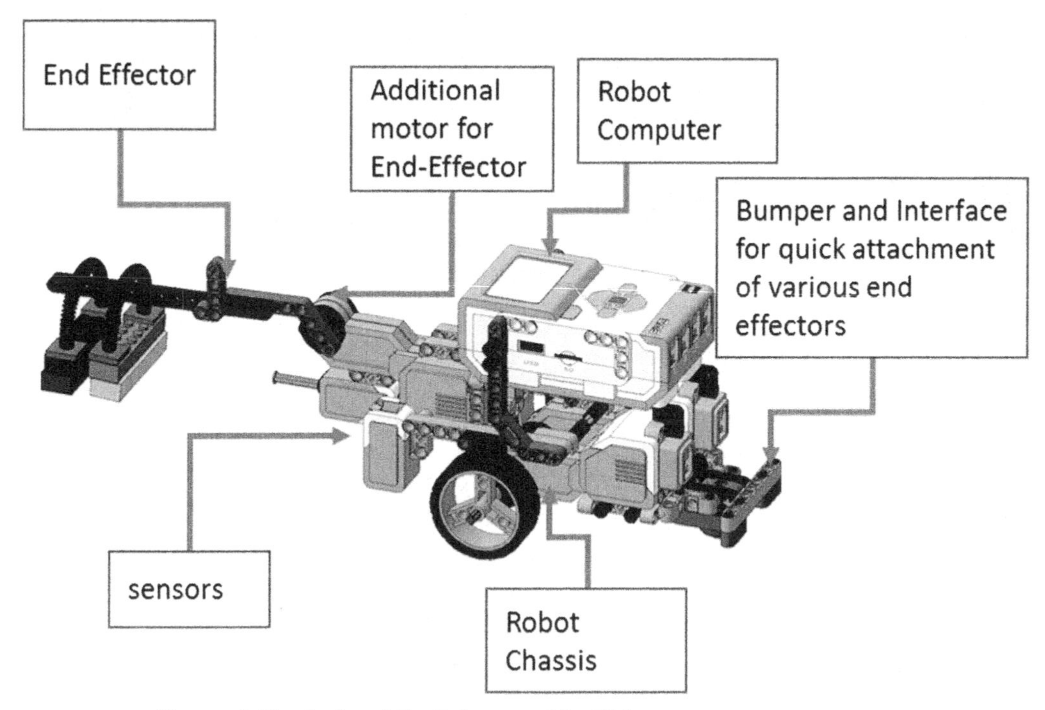

Figure 6: Typical mobile Robot used in FLL

A *configuration* refers to the arrangement of various elements of a mechanism or mechanical system. In context of FLL, let us refer to the examples shown in the Figure 6. The core mobile Robot comprises of the Chassis and motors for the Robot to travel. Additional motors are attached to the robot for attaching the End Effectors.

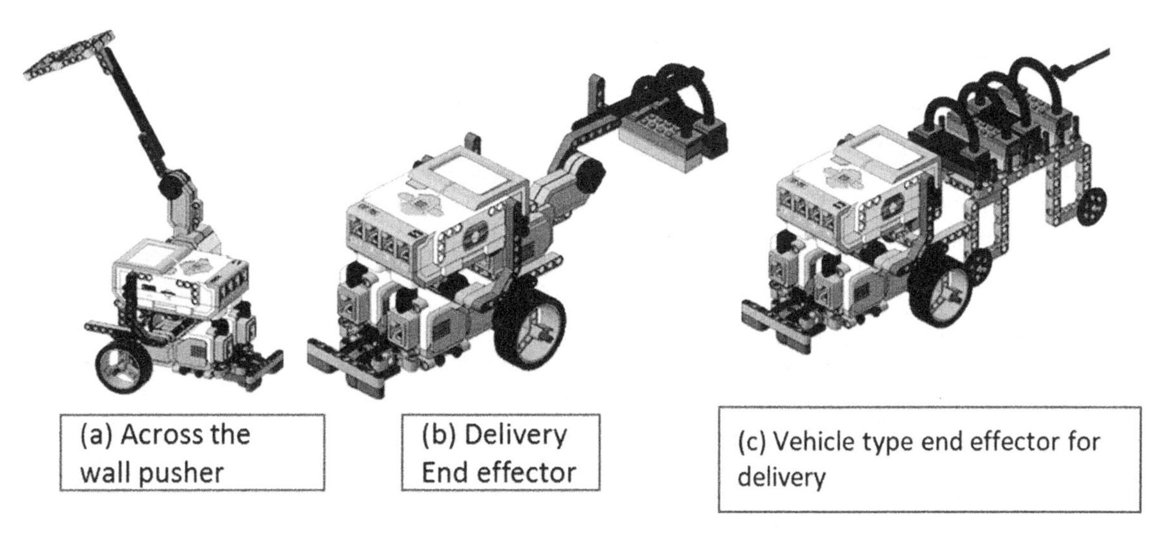

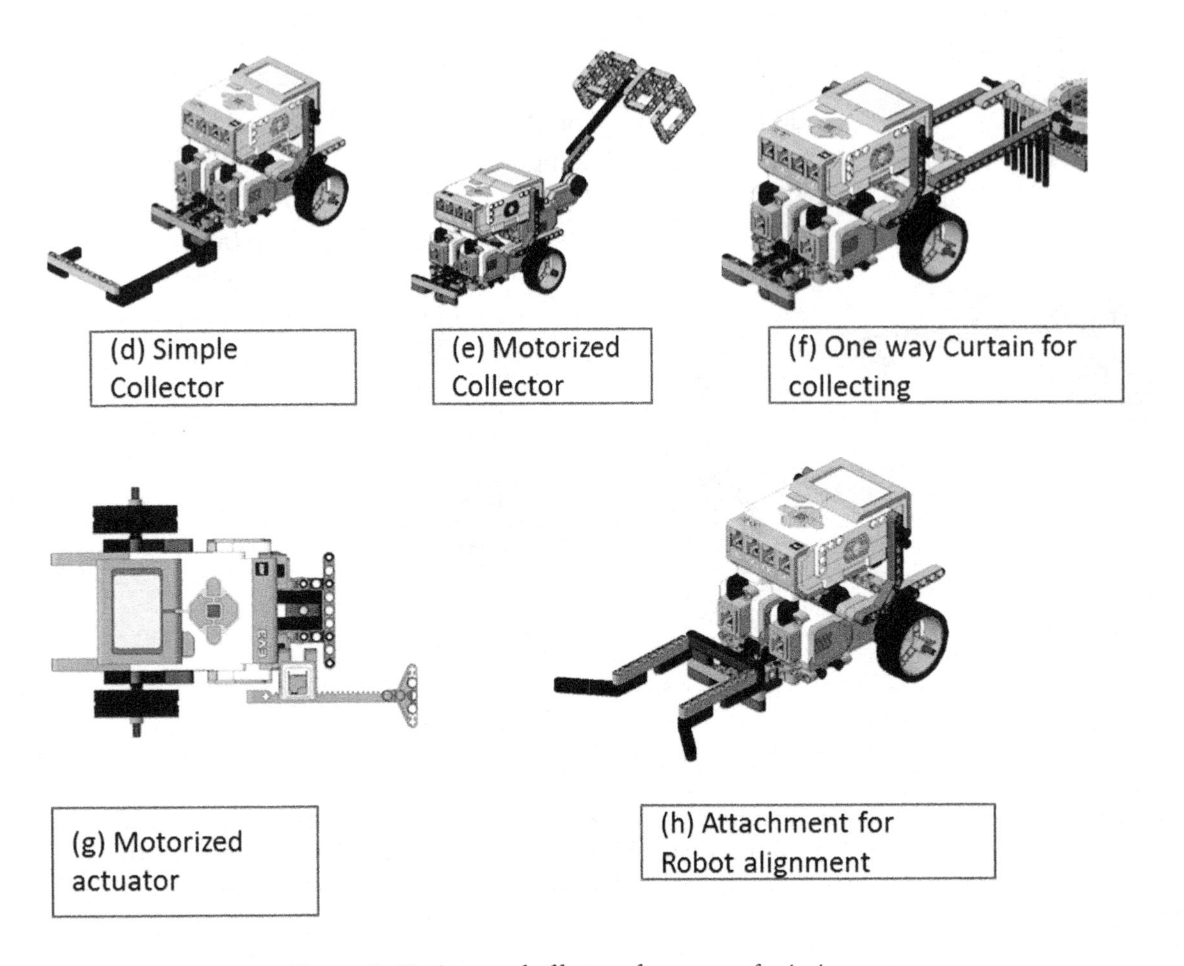

Figure 7: Various end effectors for range of missions

Figure 7(a)-(h) show different end effectors ranging in complexity and configurations. Each end effector is unique and corresponds to a mission. Figure 7(a) is a pusher mechanism that allows access across a tall structure. Figure 7(b) and (c) show mechanisms for delivery of items above the ground level. Figure 3(d)-(f) show end effectors for collecting items. Figure 7 (g) shows a rack and pinion based actuator and Figure 3(h) shows a robot with attachment for aligning a Robot. While an end-effector may be used for a specific task, many end effectors may be configured to perform multiple tasks.

When designing an end effector, it is highly recommended that the designer consider using it for multiple tasks. One may also design an end effector such that with minimal changes and adjustments, the same end effector can be used for multiple missions. As illustrated in Figure 8, a scoop like attachment, used for collecting objects, can be made multifunctional by attaching an axle to hang the mission pieces with loops. Similarly, as described in Figure 9 an alignment end effector can be easily converted to a collection end effector using a simple drop-in and remove wall.

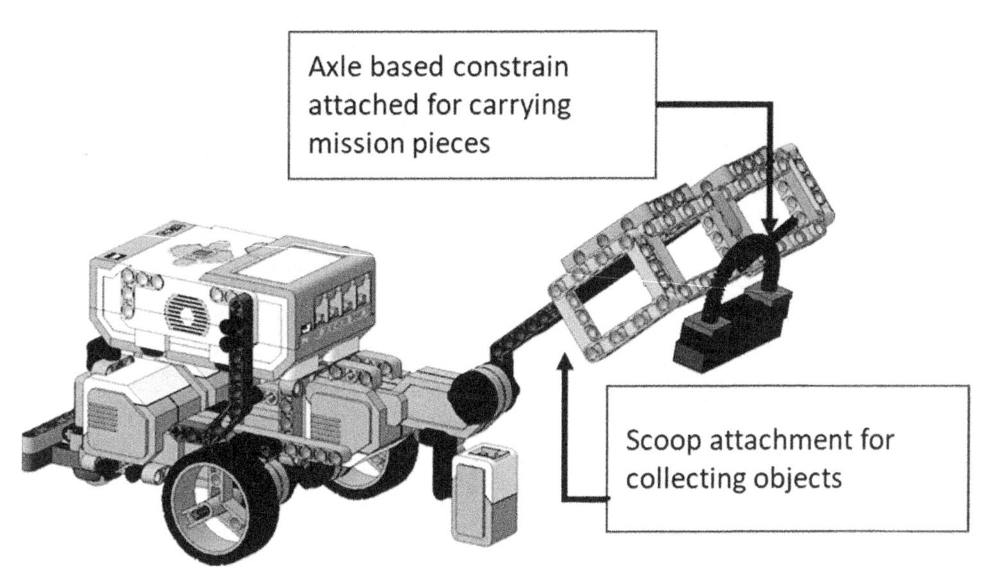

Figure 8: Configuring a scoop for multiple usage by attaching an axle

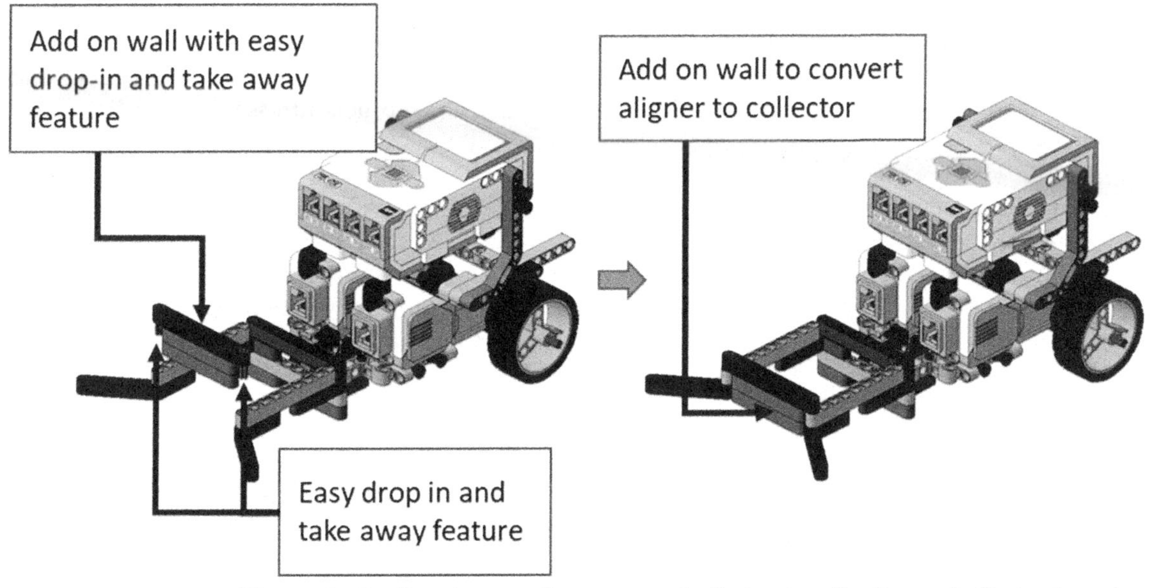

Figure 9: Configuring an alignment end effector to collection end effector by using a simple add on wall

Subcomponents of a Mobile Robot

As described in the Figure 5, typical mobile Robot, in addition to the onboard EV3 computer, comprises of following units:

1. *Vehicle (the robot body)* - this is what most coaches and teams will term as the base robot. This robot has the basic chassis with two large motors for moving the robot as well as one or two other motors to operate the end effectors or attachments.

2. *End effector (the attachments)* - these are detachable mechanisms that will be attached/mounted on the robot to perform one or more actions e.g. pick a piece, drop a piece. You would want to design a mechanism on the robot body so that multiple attachments can be attached and detached with relative ease and minimal time involvement since all FLL competitions have multiple challenges which can usually not be solved by a single attachment. Additionally, the challenges are time constrained to 2 minutes and 30 seconds, so any time spent changing the attachments is down time where your robot is not scoring points.
3. *Sensing unit (the sensors)* - even though we list this as a separate item, the sensing unit is the various sensors that you will attach to your robot. In general, the sensing unit is a permanent part of the robot and is attached rigidly to the body of the robot. The sensing units are attached to the robot with structural components such as pegs and axles and attached to the brick with a cable. In EV3 Mindstorms world, the sensing unit consists of up to 4 sensors of type *touch, ultrasonic (for distance), color, Infra-Red distance measurement sensor, and gyroscope* - for measuring angles.

Robot Design as a System

First of all, it must be emphasized that every design has opportunity to improve and secondly, a design is perfectly acceptable as long as it performs a task with reliability and within specified constraints of time and space. However, there are some fundamental rules of design that need to be addressed for a reliable and efficient performance.

Let us discus few fundamental of physics and geometry that will help us understand as well as implement the robot kinematics.

1. A minimum of three-point support is needed to fully constrain the motion of an object.
2. The Robot support points at the wheels must work in a synchronized, well defined manner.
3. The lesser the redundancy, the higher the accuracy of the Robot navigation.

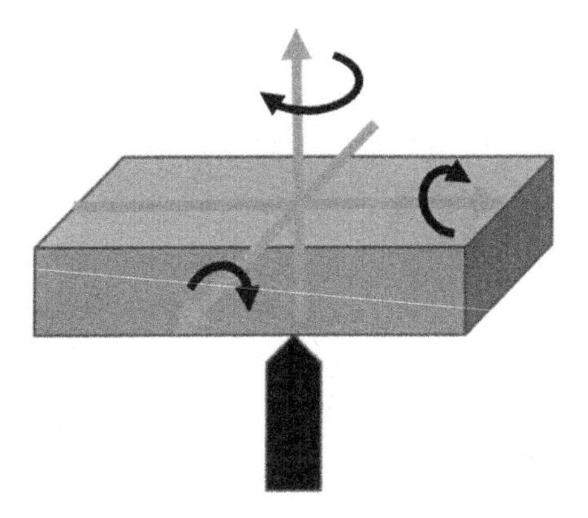

(a) Object supported at one point may tilt about a plane (Multiple-Axes) and spin

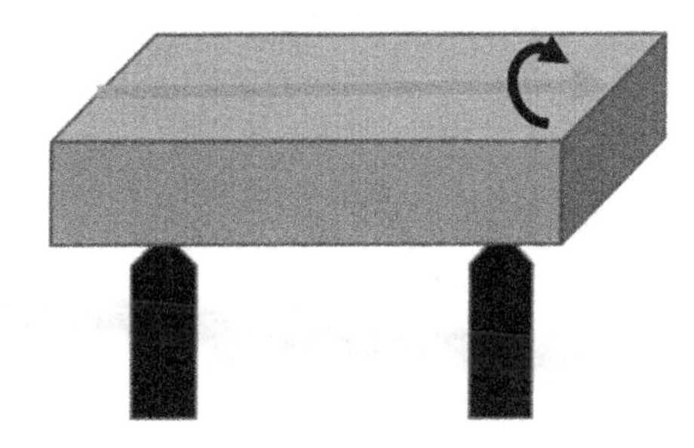

(b) Object supported at two points may tilt about the axis made by two points only

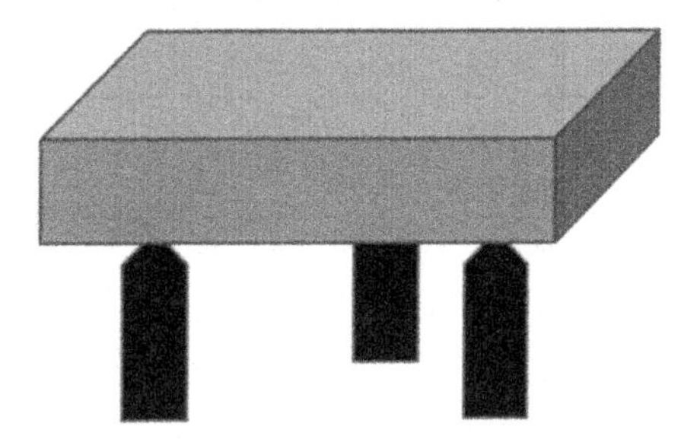

(c) Object supported at three points is fully constrained

Figure 10: Physics of object support points

To elaborate the first idea, let us refer to an example shown in Figure 10. A 3D object, when supported on one point may tilt along any of the multiple axes passing through the support point. The same object, when supported on two points, may tilt about the axis passing through the two points. By contrast, the object when supported on three points is fully constrained. The reason behind this is that at any given point of time there are only three points on which the object is supported, and any additional support introduces redundancy.

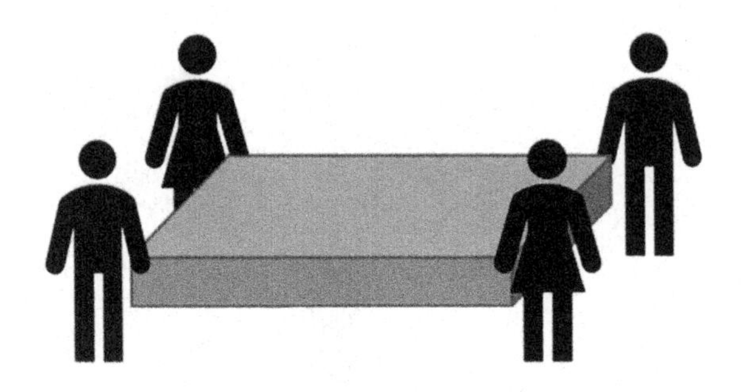

(a) Analogy – A heavy object being moved by 4 people.

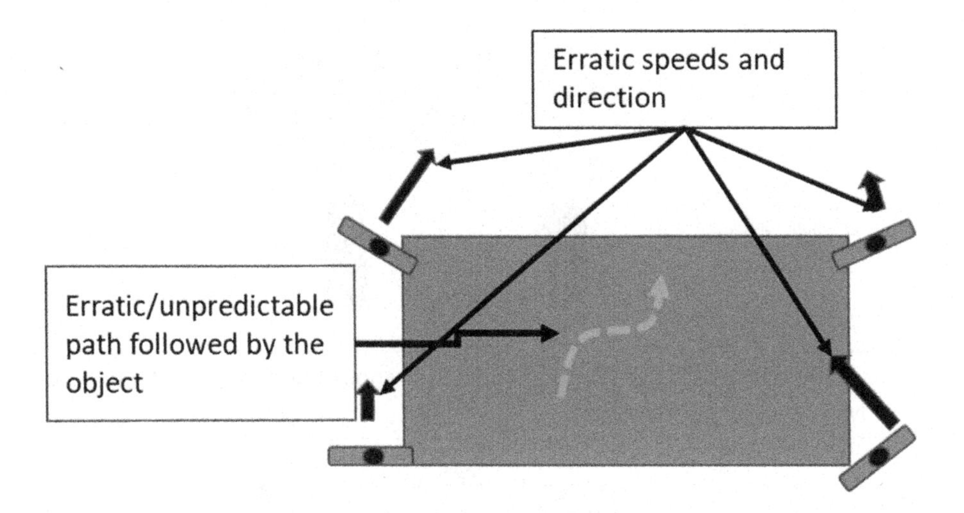

(b) People walking in random directions and speeds causes very unpredictable movement of the object.

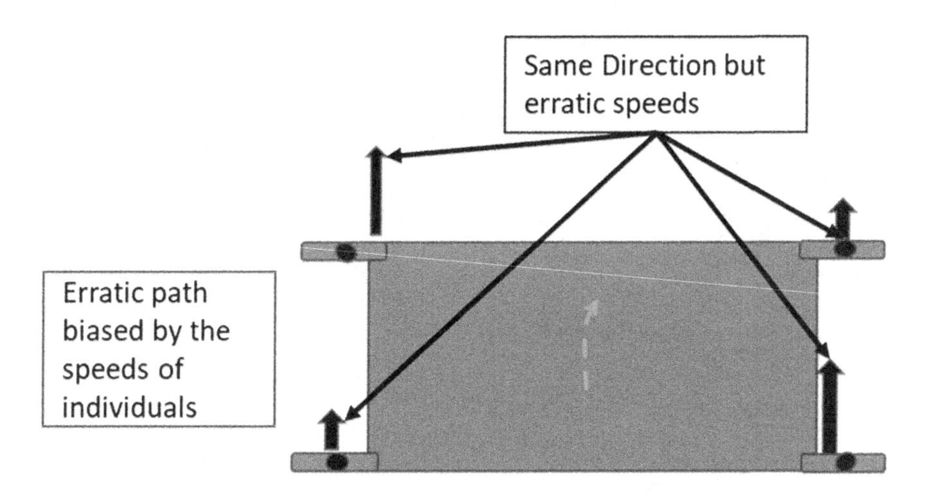

(c) People walking in same directions but different speeds causes unpredictable movement of the object.

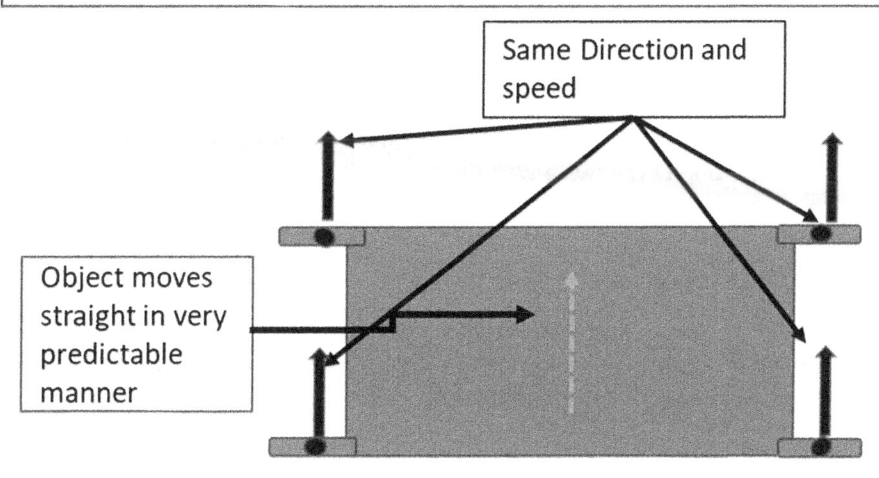

(d) When people walk in same directions and same speed, the object moves straight, in very predictable manner

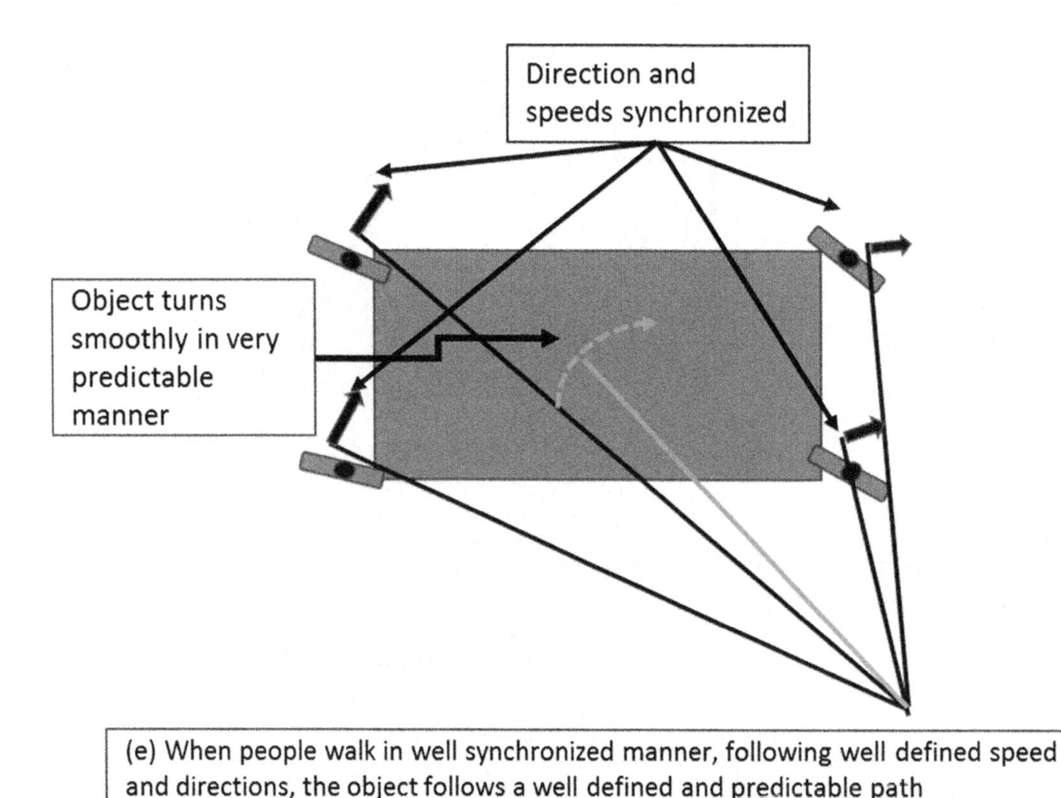

Figure 11: Analogy of four people carrying an object

For a Robot to be better controlled and exhibit repeatable performance, the number of wheels or support points on the robot that contact the ground should be kept to minimal. Another key point to be emphasized is the role of each support point. Let us consider and analogy of four people moving an object (Figure 11). If the people carrying the object walk in random directions and with different speeds, the movement of the object will be very erratic and unpredictable. Even with each person walking in same direction but different speed, the movement of object will be erratic and unpredictable. For the object to move in predictable manner, it is important that not only the speeds but also the direction of each person is well synchronized.

Similarly, for a Robot to move in a predictable manner, the movement of each wheel on the robot should be well synchronized with other wheels. Variability in a robot is introduced by multiple factors e.g. for a robot, there may be small geometrical differences in the shape and size of wheels. One may often see that for similar displacement inputs, that is, the speed or the total number of turns, two similar looking wheels may not behave identically. Despite being similar, prolonged usage, mechanical loads, wear and tear on individual wheels may cause wheels to behave differently. Certain measures (discussed in later sections) may be taken to overcome these uncertainties.

Now, let us reconsider the problem of an object being moved by multiple persons (Figure 11). Imagine if we could reduce the number of people handling the object. The predictability of object handling will improve. Similarly, for a Robot, where each wheel has some manner of variation, with increase in number of wheels and support points the overall unpredictably increases. Most of the popular mobile robots, AGVs (Automated Guided Vehicles) and research Robotic vehicles have the minimum number of needed, three wheels only. We do not imply that an FLL robot must be a three-wheeled robot. User may choose a design that serves the purpose; however, understand the principle of three-point support and take extra care when designing a mobile Robot with more than three wheels to ensure navigational accuracy.

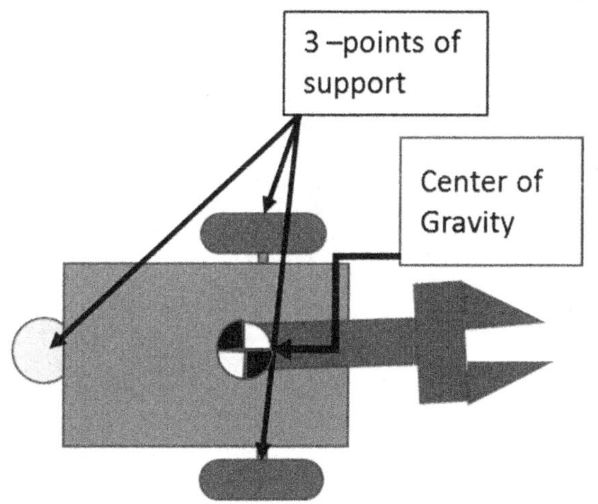

(a) Three wheel Robot in Optimal configuration

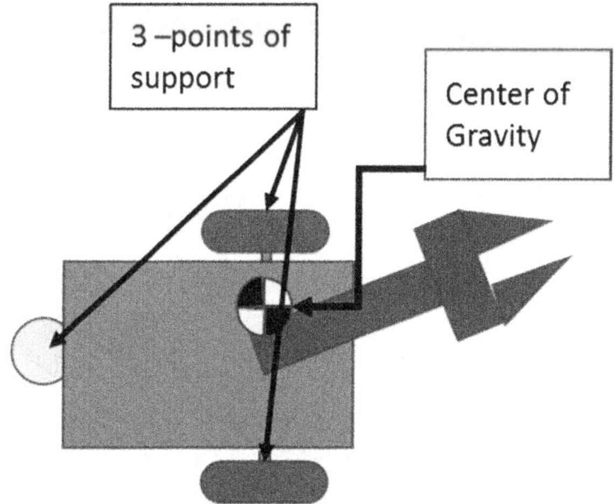

(b) Three wheel Robot configuration with Center of gravity shifted to left

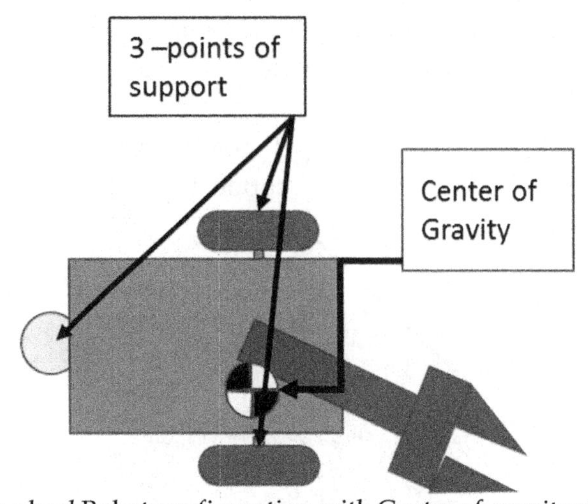

(c) Three wheel Robot configuration with Center of gravity shifted to Right

Figure 12: Three-wheel Robot with shifting center of gravity due to multiple configuration

Figure 12 above describes a three-wheeled mobile robot. Two of the wheels are motor controlled whereas one wheel is a passive non-motorized wheel. The robot is supported along the three wheels. Robot also has an arm attached to it. As the arm moves, the load distribution changes. The movement of the arm attached to the robot may shift the center of gravity of the robot. However, the points supporting the robot stay same and therefore performance of the robot will be more predictable and repeatable.

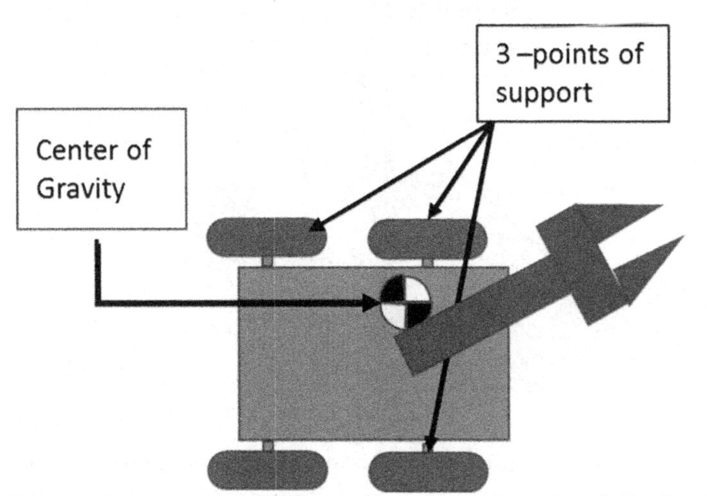

(a) 4-point supported Robot with center of gravity shifted to left

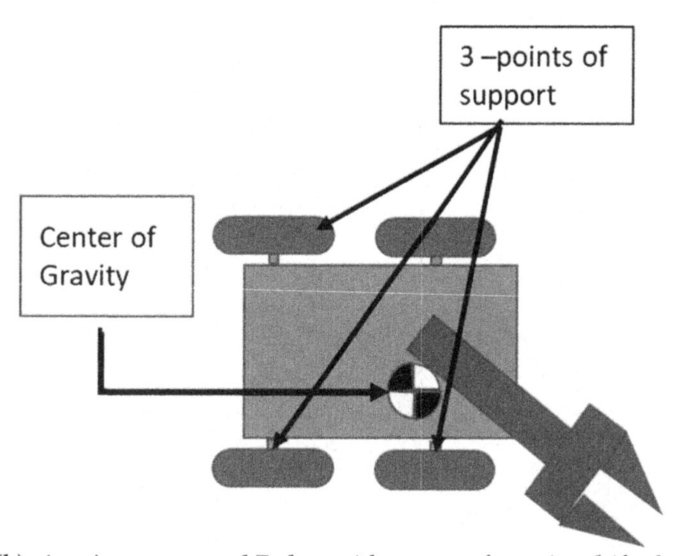

(b) 4-point supported Robot with center of gravity shifted to left

Figure 13: Shift in support points for a 4-wheeled robot

In contrast with a 3-point supported robot, if we consider a Robot with four wheels as illustrated in Figure 13, then we note that at any given point of time, the robot is actually supported by three points only. The reason this happens is that with the change in arm configuration, the center of gravity of the robot changes. As shown in Figure 13 (a) when the arm moves to the left side, the load distribution shifts to the left side and it shifts to right when the arm shifts to the right (Figure 13 (b).) Shifting loads inherently changes the points of support. The robot motion therefore has inherent unpredictability.

For four or more points of support, in addition to the shift in the center of gravity, other factors such as bumps and recessed areas on the surface alter the effective point of support for the Robot. This makes the robot motion unpredictable.

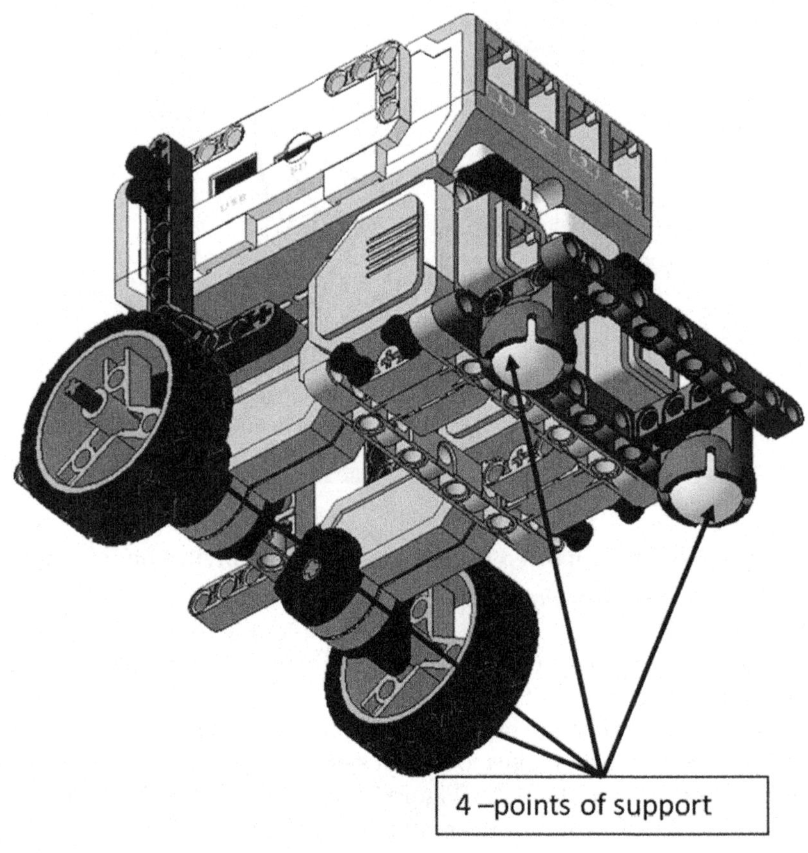

Figure 14: A popular 4 wheeled robot architecture

One of the common 4-point supported robot architecture (Figure 14) comprises of two motorized and two passive caster (another word for a free spinning roller wheel) wheels. The user must ensure that behavior of both the caster wheels must be smooth. If one of the wheels is stiffer than the other, or has higher friction, then the robot movement will be biased and inconsistent (Figure 15). The difference in the friction in each wheel may be caused by accumulation of dirt and debris. Other possibilities include scratches on the spherical steel caster ball. Prior to any critical runs, the user must test the wheels for similar mechanical performance.

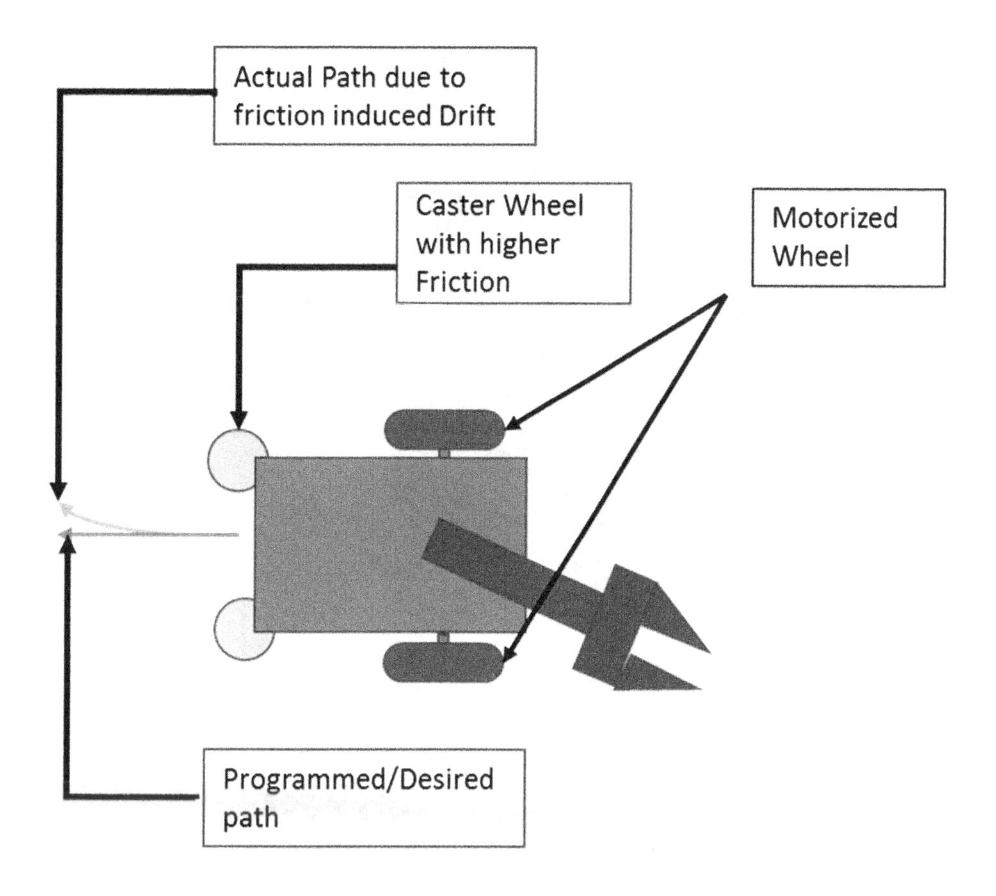

Figure 15: Robot with unequal friction in 4-point support robot

Guidelines for Robot Design

Design of the robot is very important for repeatable and predictable performance. Over the long-term building, testing and trial runs, the robot undergoes many changes as well as wear and tear. Additionally, in order to add multiple design features for various missions, robot continues to grow both in size and complexity. Many a times, designers tend to focus on immediate need rather than overall impact of the design modifications. Not only that, it may also happen that a design suitable for one mission may limit the ability to solve other challenges. When building a Robot for challenges such as FLL here are few recommended guidelines:

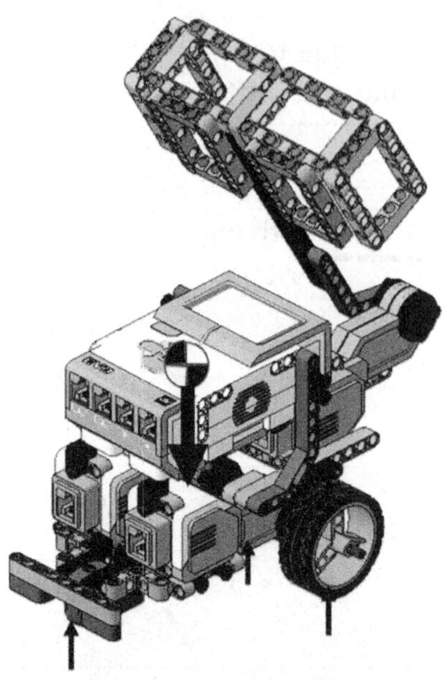

Figure 16: Robot with attachment in initial configuration so that center of gravity is well within the support points

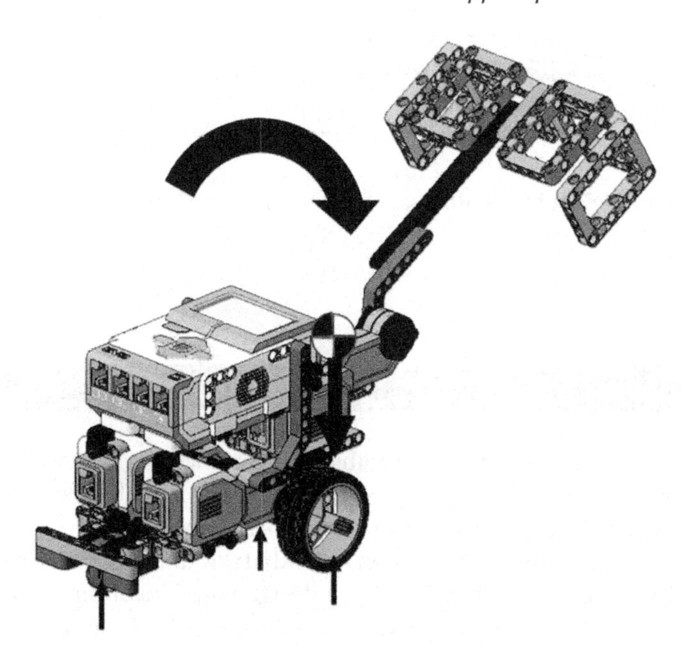

Figure 17:Robot the extended arm configuration. Center of gravity is critically located about the support points causing Robot to tip over at higher speeds

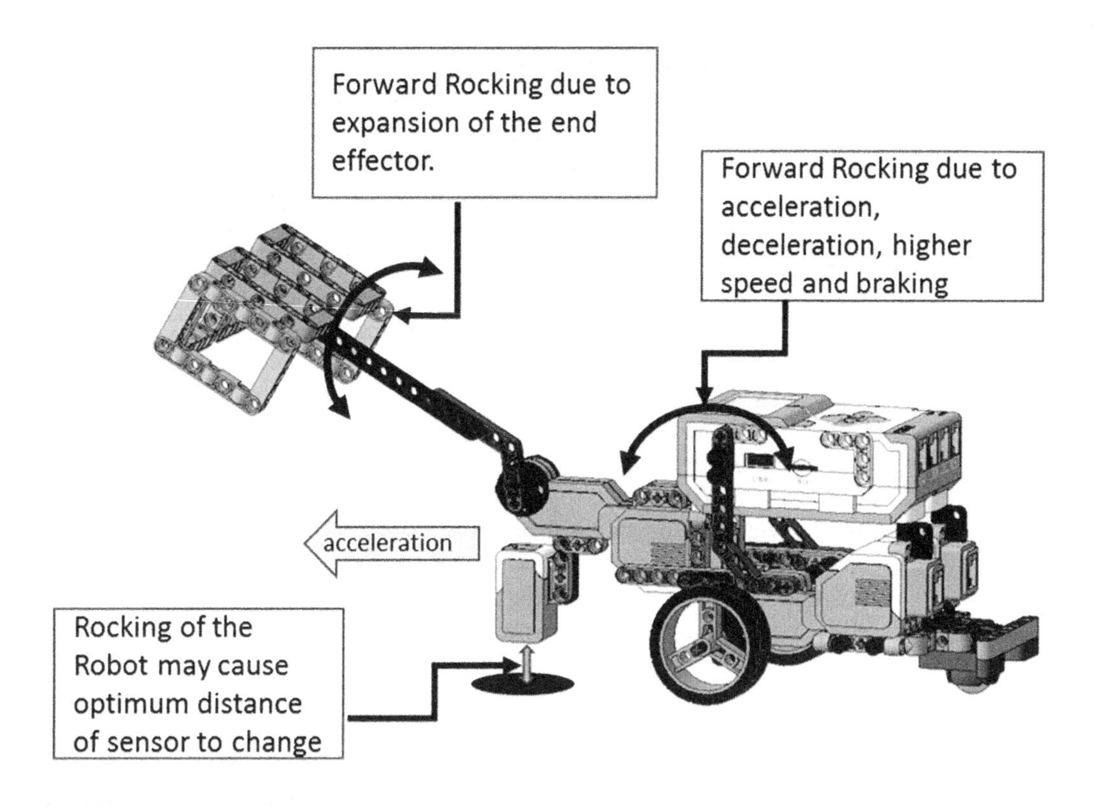

Figure 18: The rocking of the robot due to dynamic loads may cause the sensor distance to change causing erroneous readings from the sensor.

1. *Keep it simple* - The design provided by the Lego education instruction booklet that comes with the Lego Education EV3 Core Kit is a pretty good design for building the chassis. It keeps the robot small, well balanced and provides a sturdy chassis. It would be a great idea to keep the central elements of the chassis intact and then remove everything else before you start building your robot. Quite often, designers will be overexcited when modifying the robot and keep adding pieces here and there. When a structural problem is pointed out to them, their response is usually to add *more* pieces to the robot instead of rethinking the design and potentially removing a few pieces to rebuild that portion with greater simplicity. In general, the fewer pieces you use to build your robot, the better it is since you will be able to inspect, diagnose and rebuild the robot in case of issues. Pretty much every challenge in the competition can be solved with a really simple and straightforward attachment. If an attachment seems too complex, it likely is and will likely fail in incredibly frustrating ways.
2. *Make it stable* -(Figure 16, Figure 17 and Figure 18) Essentially, what this translates to in layman terms would be that the length and width of your robot should ideally be more than the height of your robot so that it is not tippy. A high center of gravity often results in an overturned robot when it hits an obstacle. The reason pretty much all EV3 robots have three points of support i.e. two driving wheels and a caster is that on such a small robot or vehicle in general, at any point of time, only three points of support are engaged so the 4th support point becomes pointless. As a general rule, ensure that all motors, beams, the brick and other structural pieces are horizontal or vertical to the ground in the base chassis of the robot. Pieces that are at an angle to the horizontal or vertical will likely cause weight to be shifted to the back or the front.

a. Another way to extend the tenet would be to ensure that any attachments that you add to the robot should not significantly shift the robot on any side as this will cause your robot to become unstable during movement and the robot will drift. In one case, one of our teams went with a forklift design with everything loaded towards the front of the robot. This caused the entire robot's rear to lift up whenever the robot would apply brakes and cause the robot to slightly shift its position on every braking. To avoid this, we used some heavy Lego bricks", (Lego item number 73843, 237826, 6006011, 6094053) in the rear of the robot to balance it out and this helped tremendously. Using caster wheel as counter load instead of *heavy Lego bricks* is another option since steel ball in the caster wheel makes it very heavy.
b. The dynamic effect (Figure 17) or the effects arising due to movement of Robot and (or) robot attachments must be taken into consideration when designing the Robot. A Robot design that is not well thought of may be stable when moving at very slow speeds; however, when moving at a faster speed the Robot my tip, slip or jump causing it to drift.
c. As shown in Figure 18, performance of sensors such as color sensor is yet another issue that may arise due to dynamic loads. The distance between the surface and color sensor is critical for line following and color detection of regions on the FLL mat. The Rocking and tipping over may cause the sensor to measure erroneous values leading to poor robot performance.
d. In case attachments shift the center or gravity of the robot well above the ground or, dynamic forces become a reason for concern, you may run the robot at a slower speed. You may also choose to increase the robot speed gradually using sophisticated programming.

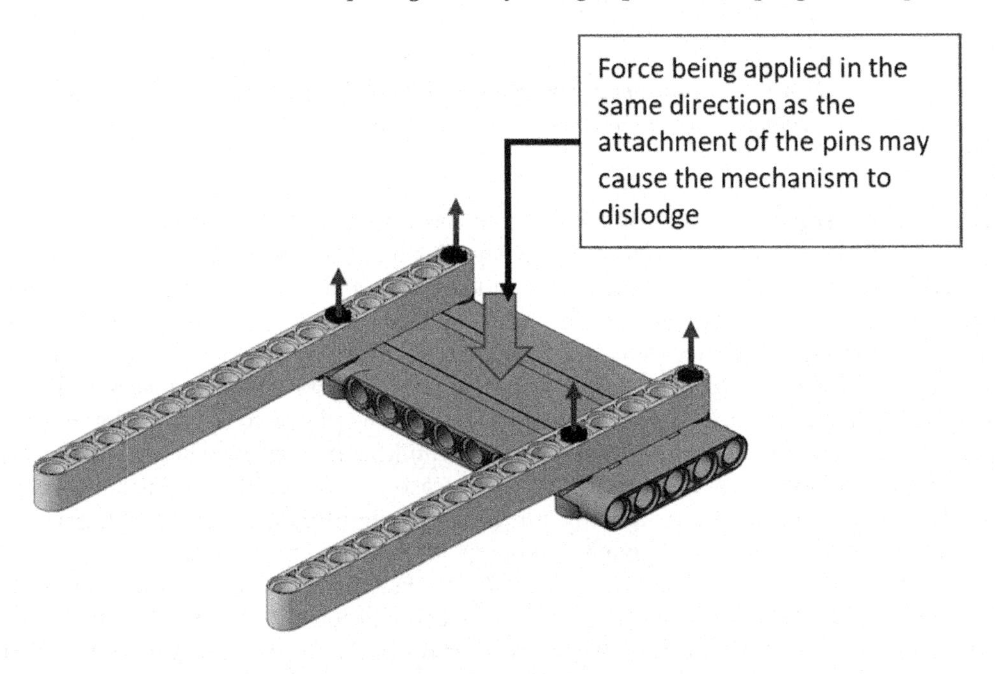

(a) Avoid attaching constraining pin along the same direction as the load

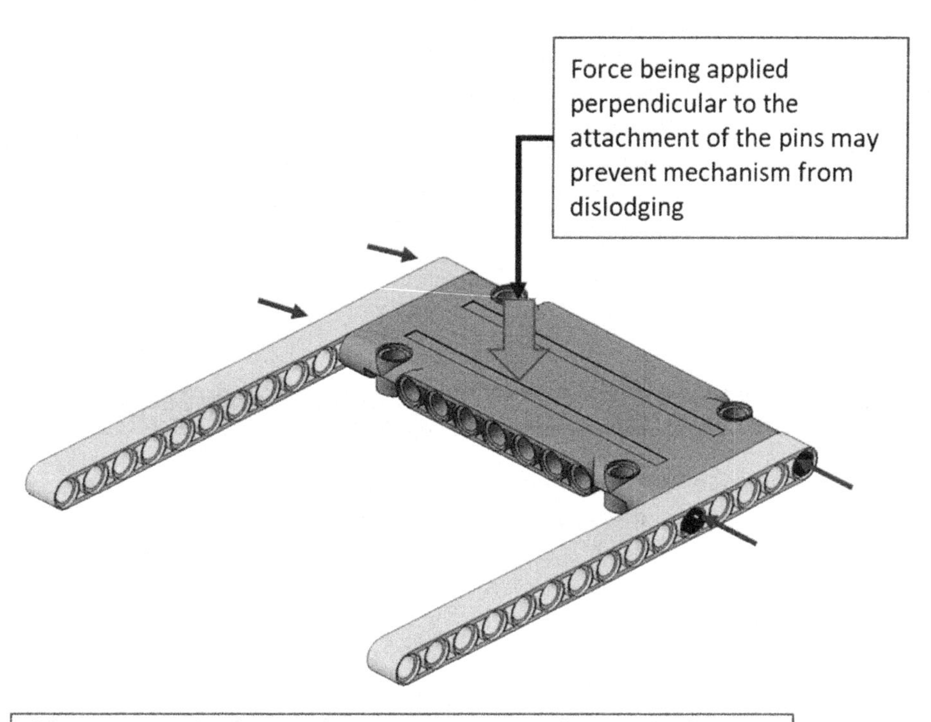

(b) A structure that has constraining pins connected perpendicular to the direction of load applied would not dislodge easily.

Figure 19: Constrain a structure such that constraining pins are perpendicular to the direction of load instead of in-line with load.

3. *Make it structurally sound* - The robot as well as the attachments undergo loads, shock and impact during the actual challenge. Despite careful handling, Robot may accidentally sustain a fall from small as well as large heights. The structural design of the Robot must ensure that it is able to endure such loads and impact with minimal damage. We tend to think that the thicker and heavier the support is, the stronger a structure is. This is a fallacy. In fact, it is the constraining of failure points that should be accounted for. Some of the principles to keep in mind when designing supports is -
    a. *Avoid cantilevers* - Cantilevers are the structures which are supported at a point but extend out enough that the load is transferred to non-supporting elements.
    b. Distribute the structure into triangular elements.
    c. Do not rely on the connecting Pins for support when pin is placed in line/in plane of the load (Figure 19). If it is inevitable, add constraints so that connecting pins are perpendicular to the plane/direction of force.
    d. Add bumpers in regions prone to load - The bumpers distribute the loads and shocks.
    e. Distribute the loads uniformly.
4. *Eliminate any plays and mechanical shift* - Whenever you are building a robot or an attachment, ensure that the mechanism is sturdy and rigid. Any unnecessary flexing of the parts or loose connections will give unreliable results.
5. *Tune your Robot frequently* - as team develops the robot or uses it during the practice run, sooner or later, various pegs and joints will become loose. This usually results in *the "but it was working just now!"* behavior. It is good idea to tighten all the pegs and visually inspect the robot frequently to correct any loose portions and more importantly all the cables. The cables quite often get loose over time and here are common ways a loose cable manifests itself:

a. Loose driving motor cables will cause the robot to stall and not move or go in a circle depending on what you have been trying to do. The reason behind this is that the Lego Move Steering and Move Tank blocks use a PID algorithm in them to make sure that the motors run in synchronization with each other. So, they will periodically, for very short periods - sub second - of time, switch off the motor that is ahead in total number of degrees to let the other motor catch up. If the *other* motor cable is disconnected, obviously, that motor is not turning and thus the software will wait. Subsequently you will see the program running but the robot stalled.

Other important considerations for overall Robot design

Here are Some of the things to keep in mind when building a robot for FLL. Some of these might not make sense to you immediately but we are providing them here for sake of completeness. These topics are covered in greater detail in later chapters, but it is good to keep these in mind from the get-go.

- Make sure that you DO use all 4 motors. Most teams that score high in the robot games usually have 4 motors. This does mean that you need to purchase another motor. We recommend a medium motor as the spare motor because it is a lot easier to fit on the robot. Comparatively, although the large motor has more torque, it is shaped in an awkward fashion and hard to place. Most FLL tasks can easily be handled by the medium motor.
- Move the color sensors right in front of the driving motors so they are (a) close to the body so that the distance between the wheels, which take action based on the color sensors and the color sensors is minimal. If this is not done, your robot behaves like a person wearing shoes larger than the size they should be wearing and keep stumbling. In robot terms, the place where the sensors are reading the value and the place where the corrective action is being taken i.e. the wheels, are far apart and the correction does not correlate with what the sensor is sensing. This decreases the reliability and precision of the color following algorithms. Additionally, the color sensors should be placed no more than the height of 1-2 FLU (Figure 169) from the ground for the best precision.
- Ultrasonic sensors should be mounted as low as possible and in a place, that will be free of robot mechanisms coming in front of it. At the same time, it should not be tilted or too close to the ground to start picking up the ground. Remember that the competition tables walls are only 2.5 inches tall.
- Touch sensor - Touch sensor may be used for accurately locating the Robot or a mission. Touch sensor may also be used for controlling the extent of motion of an arm. Unlike other sensors, the touch sensor requires physical contact with an object hence care must be taken to ensure that the contact doesn't cause drift. Also, the point of contact should not apply undue torque on the sensor.
- Gyro Sensor - we would usually recommend against using this sensor as its function can be replicated using simple calibration and it has a high degree of unreliability as the sensor may keep on drifting in many cases i.e. indicating the robot is turning even when the robot is standing still. However, if you do decide to use it, you may place it on any place on the robot since all portions of the robot rotate the same degrees. Make sure to keep it upright with the turning symbol at the top or all of your readings will be inverted. Additionally, make sure to keep the top of the sensor parallel with the ground since the gyroscope is a 2d gyroscope.

- Robots with tracks/treads - In general, unless special conditions call for it, it is not a good idea to use tracks for robots in FLL. Tracks slip on smooth surfaces and in general a robot using them does not go anywhere nearly as straight as robots using wheels. As described earlier, at any given time, the number of support points for a robot is three, the tracks and treads introduce multiple potential points of contact hence the uncertainty. We do not recommend using treads on an FLL robot unless there are specific requirements that call for it.

- Type of wheels - Designers have many choices for wheel including motorcycle, large, wide ones, squishy as well as hard wheels. The wheels that are squishier are more sensitive to local bump and load from the robot cause larger drifts.
- Make sure the wheels are supported on both sides. Having wheels that are supported on one side only usually causes the axles to bend over time. This, in turn, causes the wheels to turn at angle which might rub against the body of the robot, cause uneven wear on the tires and over time cause the robot to keep losing precision till you need to replace these parts.

Summary

Before we can build a robot for FLL or begin to program, we really need to understand what goes into building a great robot. A great robot is a robot that performs actions such as following a path or performing tasks such as picking, placing, pushing etc in a consistent manner while remaining stable and not losing its bearings. In this chapter we discussed key factors required to design a great FLL robot that will perform as expected. A FLL robot that conforms to requirements listed in this chapter is presented in Appendix D: A FLL Robot.

Mechanical System

Mechanical systems lie at the core of a Robotics system. Mechanical systems also allow movement of individual components and movement of components with respect to each other. Additionally, mechanical systems are used to build the basic support structures of a robot. Finally, Mechanical systems also allow transmission of Forces and torques from one point to another. This section is dedicated to explaining the basics of the mechanical systems.

Mechanical Drives

Usually, all motion in robotics originates as a circular motion via motor rotation, however, desired motion could be straight motion, elliptical motion or some other peculiar kind of motion. In terms of handling any objects, the Robot sometimes need to simply push or pull them. In many cases, the Robot might need to propel an object, or handle it in a very secure manner using a gripping mechanism. Sometimes, the robot needs to actively engage with a mechanism to actuate. Each such requirement can be solved using a mechanical system. The mechanical system includes mechanical drives and other linkage based drives.

As mentioned earlier, all motion in robotics starts with the motor and thus to understand motion, we need to understand motors and their characteristics. A motor is characterized by its total power which, is a combination of the speed and torque. Speed simply refers to how fast the motor is turning in terms of rotations per minute (RPM) and torque is equivalent to the force applied in circular motion. To better understand torque, consider the example in linear motion where the force is applied to push and objects in a straight line. The heavier an object is, more the force needed to push or pull the object. Similarly, for rotational motion, heavier the axle, or the wheels, more the circular force/torque needed to spin it. Looking at a real life example, the engine for a truck has higher torque than the engine for a car since it needs to pull much heavier loads. A small car such as a sedan slows down significantly when fully loaded with 4 people and cannot accelerate as fast. By contrast, a pickup truck can pull very large RVs and even heavy farm equipment. This is because the engines, which turn in a circular motion in the two vehicles, produce vastly different torques.

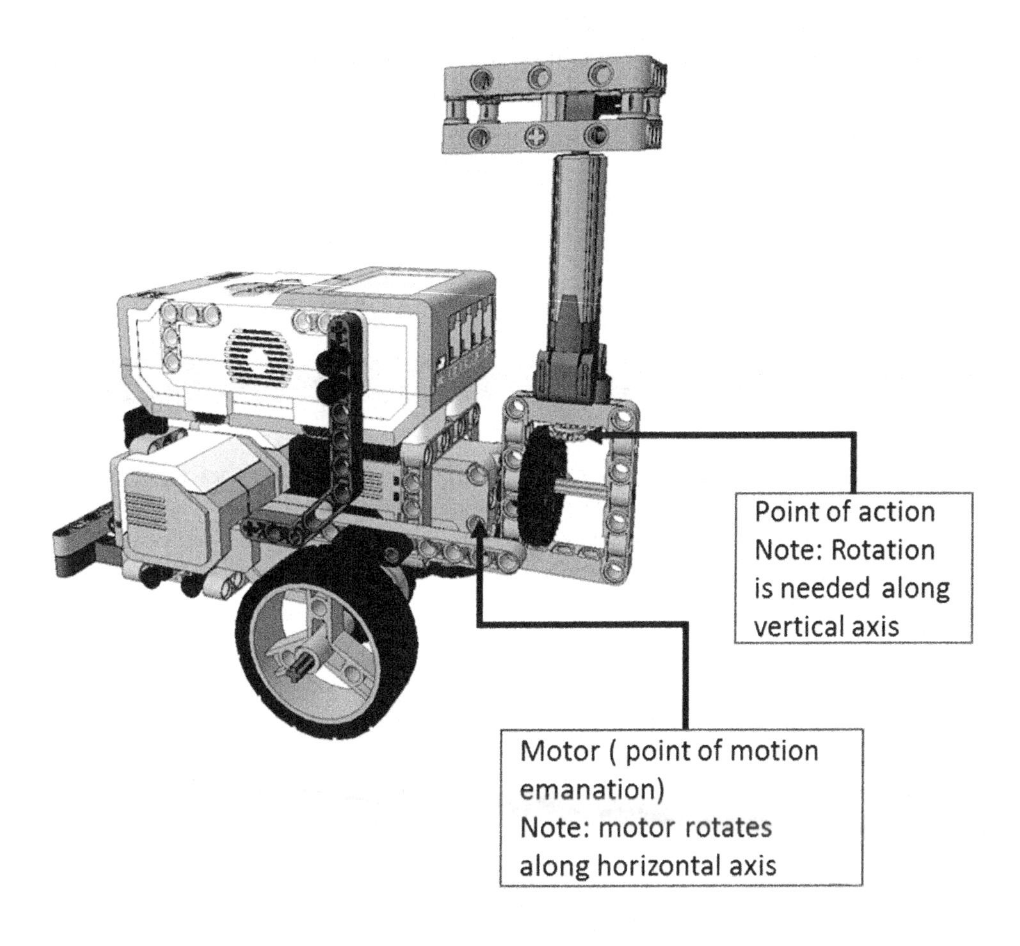

Figure 20: Mechanical drive for manipulating point of action, speeds and loads

While it is possible to programmatically change the speed, direction of rotation, and the total torque applied by the EV3 motor; there are multiple scenarios where the point of motion application as well as the axis of rotation are varied. Sometimes, we may need to perform one or more of the following actions -

1. Increase speed significantly more than what the motor setting allows
2. Decrease speed significantly lesser than what the motor setting allows
3. The effective torque needs to be amplified for load intensive tasks - such as the need to lift heavy objects which are hard for our motors to lift.
4. Change the axis of rotation e.g. instead of turning in a vertical plane, we may need to change the direction of movement to turn in the horizontal plane even though the motor is turning in the vertical plane
5. Sometimes, motor and rotating end effectors i.e. the place where the result of the action needs to take place need to be far from each other
6. Sometimes, we need to change the direction of Rotation -clockwise and anti-clockwise

As shown in the Figure 20 above, the medium motor is used as the source of motion for a linear actuator. The linear actuator has an input rotary axis that needs to spin at much higher speed than the motor speed, secondly the motor spins along the horizontal axis whereas the actual rotational input needed is along vertical direction. Gear based mechanical drives are used for both the requirements.

The manipulation of torque and speed is enabled with the help of specialized set of components or Mechanical drives. Most popular kind of mechanical drives include:

1. Gear mechanisms
2. Belt and Pulleys
3. Chain and Sprocket

Gear Drives

A gear is a rotating machine part or a circle with teeth which mesh with another gear. Further, the teeth or cogs in the gear are usually located along the circumference that mesh with the teeth or cog of another gear to transmit torque and rotational capabilities. The meshing teeth or the cogs of two gears in a gear assembly are usually of the same shape. Gears are frequently used in machines and to perform following tasks

1. Changing point of Action
2. Changing Direction
3. Changing Speed and torque
4. Single input, multiple outputs (Figure 21)

Single Input Multiple output using bevel gear arrangement

Figure 21: Using gear arrangement to convert single input to multiple outputs

Multiple gears meshing together when arranged in a sequence to transmit rotation are called as a gear train. There are a range of Gears available to us and we use one or the other depending on the application at hand. For the purpose of FLL designs we will focus on three different kind of gears-

1. Spur Gears
2. Bevel Gears
3. Worm Gears

Let's look at these gears in details in the following sections.

Spur gears

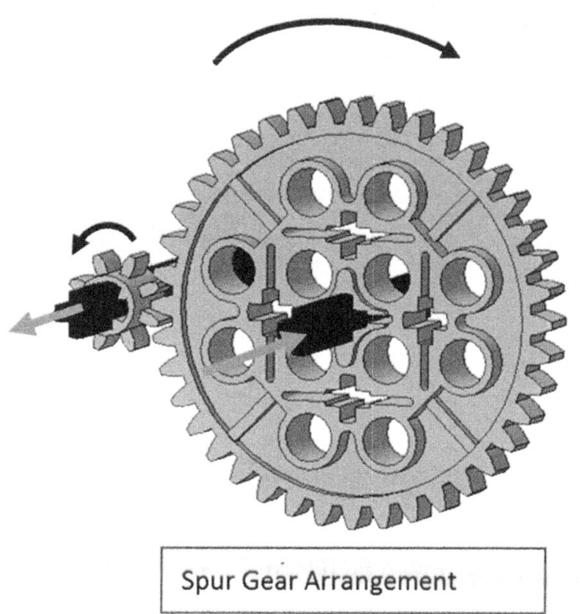

Spur Gear Arrangement

Figure 22: Spur Gear Arrangement

The spur gears (Figure 22) are simplest kind of gears with a cylindrical core and teeth projected out along the circumference of the cylinder. The shape of gears is uniform across the length of the gear. For Spur Gears, the Gear axles of gears connected with each other are parallel. One interesting property of two adjacent spur gears is that as shown in the Figure 22, the adjacent gears spin in opposite direction.

Apart from transmitting the motion and changing the direction of the movement, gears also allow you to change the speed and torque. The speed and torque scaling is governed by the number of teeth in the input (driver) and output (driven) gears. For example, if the number of teeth in the input spur gear is 48 and the number of teeth in output gear is 12 then the increase in the speed will be determined by the ratio of the (number of teeth in input gear)/(number of teeth in the output gear). This ratio is formally known as the gear ratio and for the example we just discussed, it will be 4-fold (48/12 = 4.) By contrast, if the input and output axles are interchanged in the above assembly, then the output speed will be 1/4 of the input speed.

It is important to note that torque works in inverse proportion to the speed in a gear assembly. That is if the input torque for the 48 teeth is 3Nm then the output torque will be 0.75Nm. If the 12 teeth gear is the input, then the torque advantage would be 12Nm. This is opposite of the speed as we just discussed above.

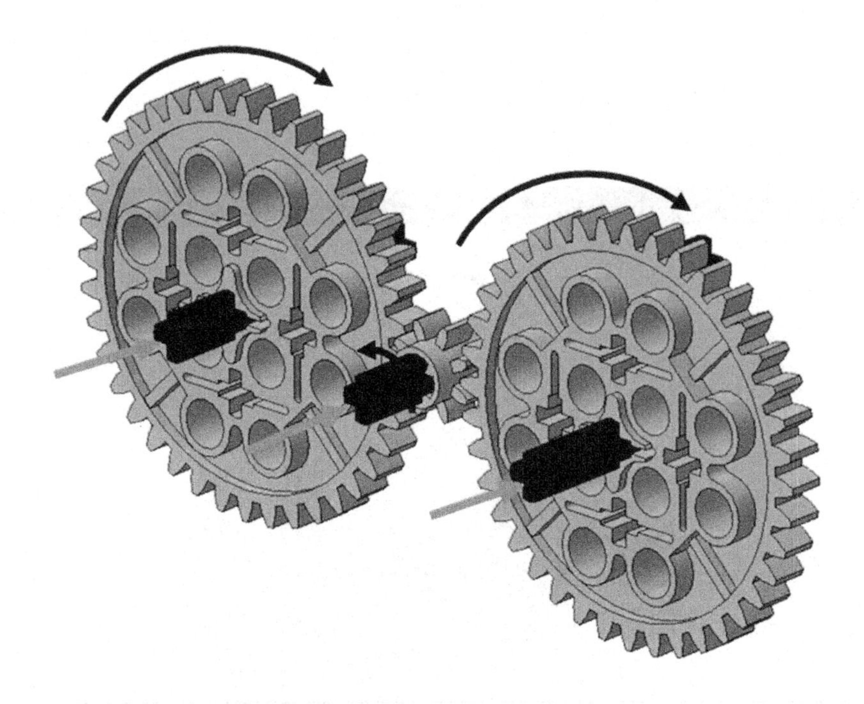

Figure 23: Spur Gear Arrangement for changing the point of action

If the input gear is rotating in clockwise direction, then the output gear would rotate in the counterclockwise direction. The primary purpose of a spur gear is to increase/decrease the speed/torque in the same plane. Additionally, it can transfer power in the same direction. The spur gear allows the input motor to be at one location whereas the output can be at a location farther away (Figure 23). Multiple spur gears may be arranged in a sequence such that the consecutive gears act as input and output gears and the point of the action of the motor is placed very far from the output. As shown in Figure 23, the middle gear, which only aids in the transmission of motion as well as change of direction of motion in the gear train, is known as an *idler gear*.

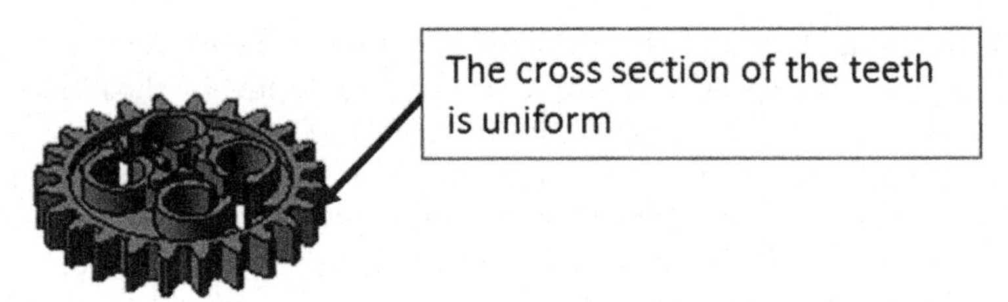

Figure 24: Exclusive Spur Gear

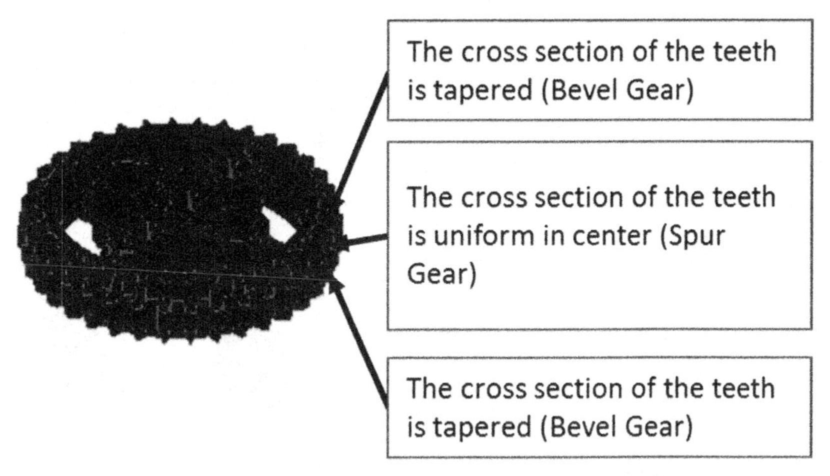

Figure 25: Gear with Both Spur gear and Bevel gear configurations

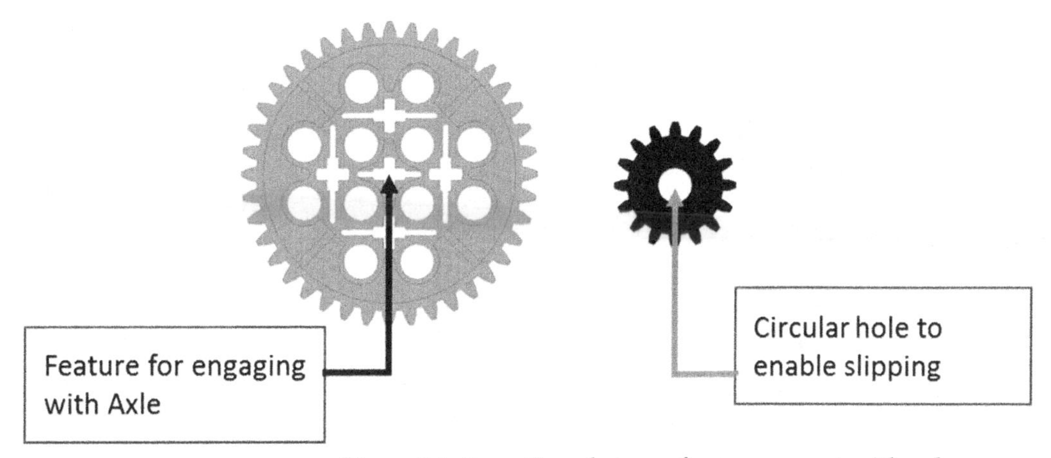

Figure 26: Spur Gear features for engagement with axle

The Lego technic catalog includes multiple spur gears. While some spur gears are exclusive spur gears (Figure 24) with the teeth having uniform cross section, other spur gears, additionally, have the teeth extended to engage as bevel gears (Figure 25).

Table 1 Spur gear groups in Lego Technic

Spur Gear Group				
8 Teeth				
16 Teeth				
24 Teeth				
20 Teeth				
12 Teeth				
40 Teeth				
36 Teeth				

As shown in Figure 26, in most of the gears, the center has a feature to allow engagement with axles whereas others have round hole in the center that prevents any engagement with the axle hence such gears can serve only as idler gears.

For reference, most popular spur gears are listed in Table 1 . The Gears are grouped according to the number of teeth. Table 2 summarizes the gear Ratio between the gears

Table 2 Output Ratio for Lego technic spur gear sets

	Output Gear							
Input Gear	**Gear Group**	**8 Teeth**	**12 Teeth**	**16 Teeth**	**20 Teeth**	**24 Teeth**	**36 Teeth**	**40 Teeth**
	8 Teeth	1	1.5	2	2.5	3	4.5	5
	12 Teeth	0.7	1	1.3	1.7	2	3	3.3
	16 Teeth	0.5	0.8	1	1.3	1.5	2.3	2.5
	20 Teeth	0.4	0.6	0.8	1	1.2	1.8	2
	24 Teeth	0.3	0.5	0.7	0.8	1	1.5	1.7
	36 Teeth	0.2	0.3	0.4	0.6	0.7	1	1.1
	40 Teeth	0.2	0.3	0.4	0.5	0.6	0.9	1

Bevel Gears

Bevel gears (Figure 27) are gears that have teeth that are slanted, and they mesh with each other at 90 degree angles.

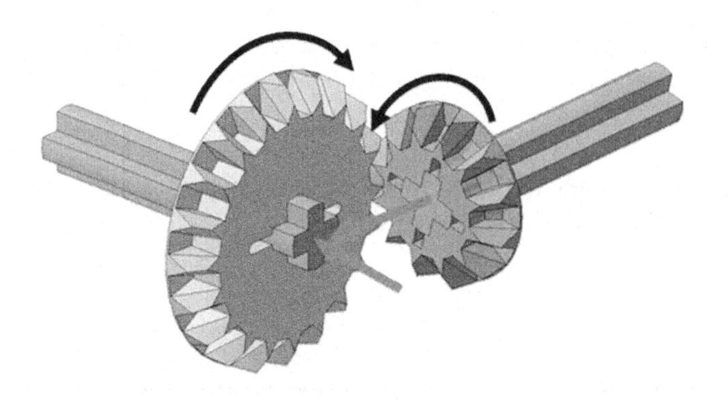

(a) Arrangement of a set of Exclusive Bevel Gears

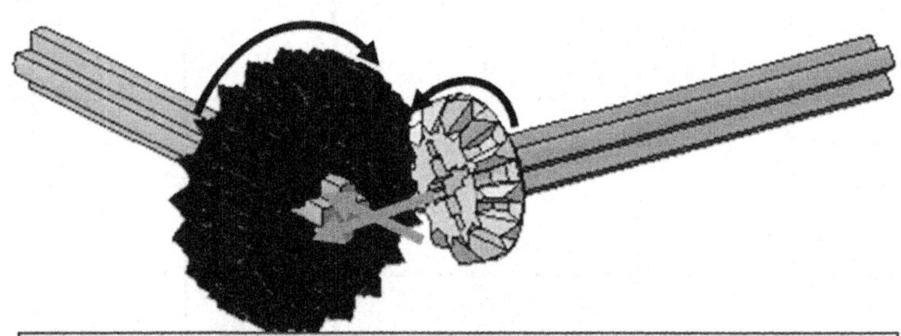

(b) Arrangement of an exclusive Bevel Gear with gear having spur as well as bevel gear features

(c) Arrangement of two gears having spur as well as bevel gear features

Figure 27: Bevel Gear Arrangement

A pair of bevel gears beaves just like a spur gears except that whenever two bevel gears mesh, the direction of the movement and transmitted force changes by 90 degrees.

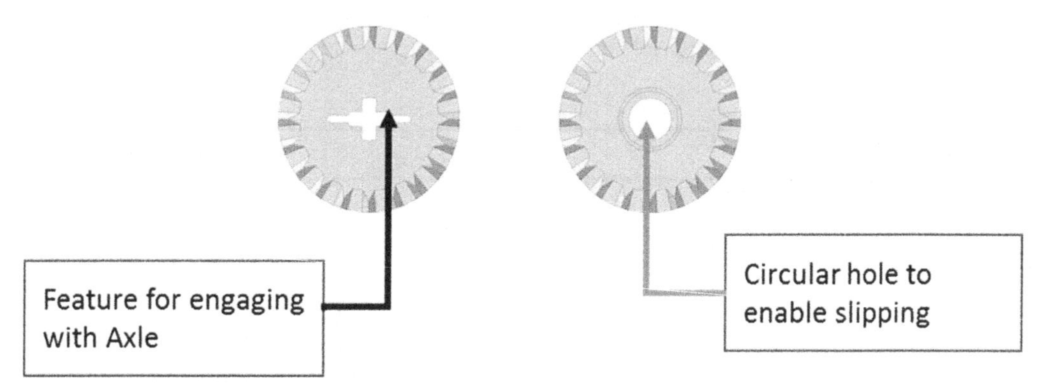

Figure 28: Bevel Gear features for engagement with axle

The primary purposes of bevel gears is to change the direction of force. In terms of gear ratio, they behave like spur gears; torque increase/decrease and speed increase decrease also follow.

As shown in Figure 28, like the Spur gears, most of the bevel gears have features to engage with the axles. Some of the gears have circular center that allow the gears to work as idler gear.

Table 3 and Table 4 below summarize different group of gears and corresponding gear ratios.

Table 3 Group of Bevel Gears

Bevel Gear Group			
12 Teeth			
20 Teeth		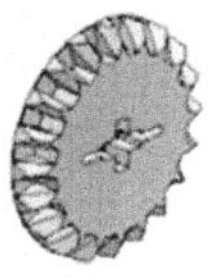	
36 Teeth			

Table 4 Bevel Gear Speed Ratio

		Output Gear		
Input Gear	Group	**12 Teeth**	**20 Teeth**	**36 Teeth**
	12 Teeth	1	1.7	3
	20 Teeth	0.6	1	1.8
	36 Teeth	0.3	0.6	1

Technic knob wheels (*Figure 29*) have a very peculiar shape. Thheir geometry is very different from the Bevel Gear; however, when engaged along two perpendicular axes they behave like two similar bevel gears and maintain gear Ratio of 1. The Technic knobs are very effective in transferring the axis of rotation by 90 degrees.

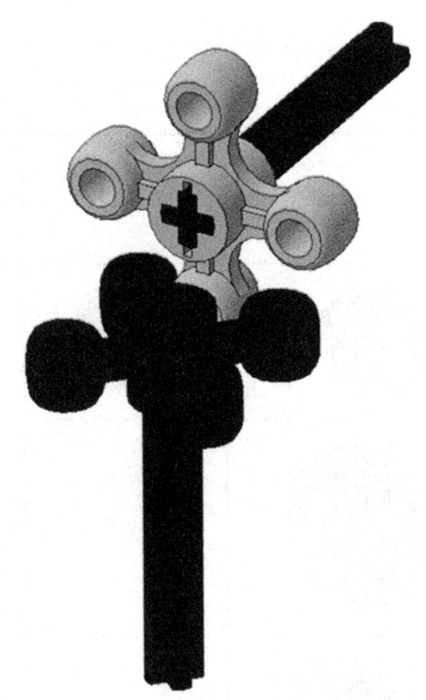

Knob arranged similar to bevel Gears

Figure 29: Knob wheels

Worm Gear

A worm gear (Figure 30) looks unique and is very different from the bevel and spur gears. In fact, it does not look much like a gear at all and instead simply looks like a screw. It has threads that wind on the surface like the threads of a screw. Unlike spur and bevel gears which can both be the input/drivers and output/driven gears, worm gear can only assume the position of a driver gear and cannot be driven by another gear. This property provides exceptional uses for the worm gear in places where self-locking mechanisms are needed, such as, in grippers (Figure 31), cranes and garage doors. Just to clarify, the word *self-locking* means that even when the power being supplied to the worm gear is lost, the mechanism will not move, even when under load. Rather, the worm gear will hold the assembly in a locked motion with no movement allowed. In contrast, if you have a gear train or assembly with only spur and bevel gears with no other explicit locking mechanisms, as soon as the input power is removed, the assembly will likely start moving under even very little load. The worm gear, due to its special construction behaves like a gear with a single tooth and provides spectacular reduction in speed and a huge increase in the torque. Given that, worm gear is quite often found in machines to decrease the speed of a very fast spinning, but relatively weak motor to have lower and manageable speeds at greatly increased torque.

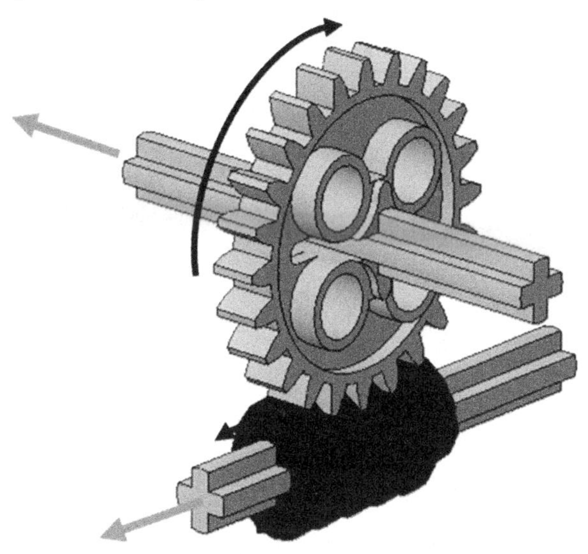

Figure 30: Worm Gear arrangement

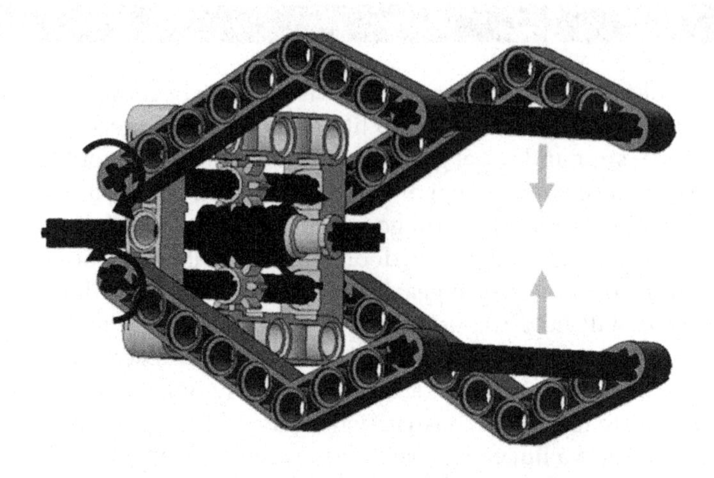

A gripper Mechanism showingWorm Gear Arrangement for self locking and larger holding force

Figure 31: A gripper mechanics based on worm gear

Other than the interface geometry, the technic worm gears have similar geometry. The most popular technic worm gears are shown in the Figure 32.

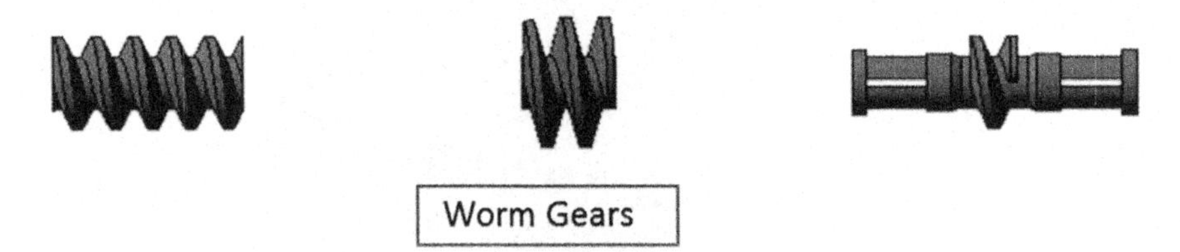

Figure 32: Popular technic worm gears

For the worm gear, each turn of the worm gear allows the progression of an attached spur gear by one teeth. Therefore, the gear ratio corresponds to the number of teeth in the gear engaged to the worm gear. Unlike spur and bevel gears, where either gear axles may be input or output; in the worm gear arrangement, the axle engaged to the worm is the always the input whereas the axles for engaging spur gear is the output gear.

Rack and Pinion

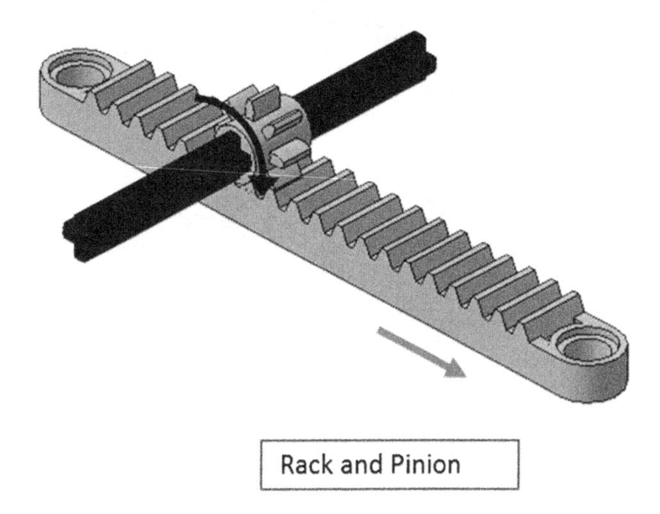

Figure 33: Rack and pinion arrangement to convert rotary motion into linear motion

A rack and pinion (Figure 33) is a type of special gear arrangement that converts rotational motion into linear motion. A circular gear also known as pinion engages teeth on a linear bar that has gear teeth is called *rack*. As the teeth for rack engages with the pinion teeth, the rotational motion applied to the pinion causes the rack to move relative to the pinion and translates rotational motion of the pinion into linear motion of the rack.

Gear Trains and Combination gear trains

A *gear train* can be built as a combination of different type of gears. For example, as shown in the Figure 34, the input axle of motor is oriented along the vertical axis, whereas the axis for *output1* axis is located perpendicular. Additionally, the output axis is offset from the motor's axis. The change in direction from vertical to horizontal direction is done with the help of bevel gear whereas the offset is accomplished with the spur gear.

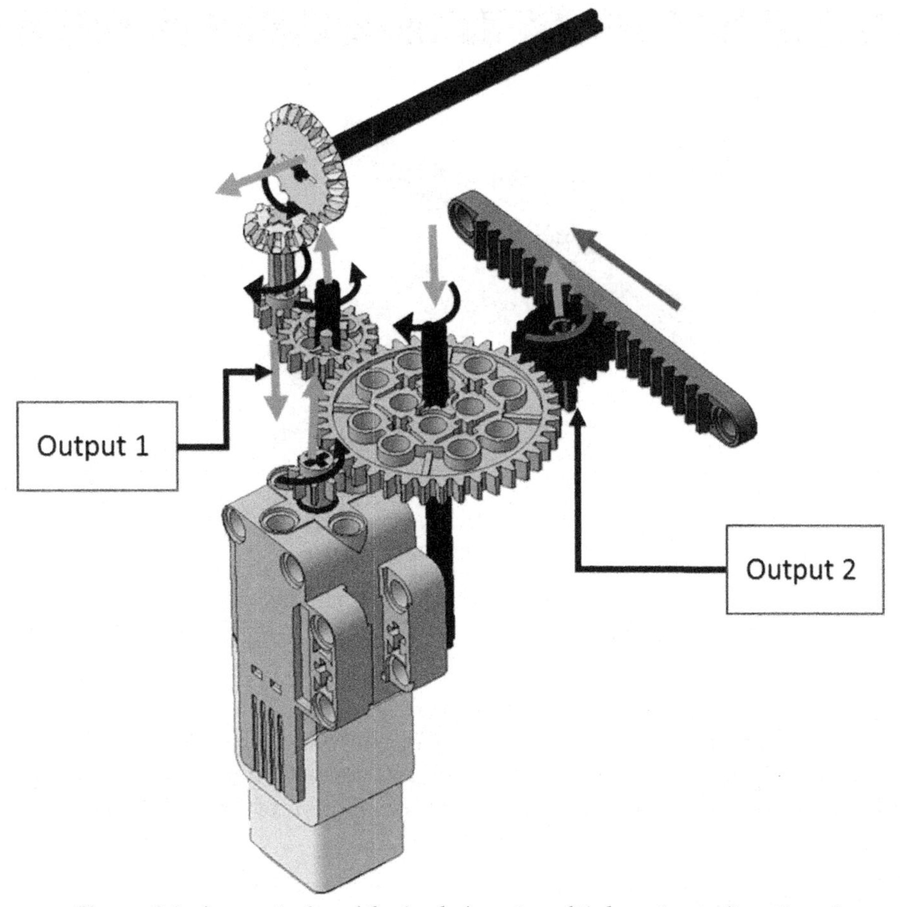

Figure 34: A gear train with single input multiple output, direction change and conversion of rotary to linear motion. The input power is supplied using an EV3 medium motor.

The *output 2* in the gear train, described in Figure 34 is a linear motion. The rotation of the gear attached to the motor in turn spins the 4o teeth gear. The 40 teeth gear spins the red idler gear. The red idler gear actuates the rack and pinion arrangement. Two gear trains driven by a single motor is very useful when time is very limited in the competition since one need not switch robot arm and attachments. Two different mechanisms attached to the same motor can perform different tasks alleviating the need to create multiple attachments.

Practical application: Worm gears are quite often used in a box configuration where they are connected to exactly one other spur gear. Such boxes are often called "Worm Gear Boxes". It is a great idea for your FLL attachments to use a worm gear for strength and precise movement.

Chain and Sprocket Drives

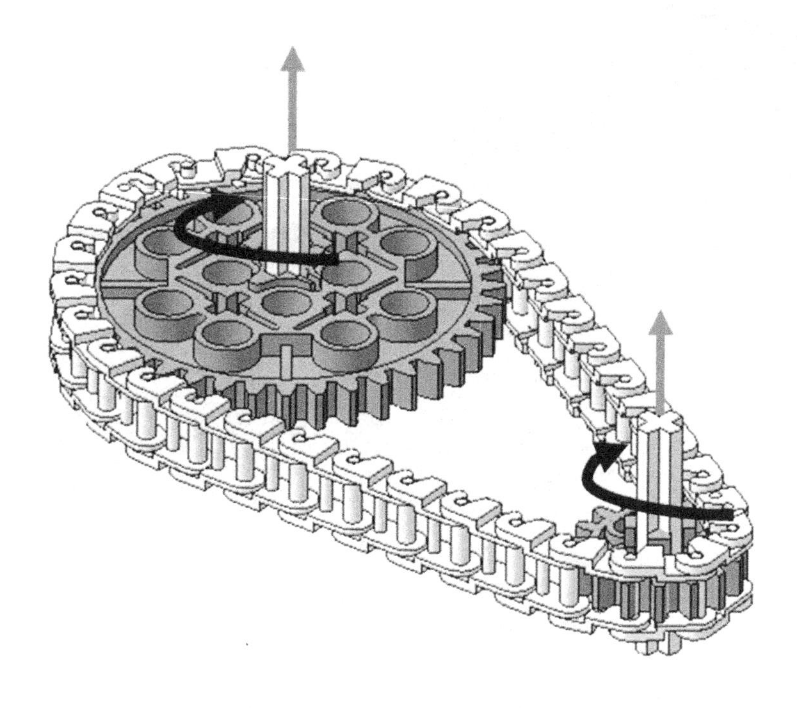

Simple Chain and Sprocket Assembly

Figure 35: Chain and Sprocket

You have likely seen a chain and sprocket arrangement in bicycles even if you did not know its technical name. We illustrate a chain and sprocket arrangement in Figure 35. To break the components down, let's look at the sprocket and chain one by one. Sprocket is a profiled wheel with radially arranged teeth like feature, a fancy way of saying that it looks like a spur gear. A *Chain* is flexible linking component with features that engage with the teeth. Similar to the Gear Drives, speed increase or decrease and corresponding torque reduction and increase is possible with chain and sprocket. Usually the chain and sprocket arrangement has the chain wrapping around the sprocket. One difference from the Gear that chain and sprocket shows is that both input and output axles spin in the same direction. An alternate configuration (Figure 36) where the sprockets are arranged externally has the axle spinning in opposite direction.

For the gear trains/drives, the distance between the axles is always fixed. In the chain and sprocket arrangement the distance between the axles can be arranged as needed. You must however, keep the axles parallel in a chain and sprocket mechanism. Chain and sprocket may be used when the motor is placed far from the point where actual action is needed. Also, chain and sprocket allows multiple axles to be driven by one input axle.

Chain and Sprocket Assembly, Internal and external engagement

Figure 36: Alternate configuration for internal as well as external engagement of sprocket and chain

One advantage of the Lego Technic chain and sprocket systems is that the chain is compatible with the spur gears with uniform cross section (Table 6: Speed ratio in chain and sprocket arrangement) and therefore the parts are reusable. The most popular gears that can be used as sprocket are described in Table 5 and the corresponding speed ratios are described in Table 6: Speed ratio in chain and sprocket arrangement

Table 5 : Spur gears used as sprocket

Gear/ Sprocket Group				
8 Teeth				
16 Teeth				
24 Teeth				
40 Teeth	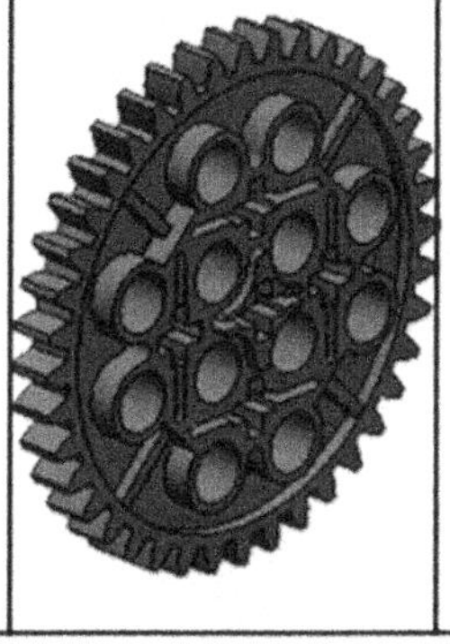			

Table 6: Speed ratio in chain and sprocket arrangement

	Output Sprocket				
Input Sprocket	Group	8	16	24	40
	8	1	2	3	5
	16	0.5	1	1.5	2.5
	24	0.3	0.7	1	1.7
	40	0.2	0.4	0.6	1

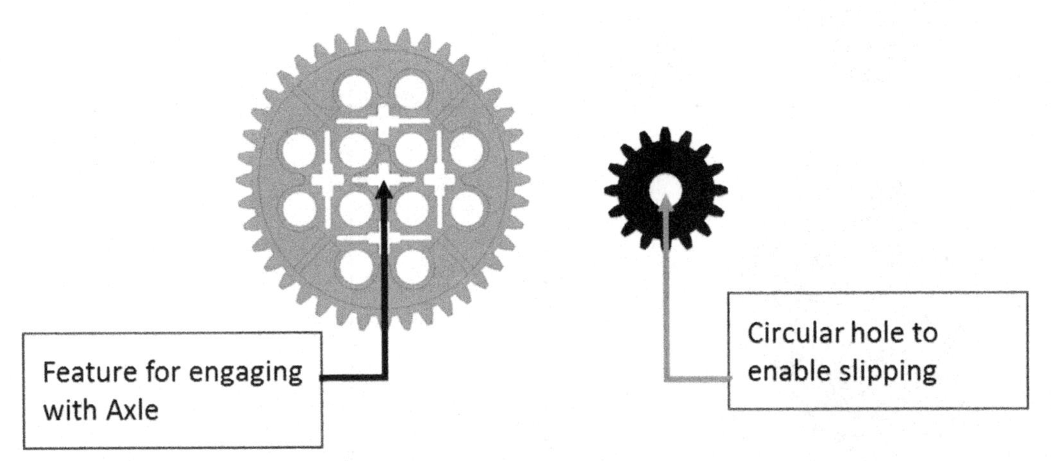

Figure 37: Features for engagement of sprocket with axle

Similar to the gear arrangement, idler sprocket (Figure 37) may be used. In fact for long chain segments, an idler sprocket is very helpful in maintaining the optimal tension in the chain.

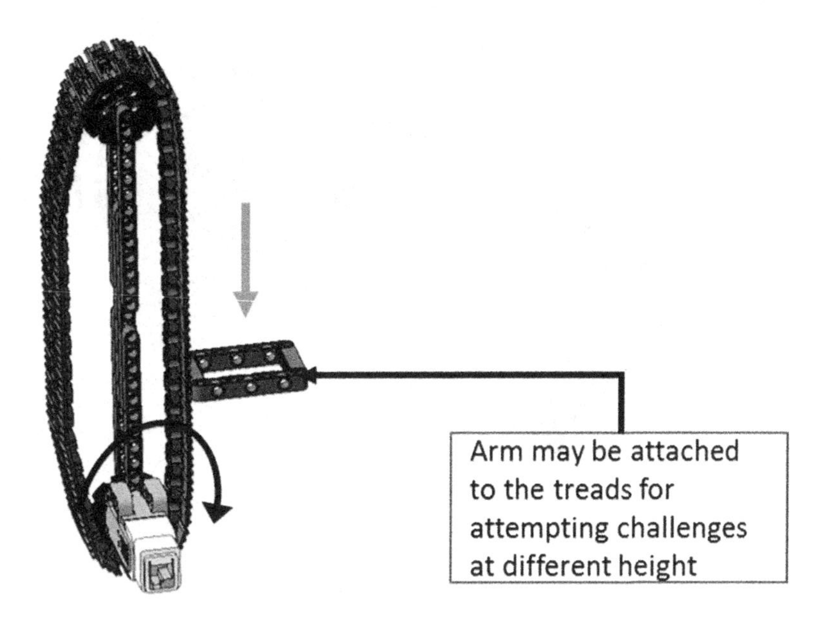

Figure 38: Tread used as Chain-and Sprocket drive for challenge needing height variation.

The Technic chain/treads can be used as chain and sprocket (Figure 38). Such an arrangement can be used to convert the motorized rotary motion into linear motion. If the missions need arms at different height then such a mechanism can be used to device a forklift mechanism.

Belt and Pulley

Similar to the chain and sprocket drive, the *belt and pulley* arrangement has a set of pulleys and a belt. As shown in the Figure 39 below, a belt wraps around the pulleys. One pulley is input pulley whereas the other pulley is output pulley. Depending upon the diameters of the input and output pulleys, it is possible to increase or decrease the speed of axle. For all the pulleys that are located along the inner side of the belt, the direction of rotation is same. The pulleys that engage along the outer side of the belt, spin in opposite direction.

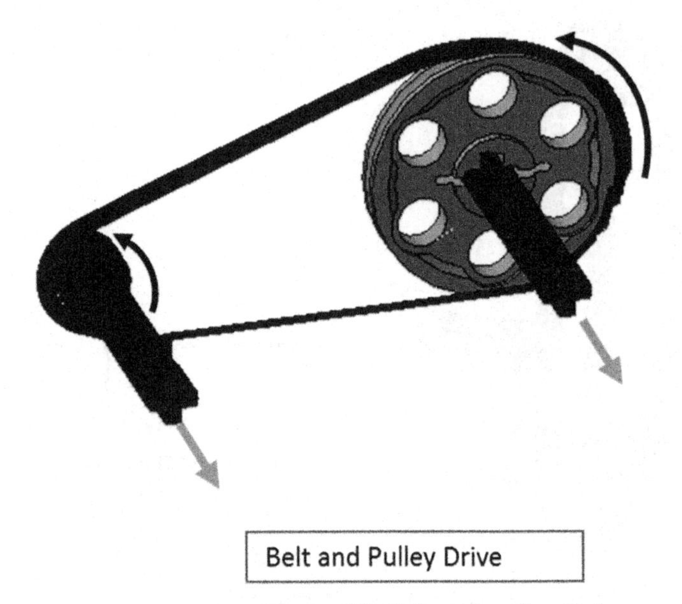

Belt and Pulley Drive

Figure 39: Belt and pulley Drive

It is important to maintain tension in the belt to prevent the pulleys from slipping. It is equally important not to put too much tension on the belts.

Some of the advantages of the pulley over gears are:

1. Distance between the input and output axis is not fixed like gears
2. It is possible to have input and output axes aligned at varied range of angles.
3. All the pulleys that contact the belt internally spin in same direction whereas he ones contacting externally spin in opposite direction.

Mechanical Linkages

Mechanical Linkages are mechanisms that are created with rigid links and joints that the rigid joints can rotate about. In the Lego world, mechanical linkages are made by connecting the beams and frames in different arrangements so that the joints are not fixed and have an ability to rotate or slide. Such an arrangement is called a *kinematic chain*. Using different kind of arrangements of the links and joints, it is possible to create a variety of unusual and useful movements and transfer force from one point to another.

Expanding on that, various kind of mechanisms can be built by connecting links of different length. Depending upon how the links are connected to each other, they move in unique ways.

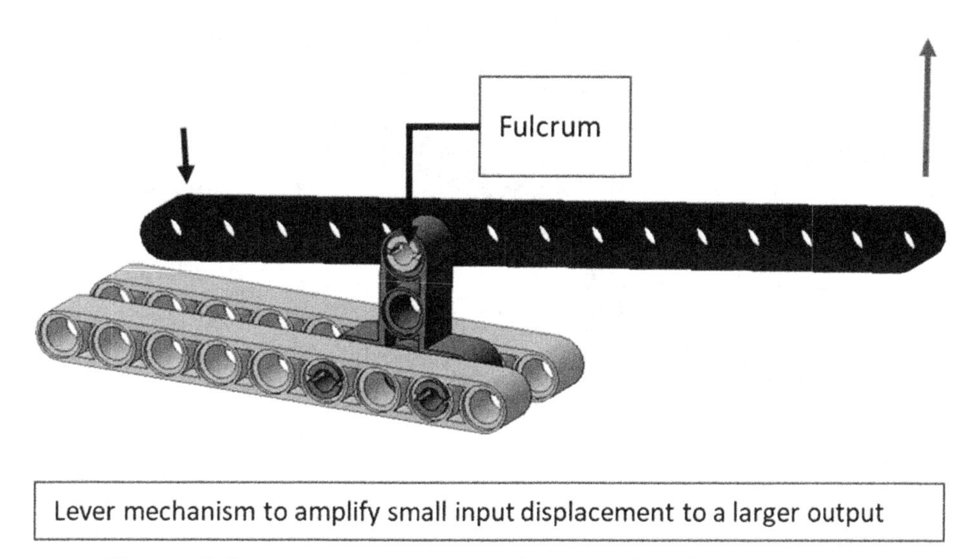

Figure 40: Lever arrangement created using technic links

The simplest mechanical linkage is a lever as shown in Figure 40. Many of you may probably have used a lever at some point. A lever is made by pivoting a straight link about a point. The point is called *fulcrum* and is fixed. Force is applied on the lever at a point away from the fulcrum and causes the lever to rotate. For a link, location of the fulcrum point with respect to the point where the force is applied, determines the force transmitted and the rotational speed of the other end of the lever. If you consider the simplest application of a lever, it is usually found in a children's see-saw in playgrounds. See-saws are usually balanced on both side to allow kids to play in harmony. If you had a see-saw where one side was much longer than the other, it would cause the kid on the longer side to raise higher than the other kid. Additionally, it would take more force for the kid on the shorter side to lift the other kid.

One of the simplest multiple linkage based mechanism is a 4-bar mechanism. Figure 41 describes a special 4-bar mechanism also referred to as *Chebyshev lambda* mechanism. The mechanism has 4 lever arrangement assembled to build a 4-bar mechanism. Of the 4 bars, one of the bars is fixed. One of the bars, also referred to as crank or the input bar is where the motor is attached. Other bars are known as output bars. By adjusting the lengths of different bars, many different motion profiles can be obtained. In a Chebyshev Lambda 4-bar mechanism, by adjusting the links to be of suitable lengths, the rotation motion of the motor is converted to special reciprocating mechanism. The output of such a mechanism has partly near-linear path.

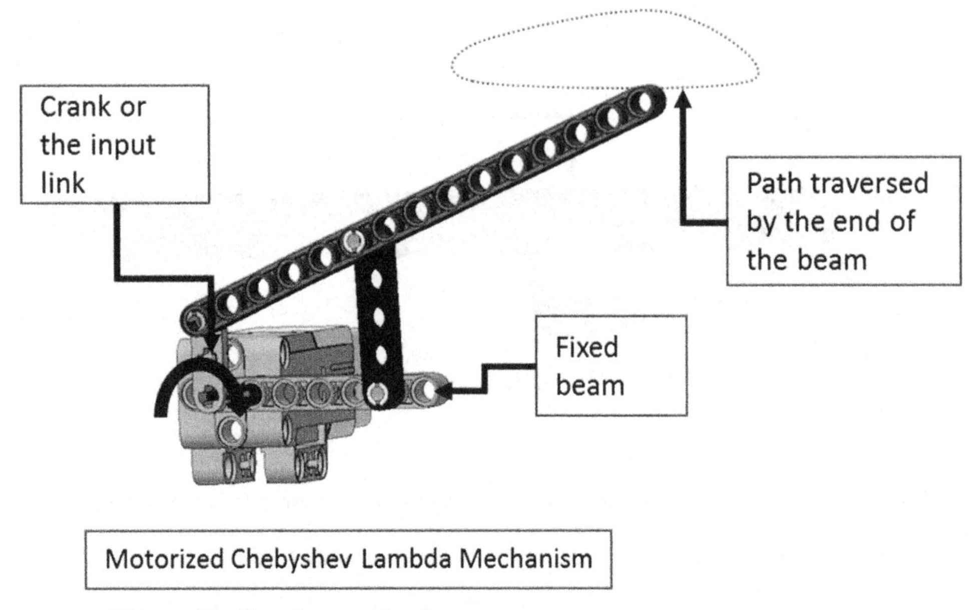

Figure 41: Four bar mechanism

Figure 42: Application of 4 bar mechanism in digger to lock the mechanism

(a) 4-Bar mechanism Parallel Bar Mechanism Unlocked-Configuration

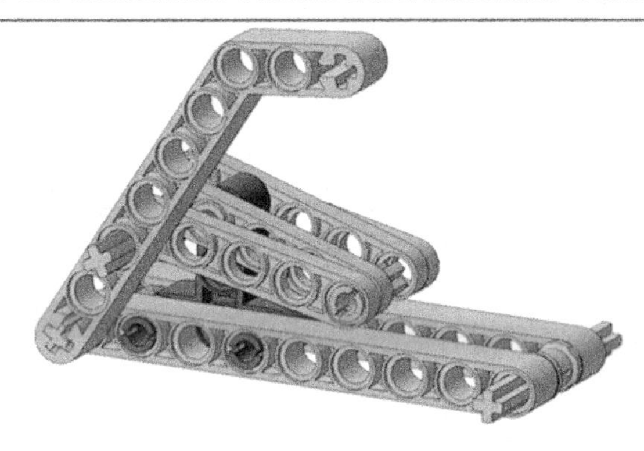

(b) 4-Bar mechanism Parallel Bar Mechanism Locked- Configuration

Figure 43: Locking 4 bar mechanism

Figure 42 and Figure 43 describe a special 4-bar mechanism where locking is enabled by establishing length ratio between the length of each link. Just like a digger when the scooper is in the configuration for collecting the load, the mechanism locks to prevent slip and accidental dropping of material.

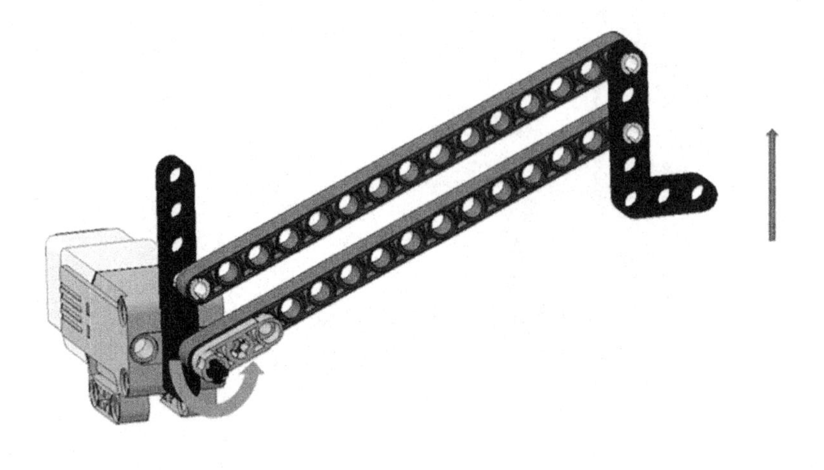

Motorized Parallel Bar Mechanism

Figure 44: a parallel bar mechanism built with technic beams

One of the popular mechanism that is built with links is called as a *parallel bar mechanism*. As shown in the Figure 44, the parallel bar mechanism has two pair of links that are connected so that the links in the pair are always parallel to each other. Such an arrangement is very useful when lifting an object in a manner where the object must stay parallel to the ground.

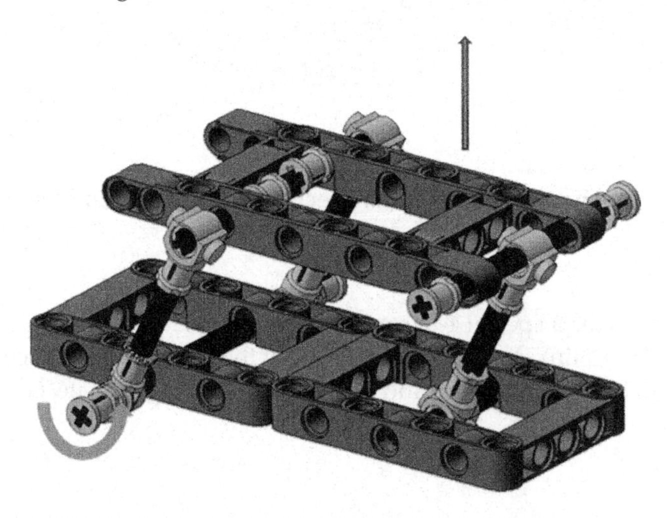

Platform that lifts parallel to the base

Figure 45: A platform based on parallel bar mechanism

As shown in the Figure 45, by arranging a set of technic connectors and axles, a platform is built so that the platform stays parallel the base. The rotation of one of the axles translates into upward linear motion of the platform.

Composite Drives

Composite drives are combination of many different kind of drives. Composite drives can be created by attaching multiple gears, pulleys, chain and sprockets along multiple axes for load transfer as well as to create multiple points of action.

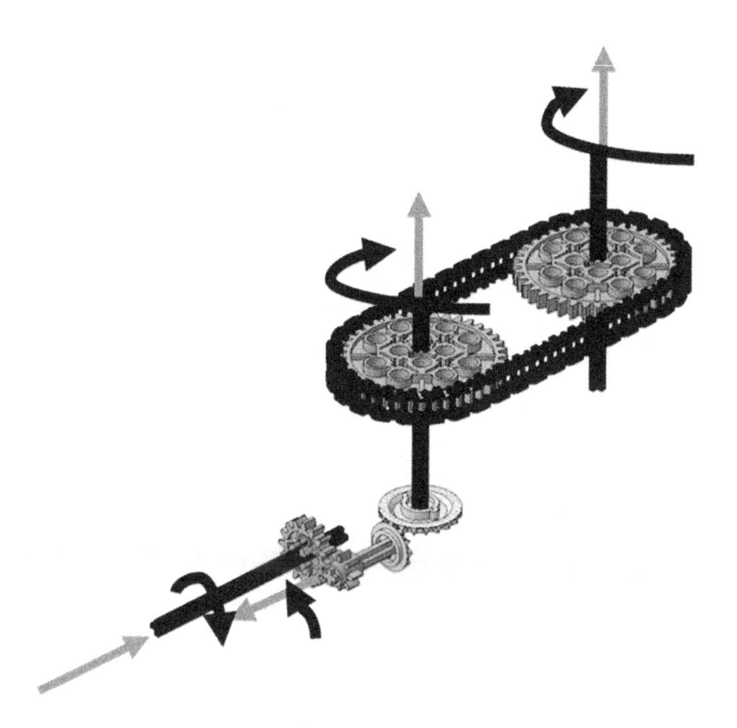

A composite Drive Comprising of spur gears, bevel gears, chain and sprocket

Figure 46: A composite drive built with Lego technic pieces

As described in the Figure 46, the composite drive is comprised of multiple mechanical drives. Spur gear is used to change the direction of rotation as well as speed. Bevel gears change the speed and the axle orientation. The chain and sprocket arrangement makes rotary motion available at two locations with the direction of rotation staying same.

With maximum 4 motor ports available on EV3. A composite drive allows the rotational motion of a motor to become available at more than one point, speed and orientation.

Summary

Any mechanical system is comprised of fundamental elements such as gears, chain and sprockets and pulleys. These elements are needed to transfer power, change the point or direction of actuation. By using gears/pulleys/sprockets of different diameters we can increase or decrease speed and change the torque to suit our purposes.

Pneumatic Systems

Pneumatic systems are based on the utilization of pressurized air for motion. You have likely seen a pneumatic system being used in nail guns where air pressure is used to drive a nail into wood. Pneumatic systems are not very popular in the FLL. However, owing to certain unique features, many teams have used them successfully. Pneumatic systems are very simple, consistent and can apply large force/torque and can do so at a very high speed.

The pneumatic system includes pumps, tubes, cylinders, valves, air tank and a manometer. Here is how the sequence of operations needed to make a pneumatic system work:

- Pressurized air is stored in a tank.
- Valve is used to control the pressurized air flow.
- Pressurized air is fed into the pneumatic cylinder to actuate the cylinder.
- The cylinder provides a fixed displacement.
- Finally, various mechanisms can be attached to amplify the linear stroke or convert the linear stroke into a variety of other type of motions. This is the key reason that pneumatic systems are popular in industrial applications.

Contrary to the motor-powered arrangement, where rotary motion is converted to linear motion with the help of specialized mechanism or Rack and Pinion; the pneumatic cylinder provides linear motion from the get go. In the following sections, we will describe the various parts of a pneumatic system in greater detail.

Pneumatic pump

Pumps (Figure 47) are the primary source of air in a pneumatic circuit. There are two kinds of pumps. One of the pumps is spring loaded and is actuated manually. Other smaller pump looks like a cylinder and is designed to be used with a motor.

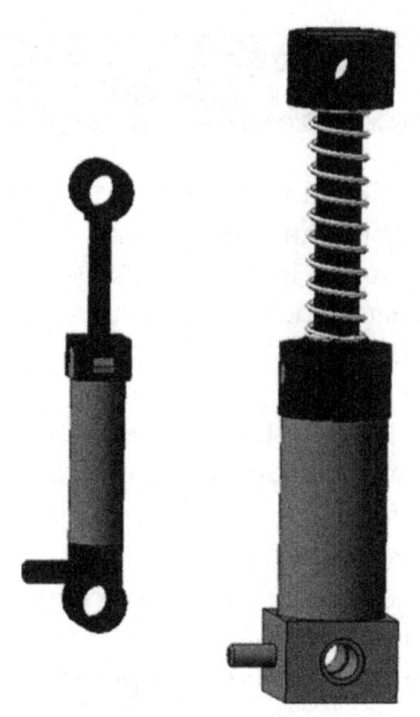

Figure 47: Lego Technic pneumatic pumps

For the FLL missions, it is recommended that the spring-loaded pump be used to fill pressurized air in the cylinder manually to save time and to ensure that the mechanism is fully loaded from the get-go. Later, a motor may be used to actuate the switch valve at the point of application.

Air Tank

Air tank (Figure 48) is the reservoir for storing pressurized air. The tank has two ports, one at either end. One port is connected to the pump whereas the other end is connected to the pneumatic circuit that is powered by the pressurized air. Both the ports are identical and hence either can be used as input or output. The tank also has features that can be used for mounting the tank onto the Robot.

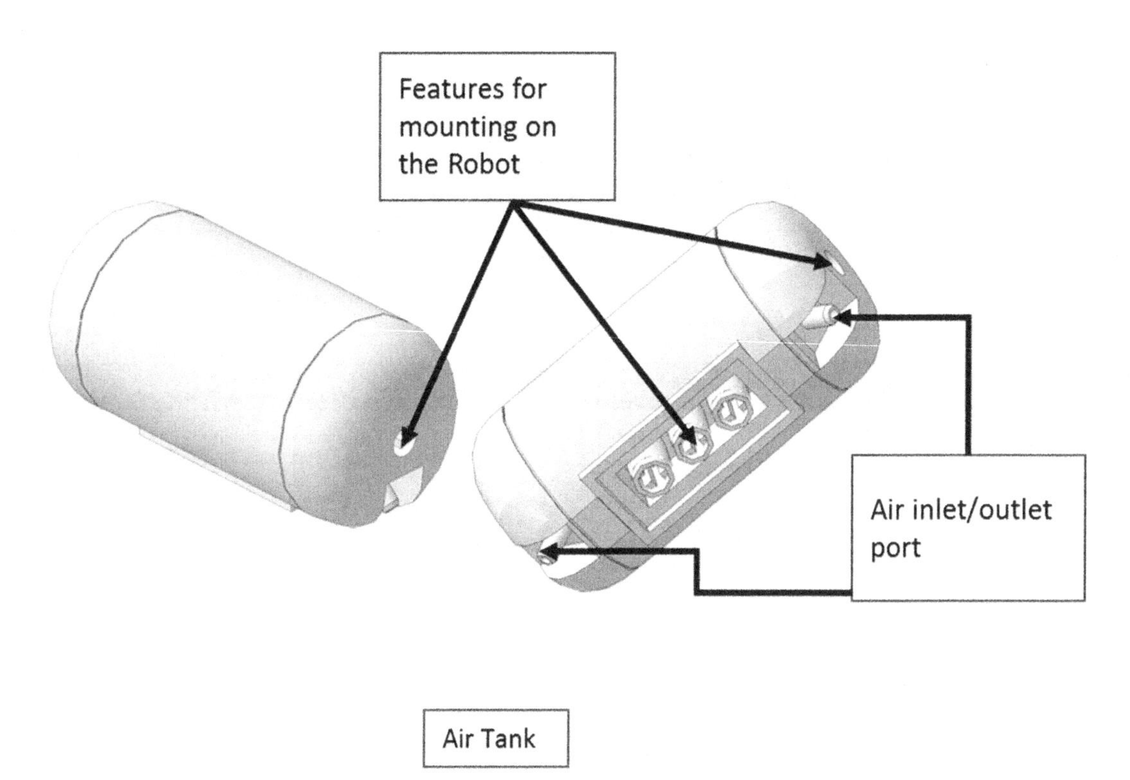

Figure 48: Lego pneumatic air tank

Flow distribution accessories

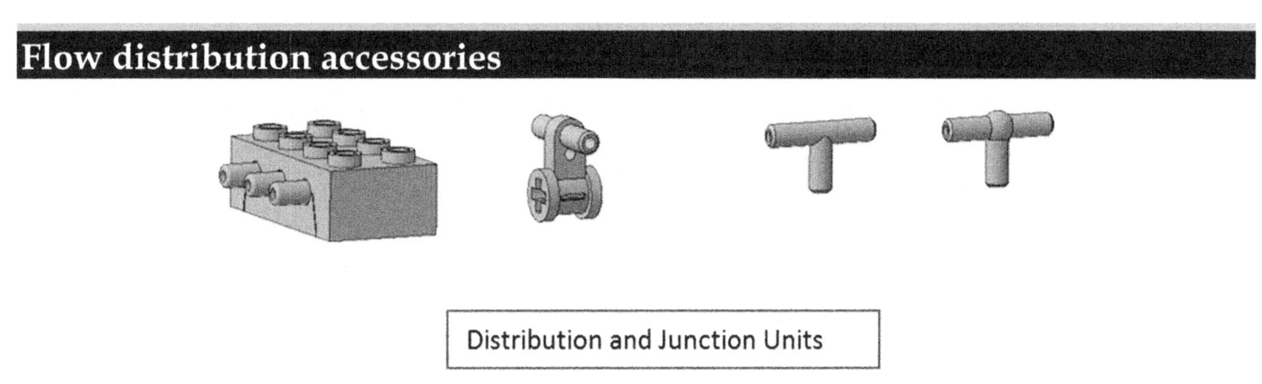

Figure 49: Lego pneumatic distribution unit

Pneumatic T-Junction and black/blue tubing are used to interconnect different pneumatic components. The T-junction allows to split the flow of air into two separate streams. Certain other connectors such as hose connector allow straight connection between two different components without splitting the flow.

You may be able to alter the length of the pneumatic hoses by cutting them but make sure to check the rules for the Robot Games. FLL rules usually prohibit the customization and mechanical alteration of the individual technic parts; however, we have seen Lego strings and tubing excepted from this rule (see 2015 FLL Trash Trek Challenge - Robot Games Rule R01.)

Manometer

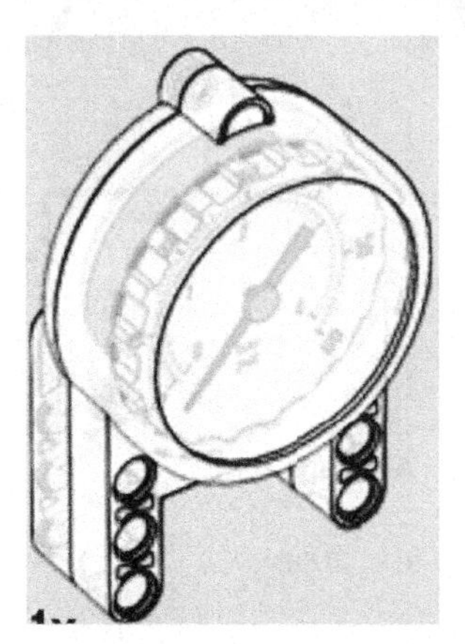

Figure 50: Lego pressure gauge (Manometer)

Air pressure gauge (manometer, Figure 50) is used to show the amount of air pressure in the pneumatic system. Manometer is just a monitoring device and is not needed for the pneumatic circuit to work. However, it is recommended to include manometer in the pneumatic circuit to monitor the air pressure and debug any problem in the circuit.

We do recommend using a manometer if you are using a pneumatic system in your FLL robot. Inflating the air tank to a high air pressure might cause the hoses to unseat themselves from the valves.

Pneumatic cylinders

Pneumatic Cylinders (Figure 51.) comprise of a piston that moves inside a cylindrical tube caused by pressurized air movement. Depending upon which port of the piston we pump the pressurized air in (Figure 52), the piston moves to extend or retract. The piston has a rod with connecting feature built into it to connect the piston to other Lego Technic parts. This is similar to the connection points with Lego motors where, direction of rotation can be controlled programmatically.

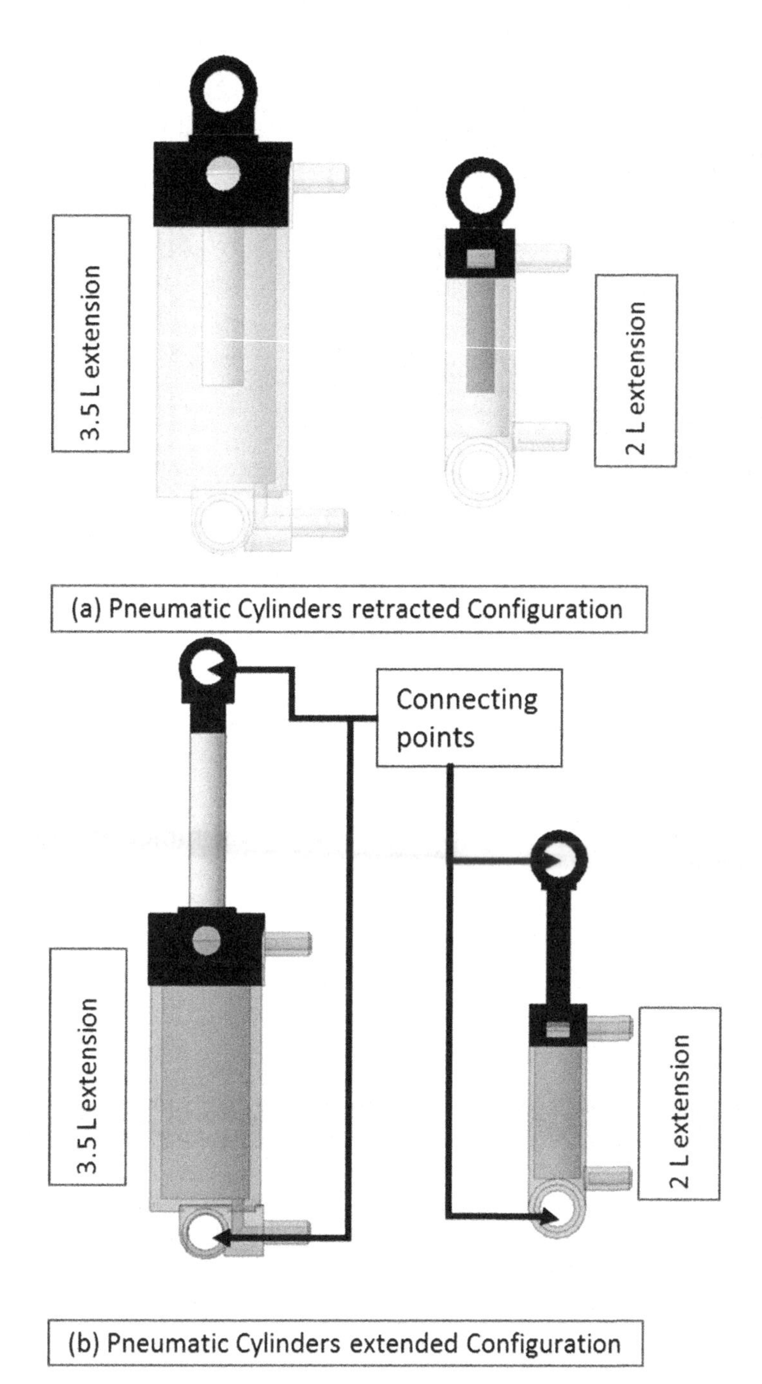

Figure 51: Lego Pneumatic cylinders

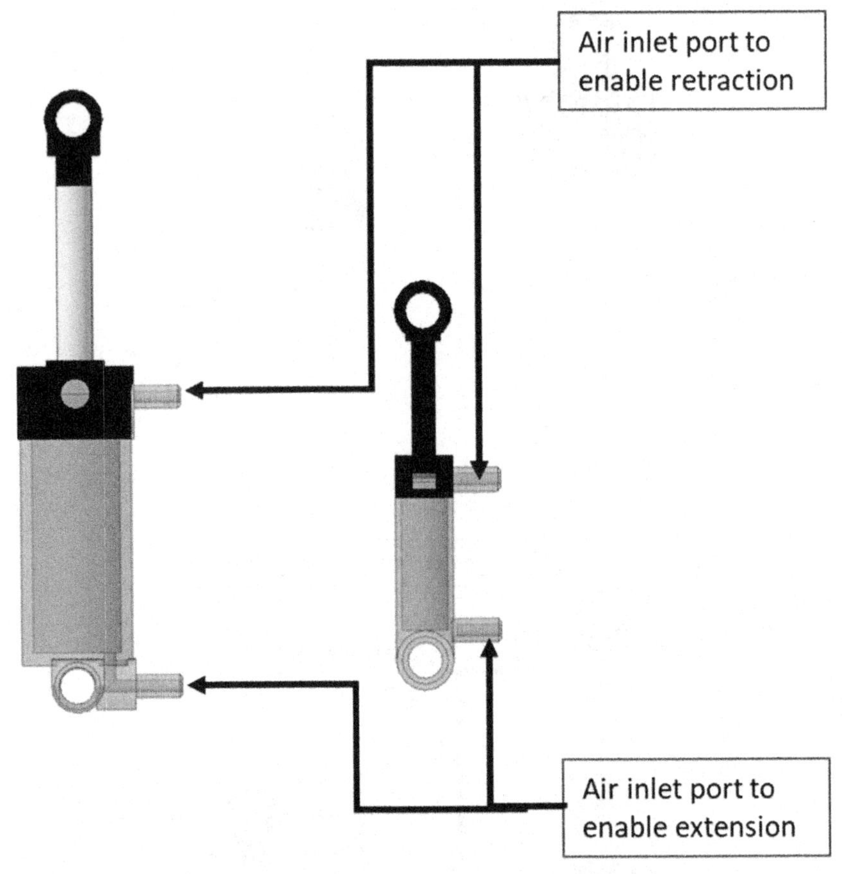

Figure 52: Ports for air inlet to enable Extension vs Retraction

Pneumatic Valve (switch)

Pneumatic valve (Figure 53) is the device used to distribute and control the direction of flow of the pressurized air. The Pneumatic valve has three ports - one input port and two output ports. The lever on the valve determines if the flow is closed, or if open, pressurized air exits from one output port or the other.

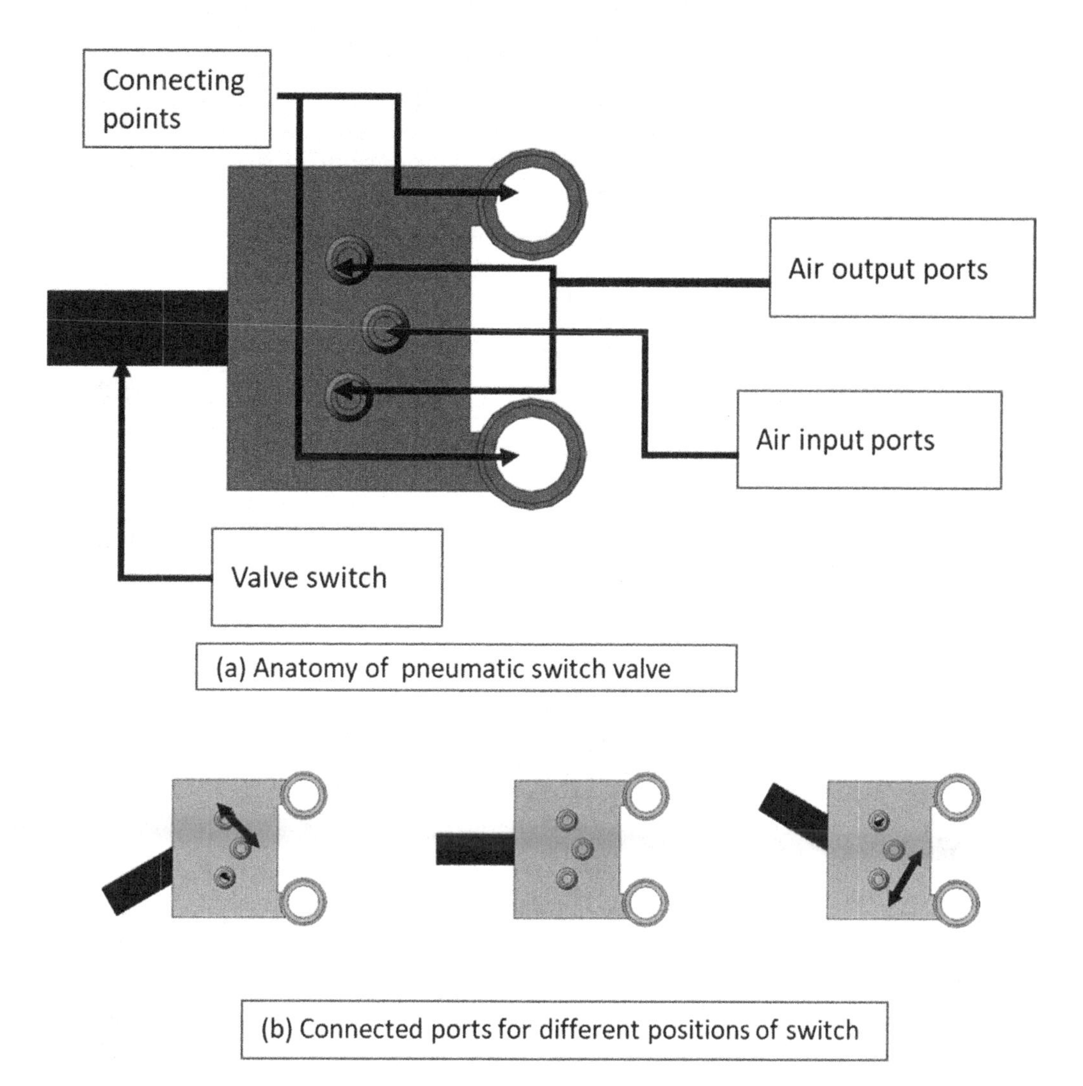

Figure 53: Pneumatic valve switch

Pneumatic mechanisms

In this section, we will showcase a few examples where you can use pneumatic systems. Pneumatic systems have great use in industry and although they are not used extensively in FLL, having a pneumatic system can be of great use in your robot depending on the problem you are trying to solve. Additionally, having a pneumatic system on your robot does make your robot unique and may score you points in the *robot design* judging portion of FLL.

Robot Design portion of FLL judging looks at a variety of factors about your robot and programming such as structural integrity, use of sensors, innovation, uniqueness as well as well documented code and programming prowess. Having a unique feature such as pneumatic system might help your team stand out from the crowd.

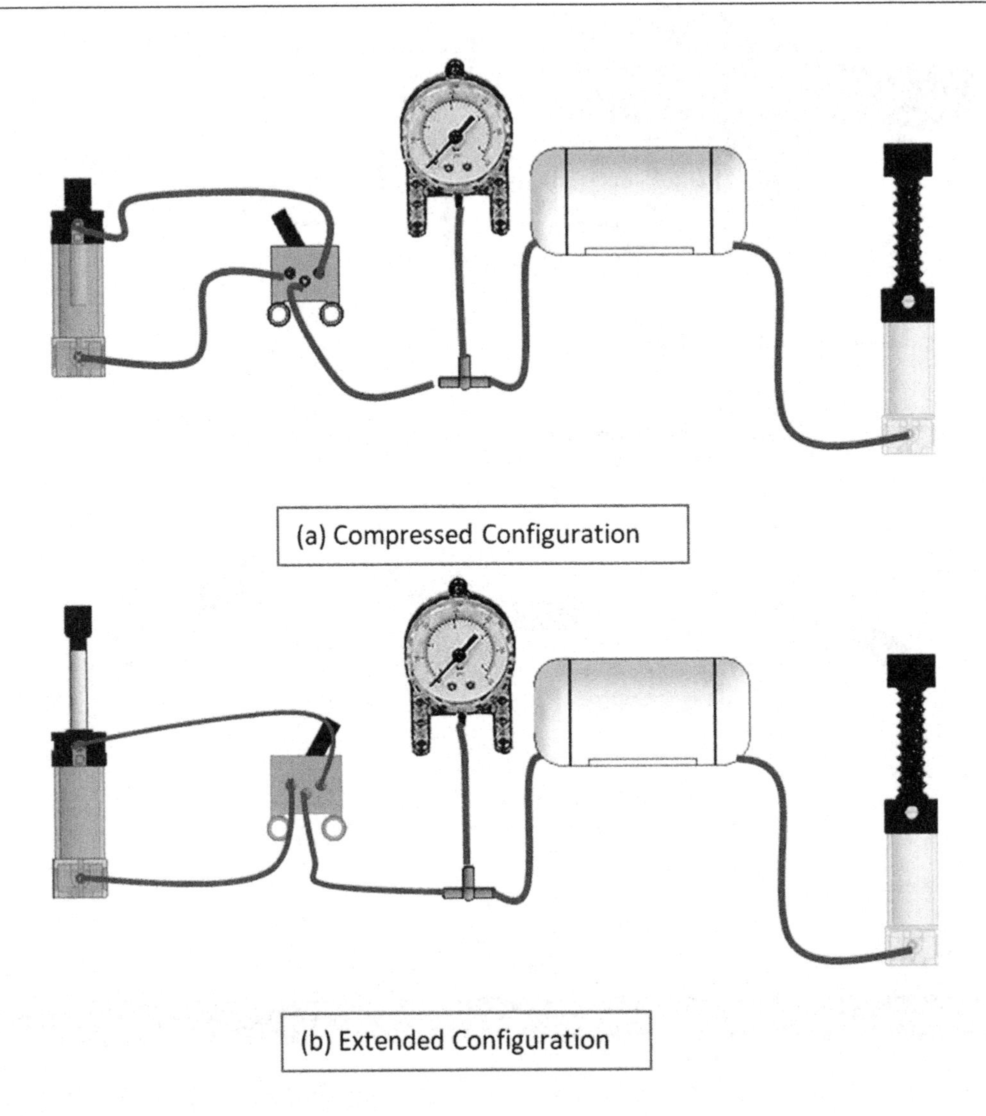

Figure 54: A basic pneumatic circuit for two-way cylinder

Figure 54 describes a basic pneumatic circuit used for actuating a pneumatic cylinder. In this circuit, the T-junction is used to connect the manometer The output from the storage tank connects to the input port of the valve and finally two outputs ports are connected to the two sides of the cylinder. When the valve lever is in middle, no air flow takes place; however, when the lever is pushed to right or left, the corresponding side of the cylinder is filled and cylinder either pushes the piston out or pulls it in.

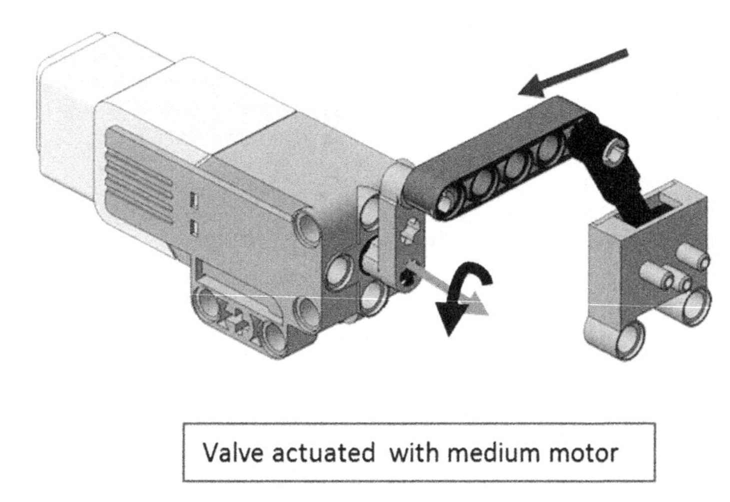

Figure 55: Motor actuated pneumatic valve

Contrary to motors that can be controlled and programmed, we cannot programmatically control the pneumatic system directly and must rely on other mechanisms to actuate the pneumatic systems. A special mechanism, as shown in Figure 55, can be built to connect the motors to the valve in order to control it with the help of EV3.

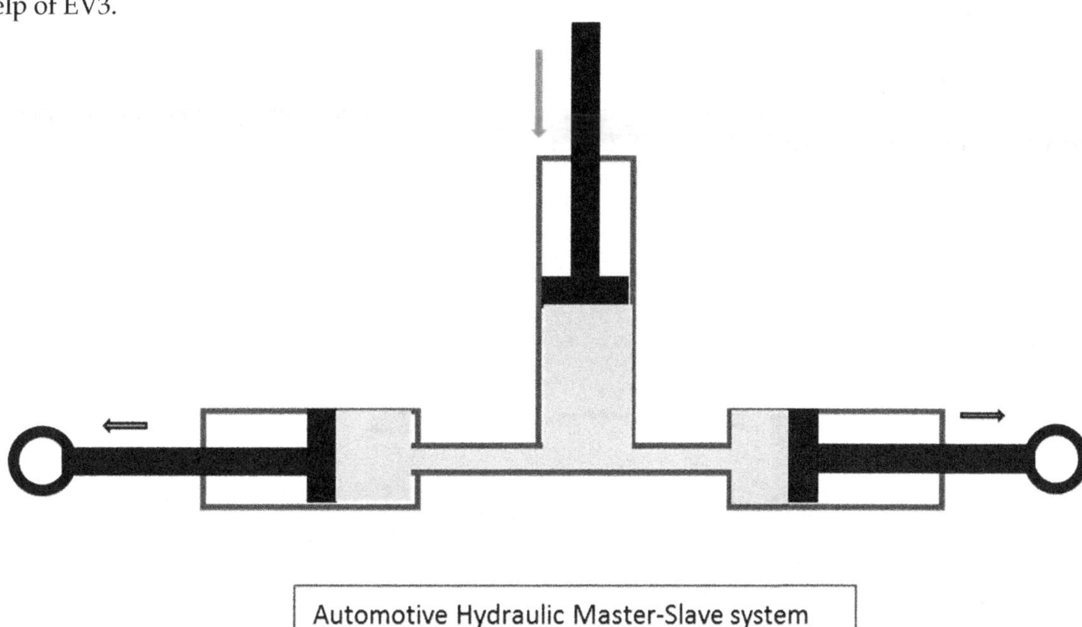

Figure 56: Automotive Hydraulic master-slave cylinder configuration

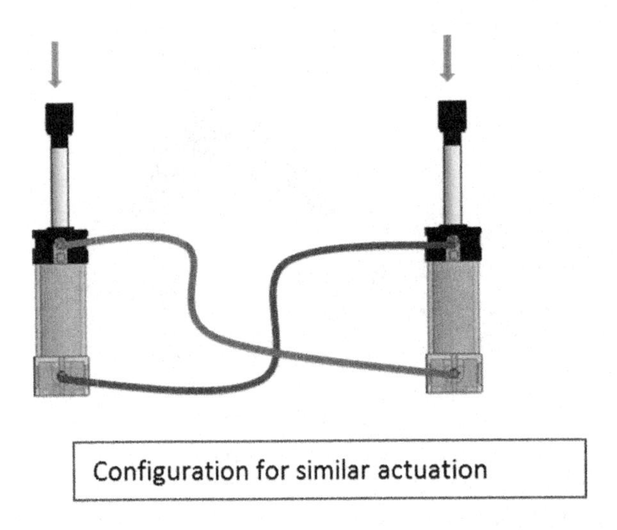

Figure 57: Identical 2-way cylinder configured in master-slave arrangement for similar actuation

Many people, who have used pneumatic systems are still not aware of a special configuration in which the pneumatic components can be operated. This configuration is known as the *master-slave* configuration. Similar to automotive systems i.e. cars where, a foot actuated master cylinder (brake pedal) transmits the pressure through hydraulic circuit to slave cylinders (the actual brakes on the wheels), the pneumatic cylinders can be connected to each other in different arrangements to directly transmit the force without employing storage tank or pump as shown in Figure 56.

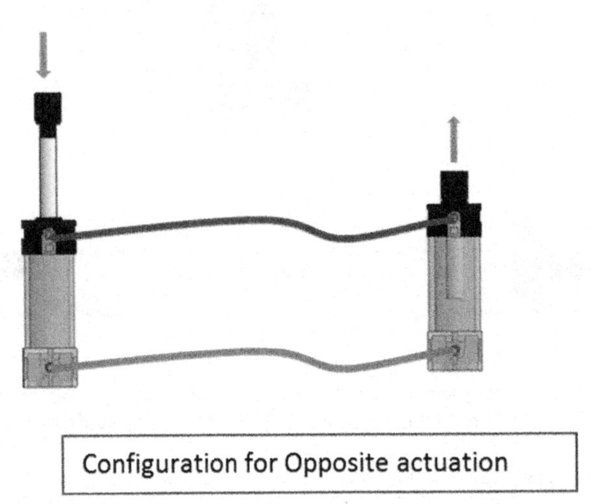

Figure 58: Identical 2-way cylinder configured in master-slave arrangement for opposite actuation

Compared to the mechanical drives, where the plane and distance between links, gears etc. is fixed, the cylinder may be arranged in any plane and are not limited by distance. In fact, one cylinder can be used to drive many other cylinder and the direction of actuation can be reversed. As illustrated in Figure 57 and Figure 58, when the corresponding sides of the cylinder are connected by a hose then, retraction of first cylinder results in expansion of the second cylinder and vice-versa. Similarly, when the air inlet port connections are inverted then the behavior of first cylinder is replicated by the second cylinder.

Pneumatically actuated Lever mechanism to amplify input displacement to a larger output

Figure 59: Pneumatically actuated cantilever arm

Figure 59 shows the examples of a lever mechanism driven by *pneumatics*. The lever is actuated by a pneumatic cylinder and small displacement of the cylinder is amplified to a larger displacement at the end of the arm. The total load supporting capability of the pneumatically controlled arm is significantly higher than that of the Lego large or medium motor.

The pneumatically actuated lever arm may be reconfigured as a parallel bar mechanism if the orientation of platform needs to sustain its original orientation (Figure 60).

Pneumatically actuated Parallel Bar mechanism

Figure 60: Cantilevered parallel bar actuated using pneumatic cylinder

Summary

Pneumatic systems operate using pressured air to actuate mechanisms. Pneumatic systems can apply amazing amount of force and can replicate or invert a motion. Although not very common in FLL, we have seen some teams use pneumatic systems to great results.

Getting familiar with the programming environment

Welcome to the programming section of the book. Even though in our experience, most participants in FLL love to build, when they get to the programming section of the robot, they get even more excited. Before we start on our journey in FLL, we need to setup the context around who is the intended audience. In this book, it is not our purpose to fully explain every detail and nuance in the EV3 programming environment because it would make the book extremely big and unwieldy. So, this book does not contain a complete step by step primer on how to program the EV3. However, we will provide an introduction as well usage of the most useful blocks and explain the situations where caution is warranted. We will also quickly touch upon blocks that are sometimes useful as we come across them in due course of programming. You should supplement the knowledge here with the Mindstorms EV3 help that is included with the EV3 programming environment. It is accessible by selecting Help->Show EV3 help and it provides a comprehensive list and functions of the programming elements and should be treated as the key reference. Duplicating it here would be unnecessary. This book contains details on programming as it relates to solving robotics problems and contains exercises to further cement the learning in the reader's mind.

Now, let's talk about the EV3 programming environment and get ourselves familiarized with it. It is worth noting that the Lego Mindstorms EV3 editor is a visual programming language editor. Figure 61 shows the EV3 editor with a sample program. EV3 programs are organized in Projects which are standalone entities like folders or binders and may contain multiple standalone programs. A *program* is a set of instructions that will execute on your robot and it will make it perform any actions of your choice. At one point of time, your robot can only be running one program. One key point to keep in mind when naming your programs is that you should avoid putting spaces or special characters such as $, <, >, (,) in the project and program names. These may fail unexpectedly and hence are not recommended.

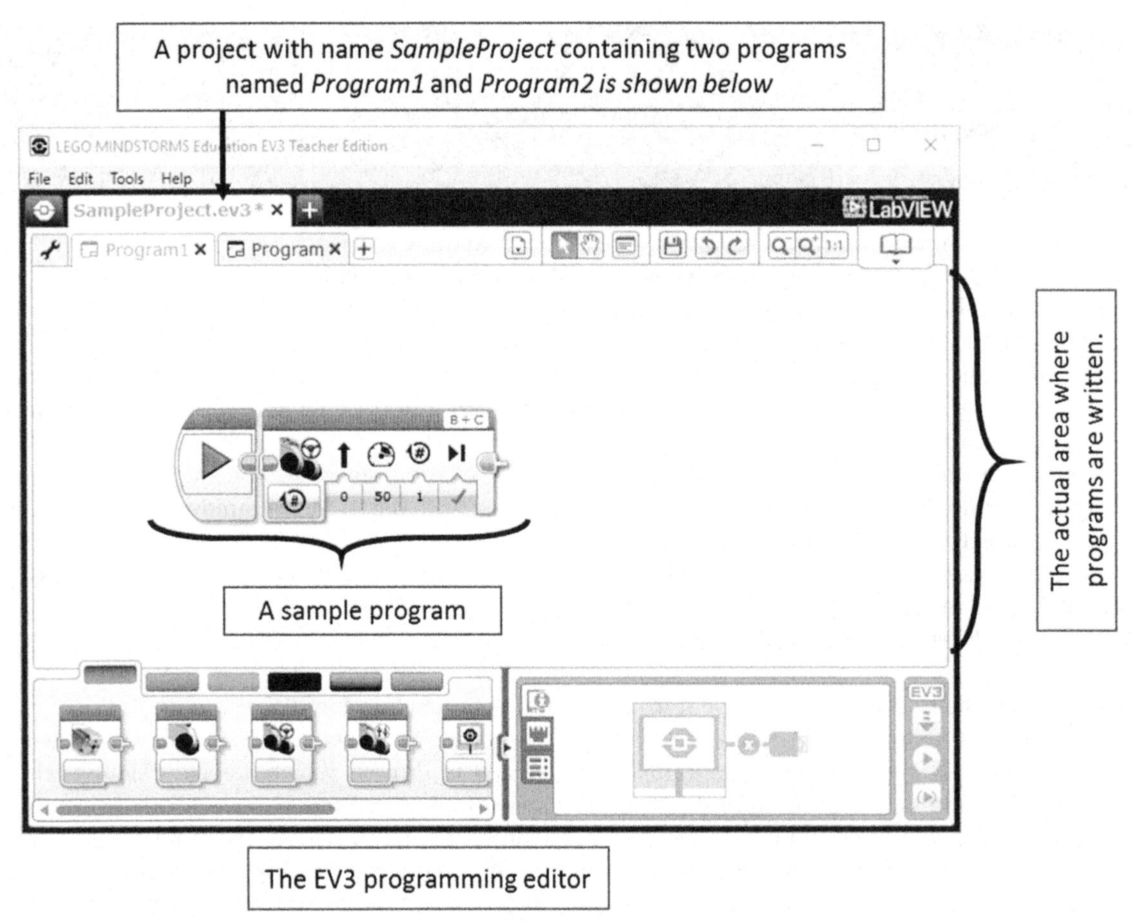

Figure 61: Getting familiar with the EV3 programming editor

Introduction to the programming

It is key to note that within the programming sections of this book, we use the Mindstorms and EV3 words interchangeably since in the Lego robotics worlds they are quite often synonyms. When we really want to refer the programmable EV3 brick, we simply refer to it as the brick or the EV3 brick.

To get started with programming, you create robot logic using visual blocks that you drag and drop into the programming window (Figure 61.) The first element in any program is the Start block (Figure 62.) Blocks that are connected to the start block, either directly or via other blocks in a chain are part of the program. Any stray blocks that are not connected to the start block in any way are not part of the program. This is illustrated in Figure 62.

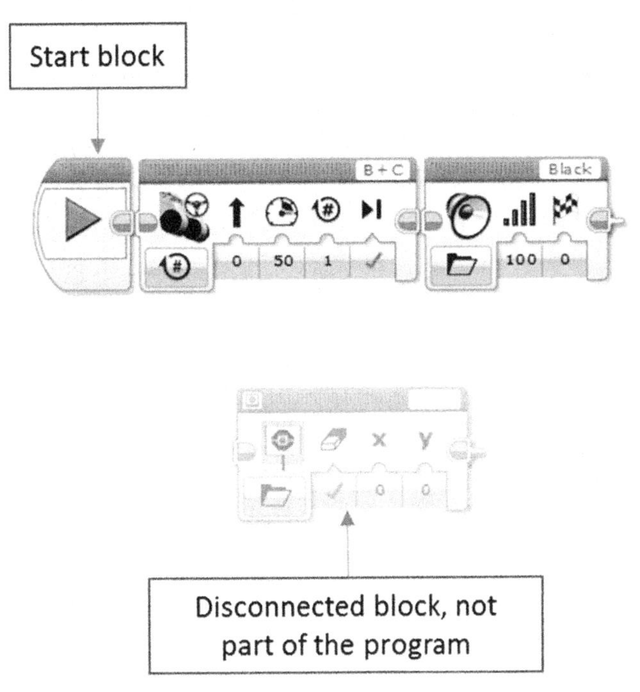

Figure 62: A sample program

A program may contain multiple Start blocks as shown in Figure 63. Unless you really know what you are doing, this is not at all recommended because of the idiosyncrasies of the editor. If a program has multiple start points and you run the program by clicking on a start block, ONLY the set of blocks connected to THAT start block will run! By contrast, if the program has been transferred to the EV3 brick and the program is run from there, the blocks connected to both the start blocks will run simultaneously. The order of operations between the blocks connected to the two start blocks is non-deterministic and will cause you all sort of anguish. It is rare to see programs with multiple start points being needed or used by any FLL teams and we discommend them.

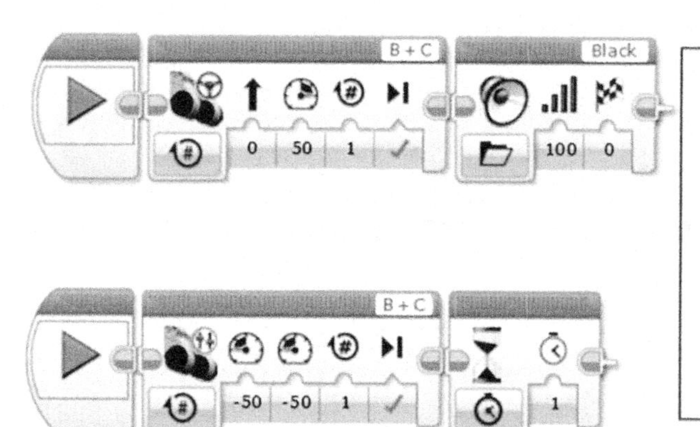

Program with multiple start blocks. The first block in each program is doing the reverse of what the other one is doing causing unpredictable results. In general, it is a bad idea to have multiple start blocks in a program.

Figure 63: A program with multiple start blocks. This is a bad idea as the results are usually unpredictable.

For folks working familiar with Software Development, the two start points create two threads and run the two parts in individual threads. Depending on the order of the execution of blocks, the order of operation in the two lines of execution may be unpredictable.

Now that you understand the basic structure of the EV3 programming editor (Figure 61), let's look at some of the essential features of the editor. Many of these features are hidden in plain sight and even experienced teams do not know all of them.

1. As you continue to add new programs in a project, eventually the space to show new programs disappears and older programs do not show in the tabs. The marker as shown in Figure 64 will give you quick access to these programs. Remember that even when you click the 'x' icon on a program to close the program, it is not deleted. Instead, it is simply closed and is still accessible using the program selector icon. To delete a program, you would have to get to the list of all programs via the wrench icon.

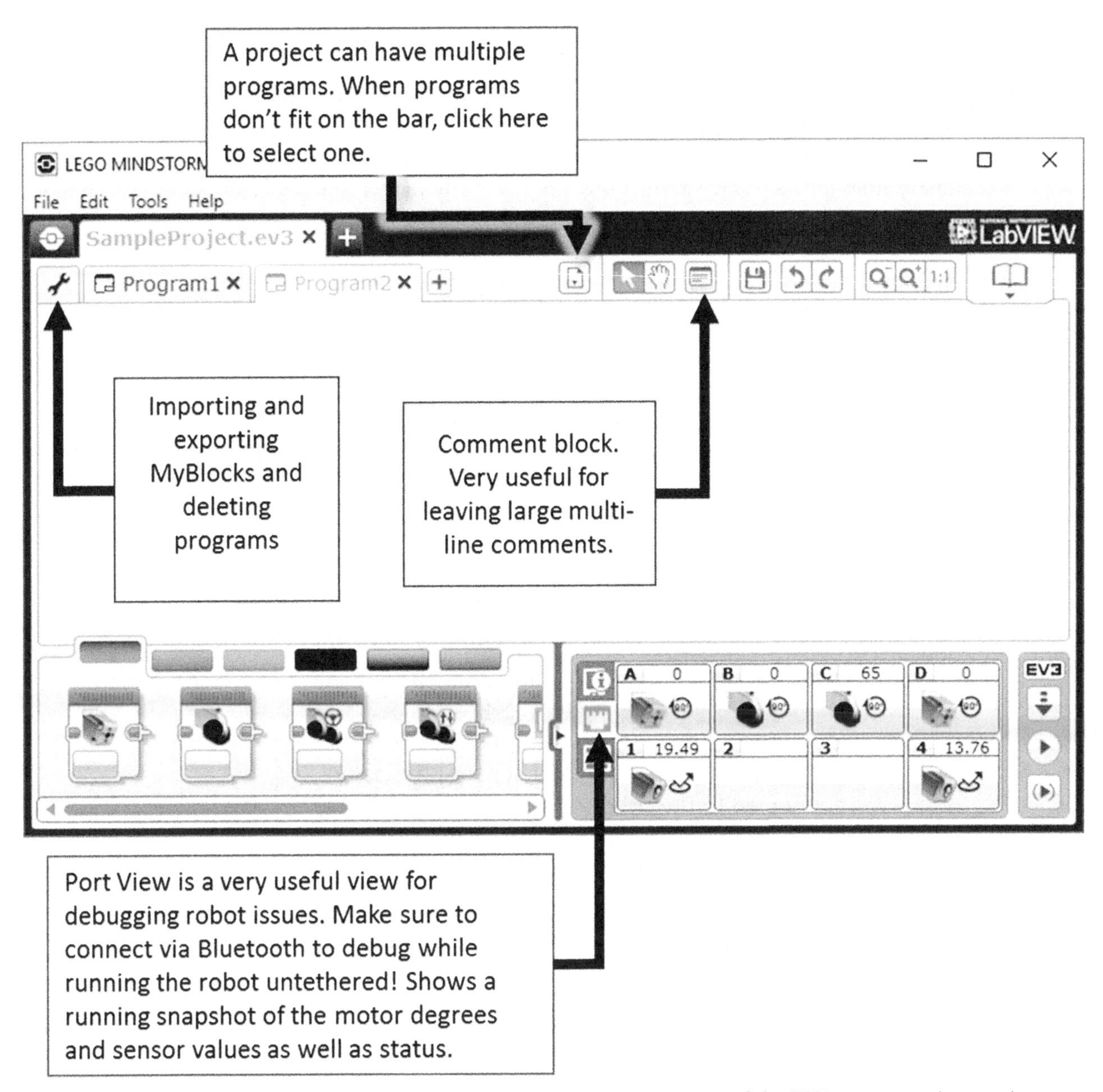

Figure 64: Some of the most important components of the EV3 programming environment

2. There is a comment block in the EV3 - we go over this in a later chapter, but we find it relatively less useful compared to the comment icon illustrated in Figure 64 which for some reason does not belong with the blocks, but is instead placed in the top bar. It can accept multi-line comments and is very useful for leaving comments and notes about the program in a descriptive fashion.
3. When you click the wrench icon to the left of all the programs, it leads to the settings for that project, including the ability to delete programs and import and export myblocks (see point 6 below) from that program. When you are done with the settings that you accessed via the wrench icon and want to get back to your program, simply click on any of the program icons next to the wrench icon and you will get right back. The wrench icon will sometimes disappear when you have a lot of open

programs. In this case, you will have to keep closing programs until the wrench icon becomes visible again.

4. The Port view in the right bottom corner is extremely useful for debugging the programs. It is identified by the fork like icon. To optimally use it, connect your robot to your computer via bluetooth which allows you to download and run programs in a wireless fashion on your robot. Additionally, this allows you to see the programs being run in real-time and you can see which block the robot is currently executing. It also shows you the current state of the sensors. When your robot misbehaves, or a sensor does not work as expected, being able to keep an eye on the current state is invaluable.
5. The EV3 programming environment has a feature called AutoID built into it. If you place a programming block into the program that is using a motor or sensor and in your program, you incorrectly specify the port that the sensor is on, you will see a warning icon in the program as shown in Figure 65.
6. MyBlocks, which are essentially groupings of EV3 blocks that you create to make your programs modular, are visible only within one project and do not carry over from one project to another unless you manually export them from one project and import them in another. MyBlocks are explained in detail in the *MyBlocks* chapter.

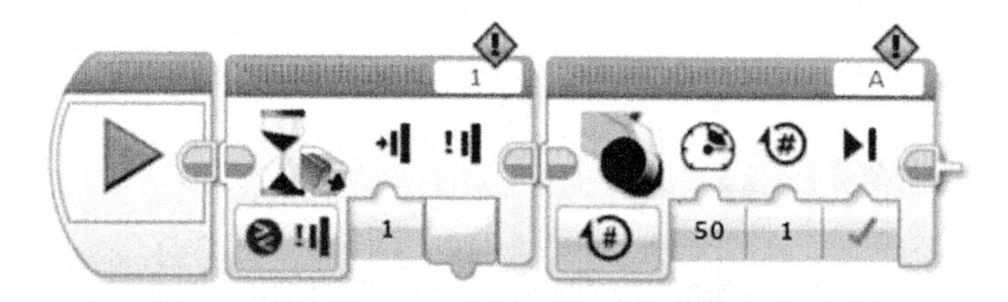

In the robot, we have a color sensor physically connected to the robot on port 1 and on port A, we have a medium motor. In the program here, we have specified the touch sensor being available on port 1 and the large motor on port A. The EV3 editor identifies the errors and shows a warning sign on the blocks.

Figure 65: EV3 AutoID identifies when the port configuration does not match what you are using in your program and shows a warning icon.

Now that you know your way around the EV3 code editor, we are listing all the programming blocks and explaining them in a short succinct manner. In some cases where something merits special explaining e.g. a block of importance in a specific scenario, we will explain in a bit more detail. Don't worry if you don't get the nuances of the block as explained in this chapter. This is an overview chapter and we will get to the blocks in details in later chapters. If a block catches your attention and you want to learn more about it, feel free to jump to the appropriate chapter in the book and then come back here to continue in a structured way. Additionally, you can also use the EV3 help to get more details on a block. The EV3 help that is included with the EV3 programming environment is accessible by selecting Help->Show EV3 help and provides a comprehensive list and functions of the programming elements and should be treated as the key reference. Duplicating it here would be unnecessary.

Green/Action Blocks

To get started, let's look at the green blocks which are called Action blocks in the EV3 editor.

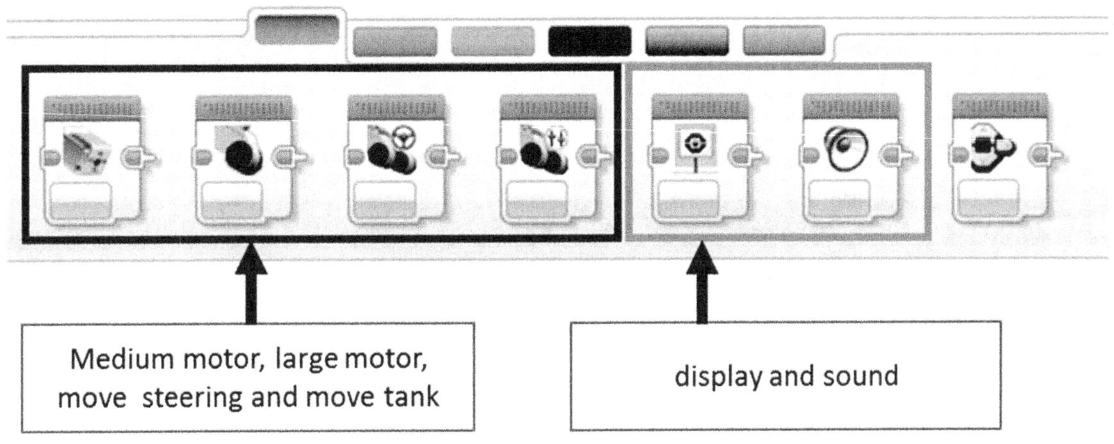

Figure 66: Green/Action Blocks

The green blocks (Figure 66) are some of the most important and fundamental blocks in the entire EV3 programming environment because they provide the means to run the motors that move the robot as well as any mechanisms that you will build on the robot. You can control motors in a pair or you can control a motor individually. The blocks highlighted in Red are the ones that we use very heavily. The ones in orange grouping are used infrequently. Here is what they do and how to use them effectively:

- *Move Steering*: The move steering block allows you to move the two large motors in a synchronized fashion which is primarily useful for robot movement. When you use the move steering block or the move tank block to turn the two large motors for robot movement, the EV3 software internally uses a software algorithm known as a PID algorithm to keep the motors running in lock-step or synchronized fashion. The PID algorithm works by supplying power to both the motors and then observing the tachometers on the two motors. If it observes a significant difference between the number of degrees turned by the two motors, it will stop the motor that is ahead and give time for the other motor to catch up. It is easy to verify this by using this block and running two large motors via the move steering and move tank blocks. If you physically stall one of the motors by holding it in your hand and not letting it move, you will realize that the second motor will stall within a short period of time.
- *Move Medium motor and Move Large motor:* They are blocks that can control only one motor at a time. Single motor control is useful for robot attachment control. Both blocks are functionally equivalent except that they control the different sized motors. One thing to keep in mind about these blocks is that they are *blocking* i.e. if you asked the motor to turn a specific amount and the motor gets stuck for any reason e.g. an attachment gets stuck, the EV3 software will keep trying to move the motor and the entire program will stall.
- *Move Tank:* Same functionality as Move Steering except that you can control the motors individually at a time. Many teams decide to use the Move Tank block because they like to be able to control the motors individually and others decide to use the Move Steering block because they feel it is simpler to use. It is a matter of personal preference on which block you decide to use to move your robot.
- *Display block:* The display block allows you to display text or images of your choice on the EV3 screen. Its primary use is in debugging programs for robot runs.

Even though the move blocks allow you to specify the amount of turn in the motors as number of turns or degrees, we strongly recommend that you ONLY ever use degrees. The reason for that is the tachometer/encoder embedded in the motors measures the turn amount only in degrees and has 1-degree resolution.

Orange/Flow Blocks

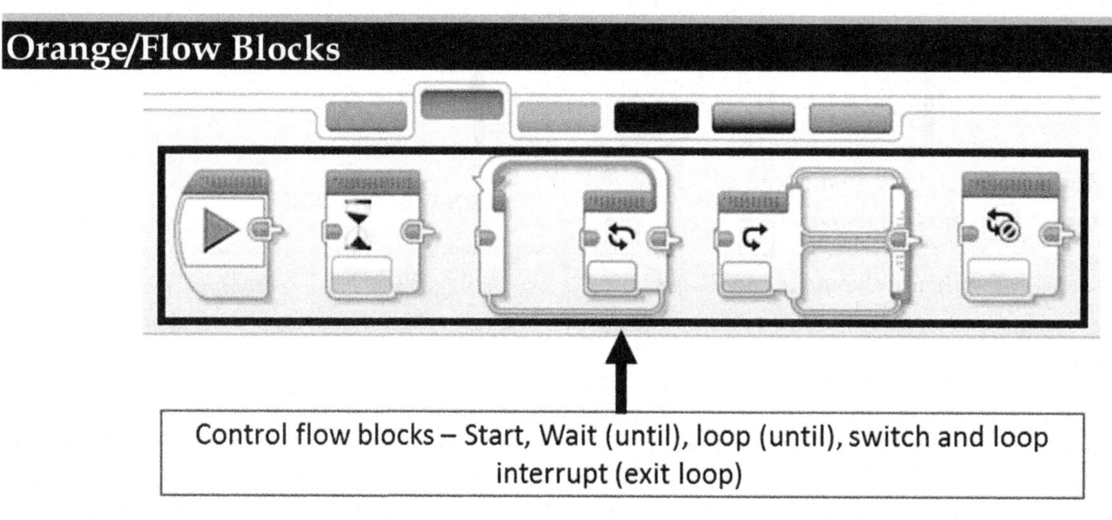

Figure 67: Orange Blocks

This section briefly goes over the blocks that allow you to change the way programs run on your robot. Usually, in a program, we keep taking deterministic actions one after another e.g. move straight, turn left, move forward, turn right and stop; however, there are times we need to program the robot with richer logic such as *wait until the traffic signal has turned green* or *if while moving you see a wall, turn left* or even *keep on pumping water until the bucket is full*. These paradigms are known as a *Wait, conditional execution* and *loop* in programming terminology.

- *Start block:* This block by itself does not perform any action, except that when you click it, blocks connected to it will be run on the robot. Additionally, any set of blocks can be part of a program only when they are connected to a *Start* block.
- *Wait block:* We like to call this block the *Wait Until* block because when running, when your program reaches this block, it will stop and wait *until* the condition specified by this block is fulfilled. Conditions that this block can wait are wait until time elapsed, wait until touch sensor detects touch, wait until color sensor detects color etc.
- *Loop block:* This is a block that repeats any other blocks placed inside it either (a) infinite number of times (b) the number of times specified or (c) until a logical condition or a sensor value meets a criterion.
- *Loop Interrupt Block:* The loop interrupt block is usually placed within a loop and when the program reaches it, the loop interrupt block makes the loop stop executing and the program instead jumps to the block immediately after the loop.

- *Switch block:* The switch block is a block that allows your code to branch into separate directions based on a variety of conditions or sensor states.

In our experience, Switch and Loop blocks, in context of FLL are primarily used within myblocks. FLL programming, excluding myblocks is surprisingly linear i.e. blocks follow blocks and it is rare to see any loops or switches in key programs excluding the Master Program.

Yellow/Sensor Blocks

The yellow tab in the programming palette includes all the stand-alone sensor blocks. These blocks allow you to get the sensor readings from the sensor without being constrained by any other blocks. In EV3 programming, in most cases, the sensors are used as part of the switch, loop or wait blocks; however, sometimes you need the value of a sensor in a standalone fashion such as the color sensor block in case of the PID line following discussed in the *PID algorithm for wall following using Ultrasonic and line following using Color Sensors* chapter. Although, a majority of the time, you are likely to use only the highlighted blocks i.e. Color and Motor Rotation Sensor blocks, below, we are describing all the blocks in this palette for sake of completeness.

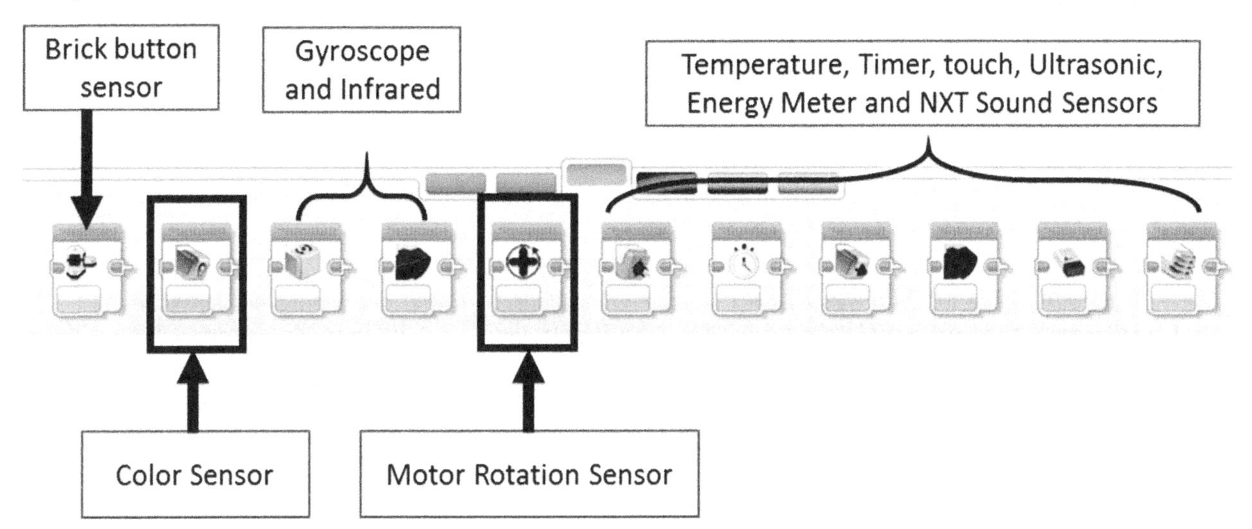

Figure 68: The sensor blocks. The highlighted Color and Motor Rotation Sensor are extremely useful and their use is showcased in later chapters.

- *Brick Button Sensor* - This block provides an ability to read the current state of the brick button or compare the state of a brick button against a known state.
- *Color Sensor* - This block allows you to calibrate the color sensor by allowing you to set the minimum and maximum values for the amount of reflected light the sensor can see in an environment. It can also measure the color the color sensor is pointing to or the reflected light intensity it is seeing.

- *Gyroscope Sensor* - This block allows you to reset the gyroscope value to zero and additionally, allows you to read the current angle the robot is facing since it last startup or last reset of the gyroscope. It can also measure the rate at which the robot is turning.
- *Infrared Sensor* - The primary use of this block is to allow you to measure distance from an obstacle. It can also detect an infrared beacon or used as the receiver end of the Lego EV3 infrared remote. This sensor is NOT ALLOWED in FLL.
- *Motor Rotation Sensor* - This block is used primarily to reset the number of degrees accumulated on a motor or to retrieve the degrees turned by a motor since the startup or last reset.
- *Temperature Sensor* - This block can provide temperature values from ambient temperature. This sensor is not available in either Home or Education EV3 kit and is not allowed in the FLL.
- *Timer Sensor* - This block allows you to create *Timers* that can measure time. A total of 8 timers can be created. The created timers are numbered 1-8 and the functionality in this block allows you to reset them to zero, read the timer value at any time or even to compare the value of the timer against a number. It is a fairly advanced sensor and it is unlikely you will have a need for it.
- *Touch sensor* - This block allows you to measure or compare the current state of the Touch sensor i.e. pressed, not pressed or pressed and then released.
- *Ultrasonic Sensor* - This block allows you to measure or compare the distance of the Ultrasonic sensor from an obstacle.
- *Energy Meter Sensor* - This block can be used to measure electrical energy i.e. current, voltage, wattage etc. using the Energy Meter Sensor. This sensor is not available in either Home or Education EV3 kit and is not allowed in FLL.
- *NXT Sound Sensors* - This block can be used to measure the intensity of the ambient sound. This sensor is not available in either Home or Education EV3 kit and is not allowed in FLL.

The sensor blocks are somewhat infrequently used but they expose some functionality that is not available in any other block. Some key example of functionality exclusive to these blocks is the ability to reset the *minimum and maximum reflected light intensity* values on the color sensor and the ability to reset the accumulated degrees on the EV3 motors. Use of both these capabilities is crucial and is discussed in later chapters.

If you are using the Mindstorms EV3 Home edition of the software, you would not have the Ultrasonic and Gyroscope blocks available by default since the Home edition of Mindstorms EV3 kit does not contain these sensors. Instead, you would need to visit https://www.lego.com/en-us/mindstorms/downloads to download these blocks and then import them in the EV3 editor using Tools->Block Import menu option. If you are using the Mindstorms EV3 Education edition of the software, these blocks are already included, and you don't need to do anything else to use these sensors.

Red/Data Operation Blocks

The Red Blocks, formally known as the *Data Operation blocks* are blocks that allow you to perform mathematical and logical operations.

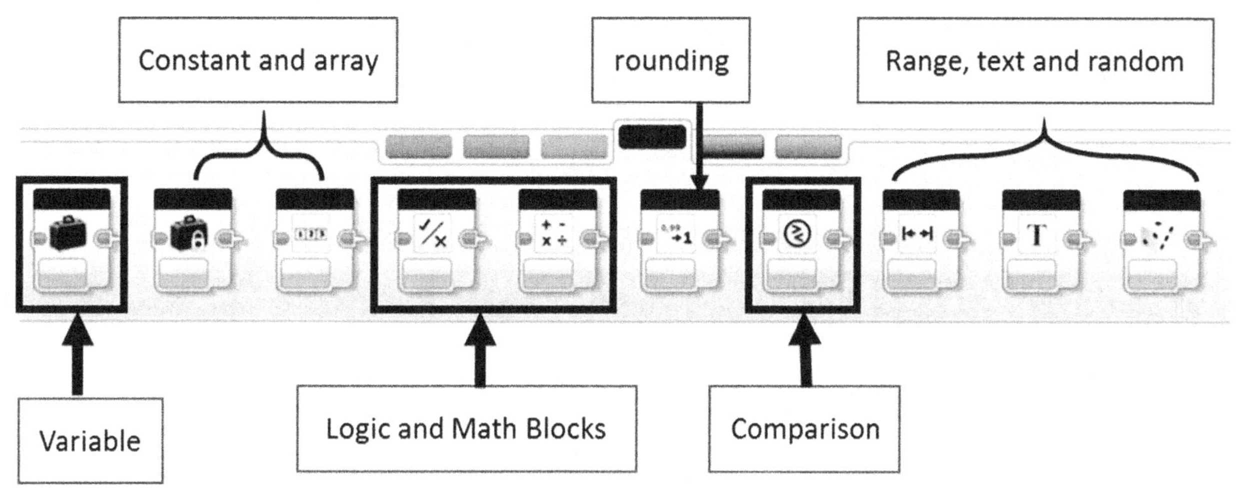

Figure 69: The Data Operation blocks

As shown in Figure 69, the Data Operation blocks contain blocks that are extremely useful for writing complex logic which requires variables, math and logical operations. The use of these blocks is almost exclusively limited to the scope of myblocks since you need these blocks for performing wall following, line following and master program. A robot run that simply uses the myblocks is not likely to use these directly in the main program. The blocks that you will likely use in your programs are highlighted in red rectangles. Below we describe the blocks for sake of completeness:

- *Variable* - This block allows you to create a variable in EV3 and assign it a name. You may read from and write to the variable. Extremely useful block and practically indispensable for most myblocks.
- *Constant* - This block denotes the notion of a Constant variable in EV3. A variable created using this block can only have a fixed value and cannot be named. We do not recommend using this block since it cannot be given a name and thus does not retain context on what that value does. A regular variable block can be named and contains all the functionality of the Constant variable.
- *Array* - This block allows you to create a list of multiple values under the same name. The use of this block is extremely rare in FLL.
- *Logic* - This block allows you to take two true/false inputs and provide a logical answer based on the two inputs e.g. (a) Are Conditions A and condition B both true? (b) Is one of Condition A or Condition C true?. Although infrequent, this block is used when writing complex logic based on variables and sensors.
- *Math* - This block is the workhorse of mathematical operations and allows you to perform math operations such as addition, subtraction, division and multiplication. It has an advanced mode that can allow you to write more complex equations in a simpler way. This block is very useful in FLL when creating myblocks.
- *Rounding* - This block allows you round fractional values to the nearest whole number. It is used relatively rarely in FLL.

- *Comparison* - This block allows you to compare two numbers. It is an extremely useful block. Although used somewhat infrequently, it possesses critical functionality not found anywhere else.
- *Range* - This block allows you to check whether an input number belongs in a range. The use of this block is very rare in FLL.
- *Text* - This block allows you to join up to three text elements into a single text block. This block is used extremely rarely in FLL.
- *Random* - This block is used to generate a random value in a numeric range or even generate a true/false answer based on a given probability of the true answer. Although, this block may sometimes be used for testing programs that require numeric input, it is exceedingly rare to see its use in FLL.

Royal Blue/Advanced blocks

The Royal Blue i.e. Advanced Blocks contain functionality that is for extremely advanced purposes and is in general not suitable for FLL. We are describing the blocks briefly below

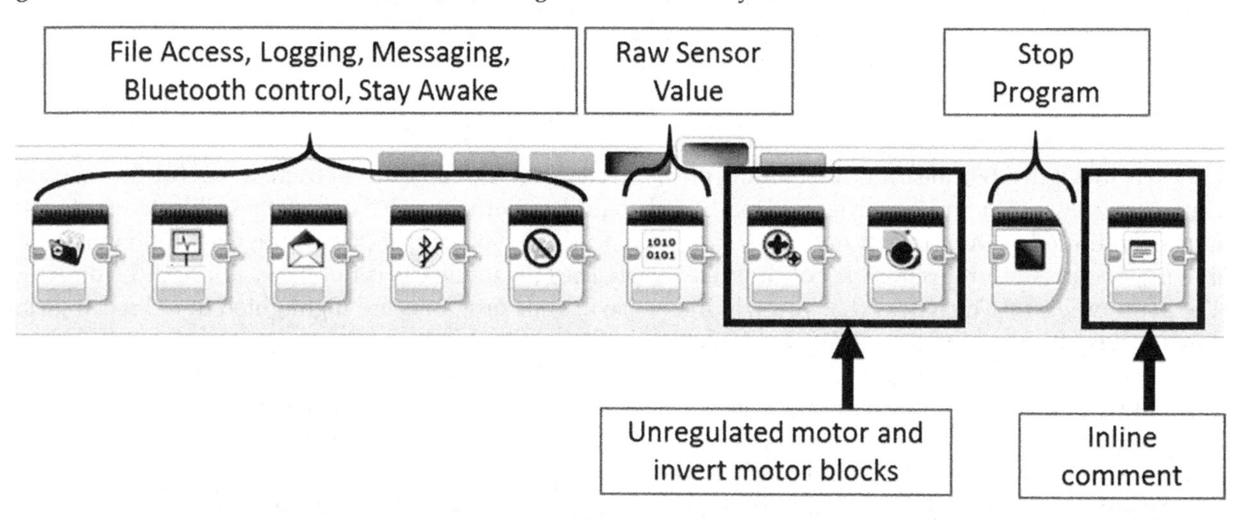

Figure 70: Advanced Blocks

- *File Access* - This block allows you to store and retrieve files on the EV3 brick. This block is extremely rarely seen in FLL.
- *Data Logging* - This block allows you to Log or store sensor dat. This block is extremely rarely seen in FLL. Potential use of this block is during debugging.
- *Messaging* - This block allows you to perform brick to brick communication between two EV3 bricks. Given that during an FLL competition, all Bluetooth communication must be switched off, this block is simply not used in FLL.
- *Bluetooth Control* - This block allows you to control the Bluetooth on the brick and to connect to other Bluetooth enabled devices. Just like the messaging, this block is not used in FLL.
- *Stay Awake* - The EV3, automatically switches off at about 30 minutes if no activity has taken place on it. This block can change that time period.

- *Raw Sensor Value* - The EV3 transforms the values from the sensor when exposing them via other blocks. This block is used to attach custom (non-Lego) sensors to EV3. In past, the NXT color sensor used to expose the color sensor raw value at a higher range than the 1-100 value that the EV3 color sensor block provides. Having a higher range allows you to create better line following algorithms. Unfortunately, on the EV3, the color sensor raw values are the same as the one the color sensor block gives out and thus, this block has little value in FLL.
- *Unregulated Motor* - The EV3 brick ensures that when a robot is moving at a certain power, it should behave nearly the same whether it is moving on flat terrain, climbing up or otherwise traversing uneven terrain. The EV3 uses a sophisticated program called PID control to ensure that no matter what the terrain, the wheels move at about the same speed. If for some reason you need to get access to the motor and control the power supplied using your own program, you may use this block.
- *Invert Motor* - This block simply has the effect of inverting the input to the motor blocks. What this means is that in the normal course of operation, the motors have a specified forward and backward directions. When the invert motor block is used for a motor, the direction of forward and backward motion get inverted. Any blocks after this block treat a directive to move the motor in an inverted way. One of the reasons we can think of using this block is if you mount your driving motors upside down and still want to keep the semantics of forward and backward as usual. This block is again, an unusual block and not commonly used in FLL.
- *Stop Program* - This block stops the program immediately even if you have more than one start block in your program. This block is rarely seen in FLL programs.
- *Inline Comment* - Unlike the multi-line comment block introduced earlier, this block becomes part of the program and can be inserted between other blocks which is its core strength. When you edit a program, you can move this block inline to insert a small comment on what a particular part of the program does. This keeps things neat and helps make sense of the program at a later time. The multi-line comment block is positional i.e. as you change the structure of the program and move blocks around, it remains in the same place and hence can quite often be in a wrong place as the blocks near it are moved around. By contrast, since this block is part of the program, when used in a reasonable manner, it keeps its logical position intact. That being said, this block can take only a few words in before the rest of the comment overflows and is rendered invisible so use your words judiciously.

Myblocks

Sometimes we need to use the same piece of code in multiple places. Instead of writing the same code again and again in different places, it is better to package it in a way so that we can write the code once and use it in the different places by referring to the single location. This functionality is available in EV3 using MyBlocks. MyBlocks are blocks that you will create for your own use. A competent FLL team is guaranteed to use MyBlocks. The most common and minimal MyBlocks that you will need are the "Move distance in Centimeters" and "Turn in degrees" blocks.

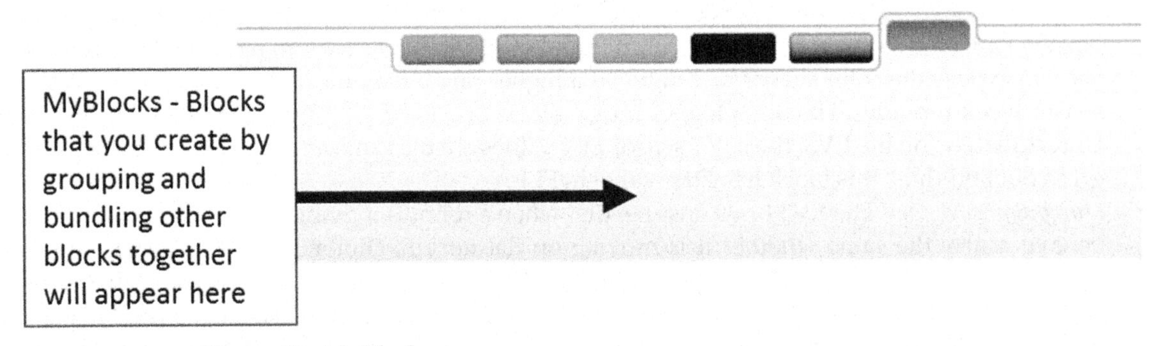

Figure 71: MyBlocks

Summary

Having a great FLL experience starts with building a robot and then having fun programming. Getting used to the programming environment and understanding the blocks that are most used, less used and rarely used sets you up for writing programs that will help you achieve your desired goals. Sometimes, even experienced teams and coaches do not know all parts of the EV3 programming environment to the team's detriment. We recommend revisiting this chapter when you are a bit more familiar with programming, so you can better understand and internalize the concepts.

Fundamentals of Robot Movement

The largest portion of the robotics in FLL is getting the robot to move in a desired fashion. Essentially, when you are having your robot run, it is critical to be able to go from the start point to the destination in a consistent, accurate fashion.

There are many factors that affect how your robot will move, some of them mechanical such as the diameter of your wheels, the distance between the wheels and whether it is using wheels or tracks or another mechanism. Additionally, there is a programming element to it which defines how to translate the specific mechanical characteristics of your robot into programming elements that are configured for your robot.

In the following examples, we are going to be looking at a robot that has two wheels and a metal caster wheel for support. This is the most common robot design used in FLL because it is very sturdy and small design that works for a very large variety of challenges. The programming techniques that we describe will apply as is to most other popular robot architectures that are based on the design that uses two motor driven wheels and a passive wheel or caster. The concepts will even apply to the robot with two caster wheels (Figure 14) although for reasons explained around that robot, you may experience variability. With that, let's look at the motion needed by a FLL robot to move around.

The two elements of robot movement in FLL are:

- Moving Straight: In this case, both the wheels are moving in the same direction at the same speed causing the robot to go straight.
- Turning - in this case, the wheels are either:
  - Turning in the same direction at different speeds (an uncommon mode of turning in FLL)
  - Turning in opposite direction at the same speed, which, although counterintuitive for many people, makes the robot spin on the spot. Note that the center point of rotation is the center point right between the two wheels during this sort of turning motion. This mode of turning is also called as *differential steering.*

Parts of the robot

To ensure that the rest of the chapter is easy to understand, we are providing the explanation of the parts and terminology around the robot in the following sections.

Large Motors

We assume that the large motors as shown in Figure 72 are the driving motors for the robot since this is pretty much the de-facto standard in all EV3 robots.

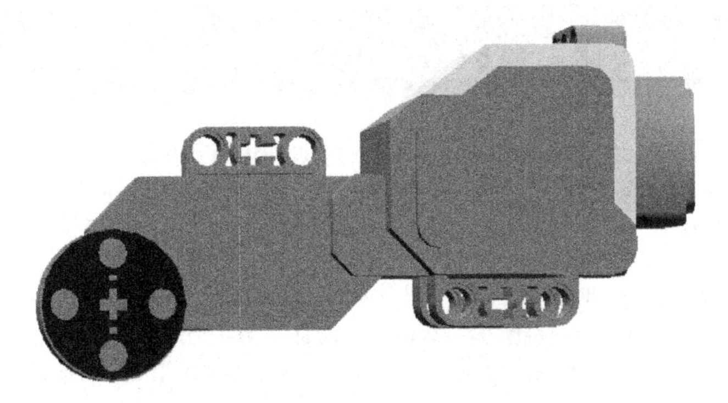

Figure 72: The large motor is the driving motor in virtually all EV3 robots

In our robot, there are two large motors, one for each wheel. Having one motor per wheel is like having a separate engine for each wheel which can control each wheel completely independent of the other. Note that this is not how real cars and trucks work in general. Usually, there is a single engine that drives both the wheels and the turning is accomplished by actively steering the wheels.

Connection to Wheels

The Wheels are directly connected to the large motors by means of axles. The implication of this design is that the motor and the wheel move in-sync with each other. One motor rotation usually means that the wheel also turns one full turn. See Figure 73 below for clarity.

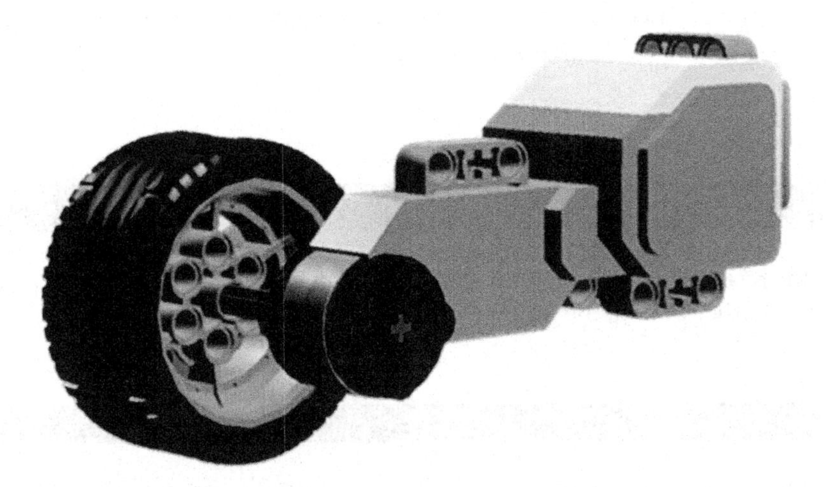

Figure 73: The large driving motor is directly connected to the wheel using an axle

Caster

The robot is supported by the omni directional metal caster (Figure 74) in the back. Casters consists of a smooth metal ball housed in a plastic casing in which it can roll freely in all directions. It is usually preferable to placing wheels in the back of the robot because it provides smooth rolling in all directions whereas wheels in the rear create friction when turning the robot.

Figure 74 - The Caster, sometimes also referred to as the marble

---

Due to its design, the space between the caster and the plastic casing tends to accumulate dirt and debris over extended use. This may increase the friction. It is advisable to clean the caster with smooth fabric frequently. Additionally, unlike other wheels, the movement for caster is governed by rolling of the caster inside plastic casing; hence, avoid any scratches on the caster wheel.

Bumper

A bumper is simply a flat Lego beam attached to the rear of the robot as the furthest point of the robot in the rear. It is similar to the bumpers in cars and trucks. This is an absolute must if you want to be able to ensure that the robot starts from the same place in every run since the bumper can be used to accurately place the robot against a line or against a flat surface to ensure repeatability of initial starting condition. See Figure 75 for an illustration of a bumper. Notice how the rear most portion of the robot comprises of two flat Lego gray beams which allow for setting the robot completely flat against lines or a wall.

The bumper can also be used as a pushing surface in the robot. Since the number of motors in an EV3 robot is limited to 4, you often need to get creative with how to perform various actions. Using the bumper to push objects is one such option. Additionally, the holes in the bumper beam allow you to attach various contraptions easily to the robot using either pegs or axles.

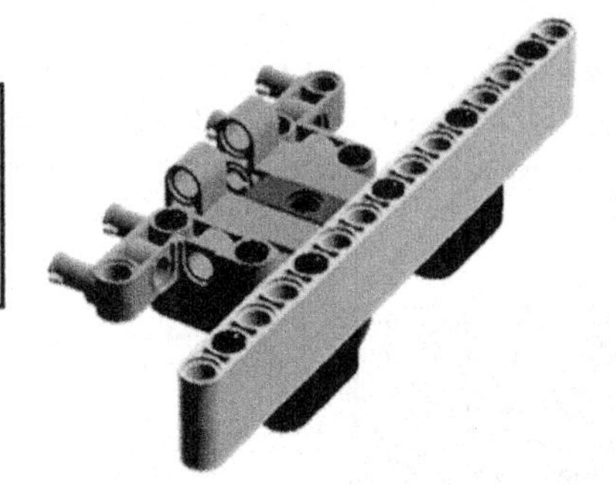

Figure 75: Bumper, the rear most part of the robot. It is a good idea to have this completely flat for easy alignment

Chassis

All the parts of the robot that form the basic driving assembly with the brick excluded. This is the stripped-down version of the robot with the minimal pieces needed for the robot to move. Having a simple, small and sturdy chassis with low center of gravity is a must to be able to create a reliable robot that behaves consistently. See a sample robot chassis in Figure 76. For the illustrations in the rest of the book, we will be using the chassis as shown in Figure 75 and the bumper shown in Figure 76 with the Znap wheel assembly (Lego Design ID# 32247)

We have found the Znap wheels (Lego Design ID# 32247) to be very reliable wheels that go fairly straight. They do wear out over the course of a season or two, but we find them well worth the cost.

Figure 76: The Robot Chassis.

Let's recap the various parts of the robot with the help of the following diagram (Figure 77). The motor and port configuration described in the figure will be used in the rest of the book for consistency.

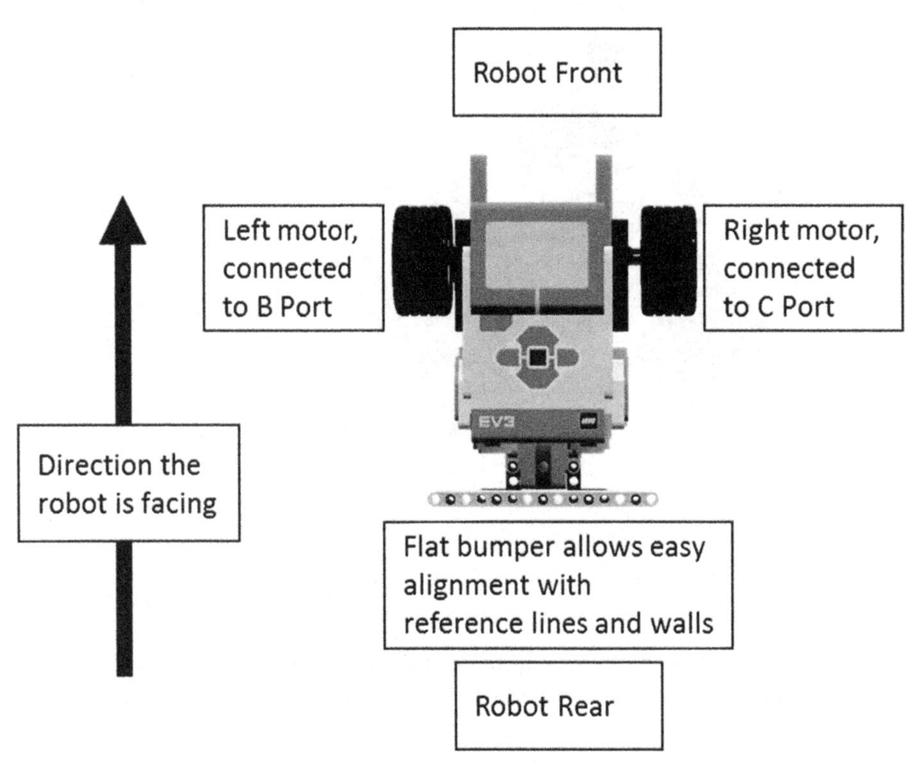

Figure 77: Robot terms and convention used in the rest of the book

Moving Straight

The first thing that you should do once you have built a robot is to calibrate the robot so that you can understand how much movement of the motors is going to cause the robot to move a certain distance or to turn a certain number of degrees.

One of the first programming blocks in the EV3 programming language that you will be using are the blocks that make the robot move. Below, we are using the *Move Steering* block to showcase how to move and turn your robot. Before we start programming using the move steering block, there is a key concept that we need to explain, which is, the concept of a *tachometer.* The Tachometer is a device that is connected to motors and counts the number of degrees turned by a motor. Both EV3 large and medium motors contain a tachometer built inside them and the functionality to measure the degrees is exposed in the EV3 programming environment. The EV3 tachometer can measure the motor turning in 1-degree increments. Thus, you can break one motor turn into 360 steps and specify the motor to spin in a multiple of 1 degrees.

Thus, given that the EV3 motors measure their movement in degrees, it is best, when programming, to specify the motor movement in degrees. Although the EV3 allows you to specify the amount of motor turn in number of rotations, internally the amount is translated to degrees and hence specifying the motor turns gives you a much better picture of what your motors are doing. In the rest of the book, we will only use the motor movement blocks in degrees.

Coming back to the move steering block, first locate the *Move Steering* block which is highlighted below:

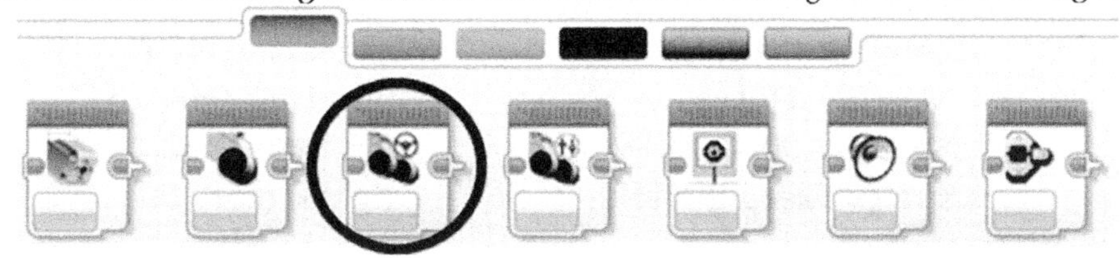

Figure 78: Move Steering block

Put a move steering block in the program immediately after the *start* block. Click on the bottom left button marked *On for rotation* and select the *On for Degrees* option. In this mode, you can enter the number of degrees that you want the motor to turn. Look at Figure 79 to better understand the various options in the *move steering* block.

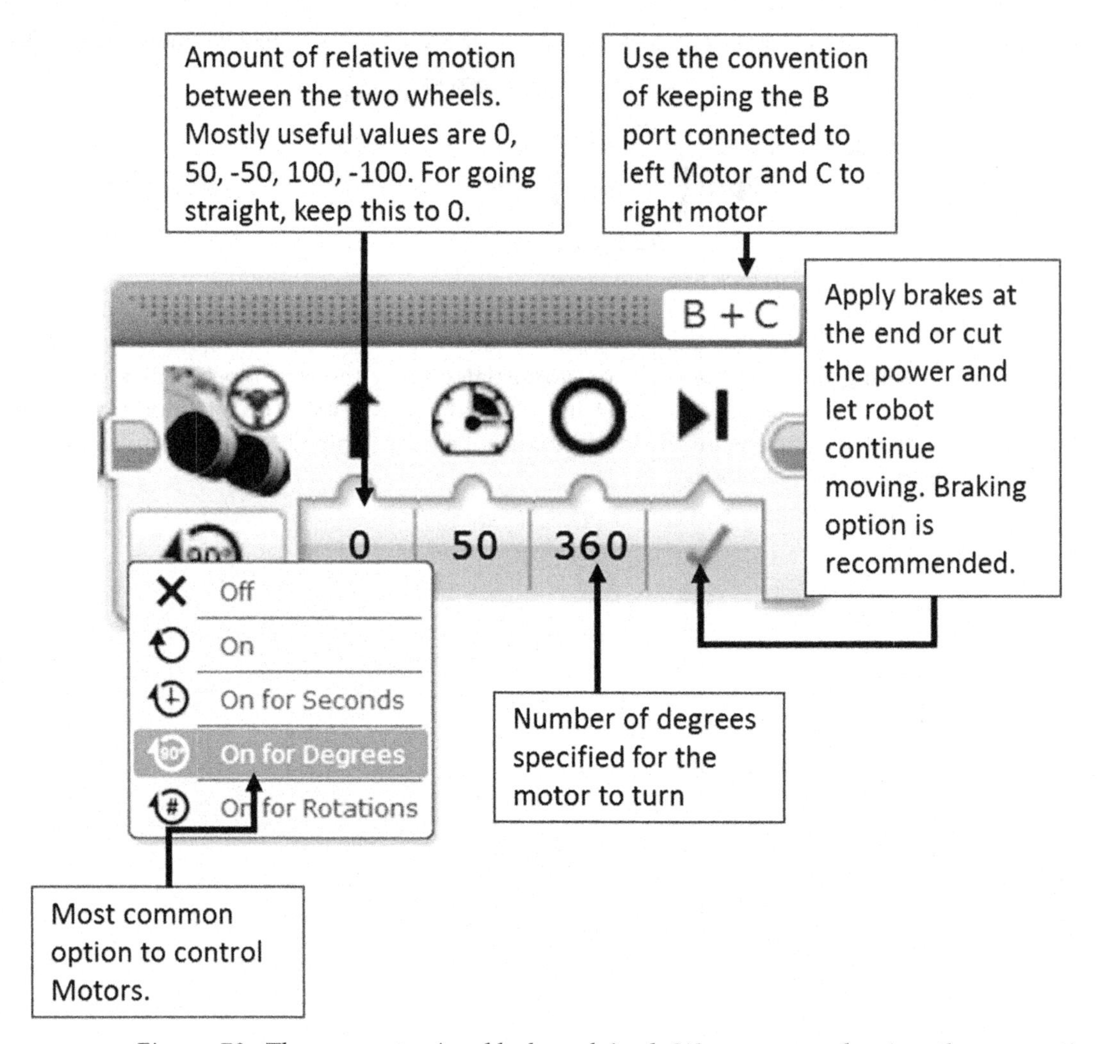

Figure 79: The move steering block explained. We recommend using the convention of keeping the B port connected to the left motor and the C port to the right motor for reasons explained later.

Although it is possible to put negative values in any of the three inputs to the move steering block, we recommend only using the degrees field to input negative values since it is easy for kids to understand that positive degrees will move the robot forward and negative will move it backwards. Additionally, later, when we discuss turning, with the B and C ports connected to left and right motors respectively, having positive degrees in this field will turn the robot right and negative will turn it left which is consistent with the integer scale taught in math class.

Armed with the above information, let's now see how to calibrate the robot to go straight. Since we will be supplying the degrees for the motor to run in the move steering block, we need to be able to calculate the number of degrees required for any distance. To go straight, set the first input in the block to 0. The second input, which specifies the power supplied to the motor can range from 0-100 and in general a value between 50-75 works quite well for moving straight. The third input is the one that we need to be calculating.

As we had discussed earlier, since one single rotation of the motor turns the wheel one single time, to understand how much the robot moves every time the motor turns one degree, we can simply set a large number of degrees in the degree field and then run the robot to measure the distance travelled and calculate the distance per degree. In this case, we can put 3600 degrees there which is essentially 10 full rotations of the motor. Note that, since the motor is directly connected with the wheel, this will also cause the wheel to spin 10 rotations. Run the program and measure the distance that the robot has moved in centimeters.

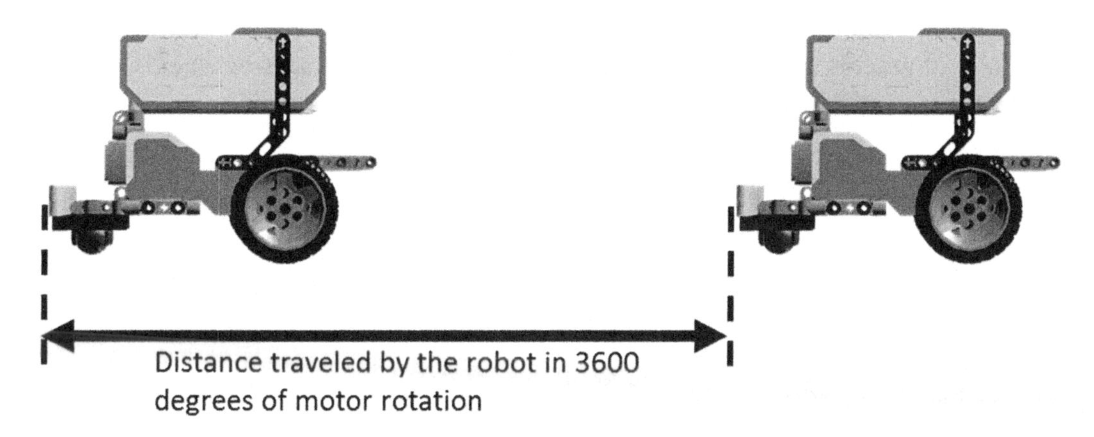

Figure 80: Measure the distance moved between the same point on robot body between the start and end positions. We recommend using a bumper as the reference point although any point may be used.

As you can make out by the tire collection shown in Figure 81, tire size dictates the distance moved in one rotation of the wheel. Depending on the size of the tire that you use, in 3600 degrees or 10 turns of motor rotation, the robot will move different distances. A smaller wheel will move the robot less distance and a larger wheel will have your robot move larger distances.

A variety of wheels can be attached to your robot. Different sets of wheels have different calibration values for both going straight and turning

Figure 81: A few Lego tires shown for illustration purposes.

Now you can calculate the calibration value as:

$$Motor\ Degrees\ per\ CM\ of\ robot\ movement = \frac{Motor\ degrees\ turned\ in\ Calibration}{Distance\ traveled\ by\ Robot\ in\ Calibration}$$

So, for example if your robot travels 200cm in 3600 degrees of motor rotation, your *Motor Degrees per CM of robot movement* = 3600/200 = 18 degrees per cm. If we simply call *Motor Degrees per CM of robot movement* as the *Moving Straight Ratio,* then whenever your robot needs to travel D centimeters, the motor in the robot needs to turn *D* x *Moving Straight Ratio.* So, using the above example values, if the robot needs to travel 100cm, the motors need to spin 18degrees/cm x 100cm = 1800 degrees. This value can be directly fed into the move steering block to make the robot travel the required 100cm as shown in Figure 82.

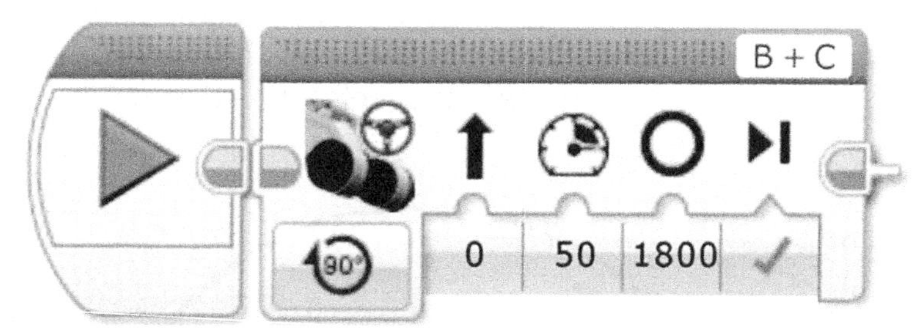

Figure 82: A sample robot configured to travel 100cm using the Moving Straight ratio of 18 degrees/cm

---

The move steering block drives the motor for at least the degrees you want to turn the motors and applies electronic brakes at the end. However, while the robot stops due to braking, your robot will still continue moving. This causes the robot to overshoot whatever degrees you were trying to travel. If your robot is connected to the computer via Bluetooth or the USB cable, you can see this overshooting in the port view. In general, in our experience, for one set of wheels and for a certain motor power, the slippage in terms of degrees is nearly the same every time no matter how far you go. Thus, to be as precise as you could, you will need to accommodate the nearly constant braking distance for the power that your robot is using e.g. if your robot always slips 5 degrees with a set of wheels and 50 power, in the above example, instead of traveling 1800 degrees, you could simply ask the robot to travel 1795 degrees, knowing that the 5 degrees of slippage will make your robot go 1800 degrees anyways.

Turning

The simplest and most common way of turning an EV3 robot is via a turning term called *Differential Steering*. Differential steering simply means that the robots spins its two wheels in *opposite directions,* at potentially different speeds. If the speed of both the wheels is the same, although in different directions, it will cause the robot to spin with the center of rotation right between the two wheels. See Figure 83 for a better explanation. For many participants, this is an unusual form of turning since pretty much all cars and trucks move using the *Active Steering,* where both the wheels turn in the same direction. *Active Steering* is rarely, if ever used in robotics competitions since it requires a much larger space to turn and requires carefully maneuvering the robot to get out of tight spaces. *Differential Steering,* on the other hand takes very little extra space to turn and can usually turn the robot in very tight spaces, a definite advantage in a robotics competition. Additionally, *differential steering* keeps the center of the robot in the same place even after the turn is complete which makes for easier *robot path planning.*

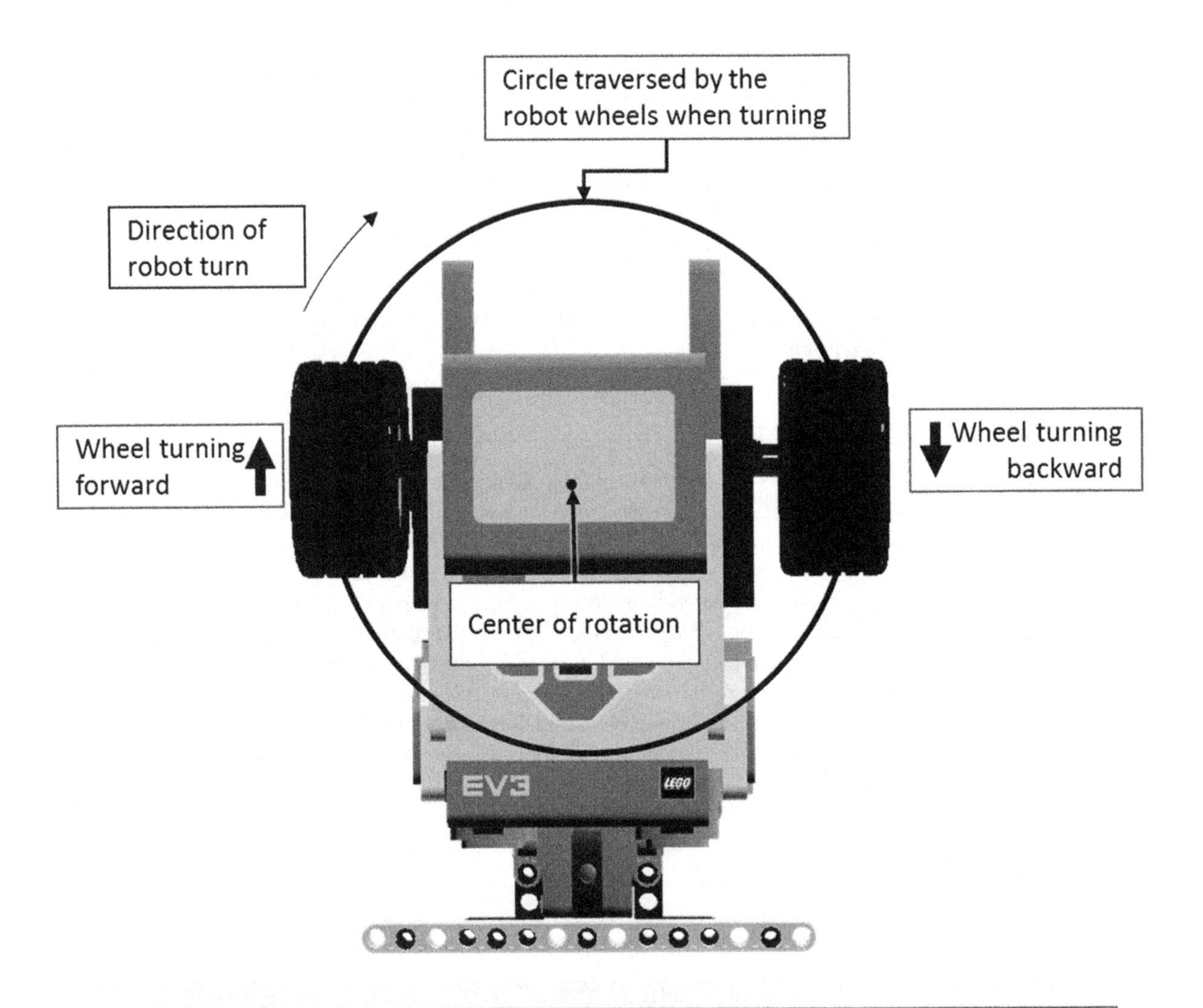

Figure 83: Differential Steering, a way of turning the robot on one spot by spinning the wheels in opposite direction. This causes the robot to move on the spot with the center of rotation right in the center of the two wheels. This is the most common way of turning the robots in competitions.

With the understanding of the *differential steering,* we can now learn how to calibrate the robot for turning. Just as we had to calibrate the robot using the calibration measurements and calculating the *Moving Straight Ratio,* we will have to once again perform the measurements and calculate a *Turning Ratio.*

Although you can use the Gyroscope sensor to perform the calibration in a simpler manner, we recommend the way described earlier. We believe that this method of calibration forces the user to understand the underlying math and mechanical aspects, rather than depending on the gyroscope which although easier, does not impart much in terms of learning. If your calibration is not consistent, you can do another calibration using the gyroscope once users are familiar with programming.

Before we get to calculating the *Turning Ratio,* we need to reaffirm some of our understanding on how the motor movement impacts the robot movement. See Figure 84 to better understand how we calculate how much the robot has turned.

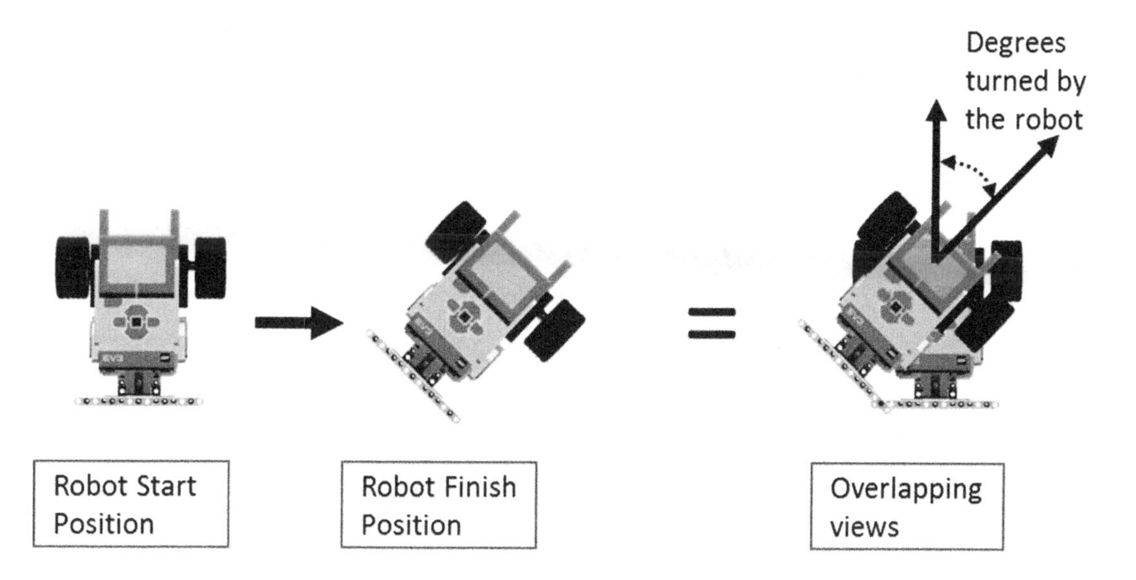

Figure 84: When the robot turns using the differential steering, the actual number of degrees that the robot has turned are calculated by measuring how much the robot has turned with reference to the center between the two wheels.

Looking at both Figure 83 and Figure 84, it should be fairly apparent that when the driving motors spin in the opposite directions, the robot wheels start traversing a tight circle as shown in Figure 83. It is critical that you understand that how much the *motors turns* is different from how much the *robot turns!* This is a common cause of confusion and the *move steering* block in the EV3 programming environment makes it more confusing.

The number of degrees the motors have turned and the number of degrees the robot has turned are two different entities and it is critical for you to understand that the degrees that motor turns impacts the number of degrees the robot turns. This is a common source of confusion!

Since the motors rotating in opposite direction causes the robot to start turning, we should calculate how much motor rotation causes how much of robot turning. To do so, first draw a straight line on a flat surface and align the rear bumper of the robot to be parallel and fully flush against the line as shown in Figure 85.

Aligning is the process of placing the robot at a start location so that the robot is placed in the same place before starting every time.

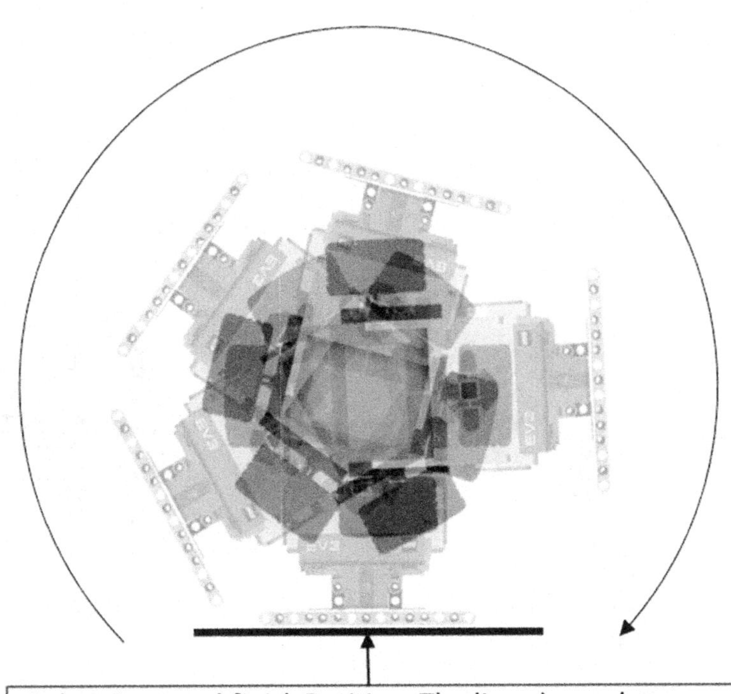

Figure 85: How to position the robot to experimentally calculate the Turning Ratio. The intermediate positions during turning are shown as is the turn direction of the robot.

After that, create a small program with the move steering block configured as shown in Figure 86. You will now start with an arbitrary number of degrees such as 500 in the degree field and keep experimenting whether these number of degrees is sufficient to turn the robot around and bring it back to the start position exactly e.g. you may find that you need to keep bumping the motor degrees from 500 to 800 for the robot to successfully complete a 360 degree turn and land back at the start position. You will likely need a few tries to get this right. To summarize the above, you keep increasing the degrees the motor is turning until the robot physically ends up turning 360 degrees.

For better accuracy, it is always a great idea to calculate the calibration multiple times and average the calibration values to reduce the error. This is a common way to reduce errors in any scientific experiment.

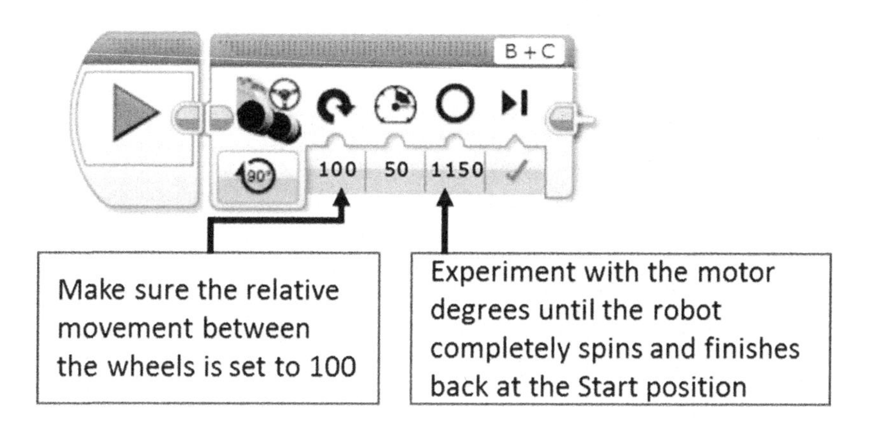

Figure 86: How to use the move steering block to calculate the Turning Ratio

Once you have experimentally figured out the total degrees the motor needs to turn to make the robot turn 360 degrees, we can calculate the *motor degrees turned per degree of robot turn i.e. the Turning Ratio* as follows:

$$Turning\ Ratio\ = \frac{Motor\ degrees\ experimentally\ found}{Robot\ degrees\ turned\ i.e.360}$$

Let's say you experimentally verify that it takes 720 degrees of motor rotation to turn the robot full 360 degrees. In that case, the turning ratio will be 720/360 or 2.0. Armed with this information, you can now turn the robot any number of degrees. For example, let's say you need the robot to turn exactly 90 degrees to the right. To do this, you need to calculate *Turning Ratio x 90* and input that value - which in the above example would be 180, as the degrees to turn in the move steering block. To turn the robot left, simply negate that value i.e. place -180 in the degrees field in the move steering block. Now, to turn the robot any amount, you can simply multiply that amount by the *Turning Ratio* and use as the input in the move steering block.

Once you have both the *Moving Straight Ratio* and the *Turning Ratio* available to you, you can easily have your robot navigate any path by breaking it down into moving straight and turn segments. Here is a simple robot path example and its corresponding robot program. We have assumed that the *Moving Straight Ratio = 18 degrees/cm* and *Turning Ratio = 2.0* in the following exercises.

Although some teams may continue doing the math and putting it in the move steering block all the time, it is traditional to wrap this logic in a MyBlock, a custom block with custom logic that you can create to simplify your programs. This is discussed later in the MyBlocks chapter.

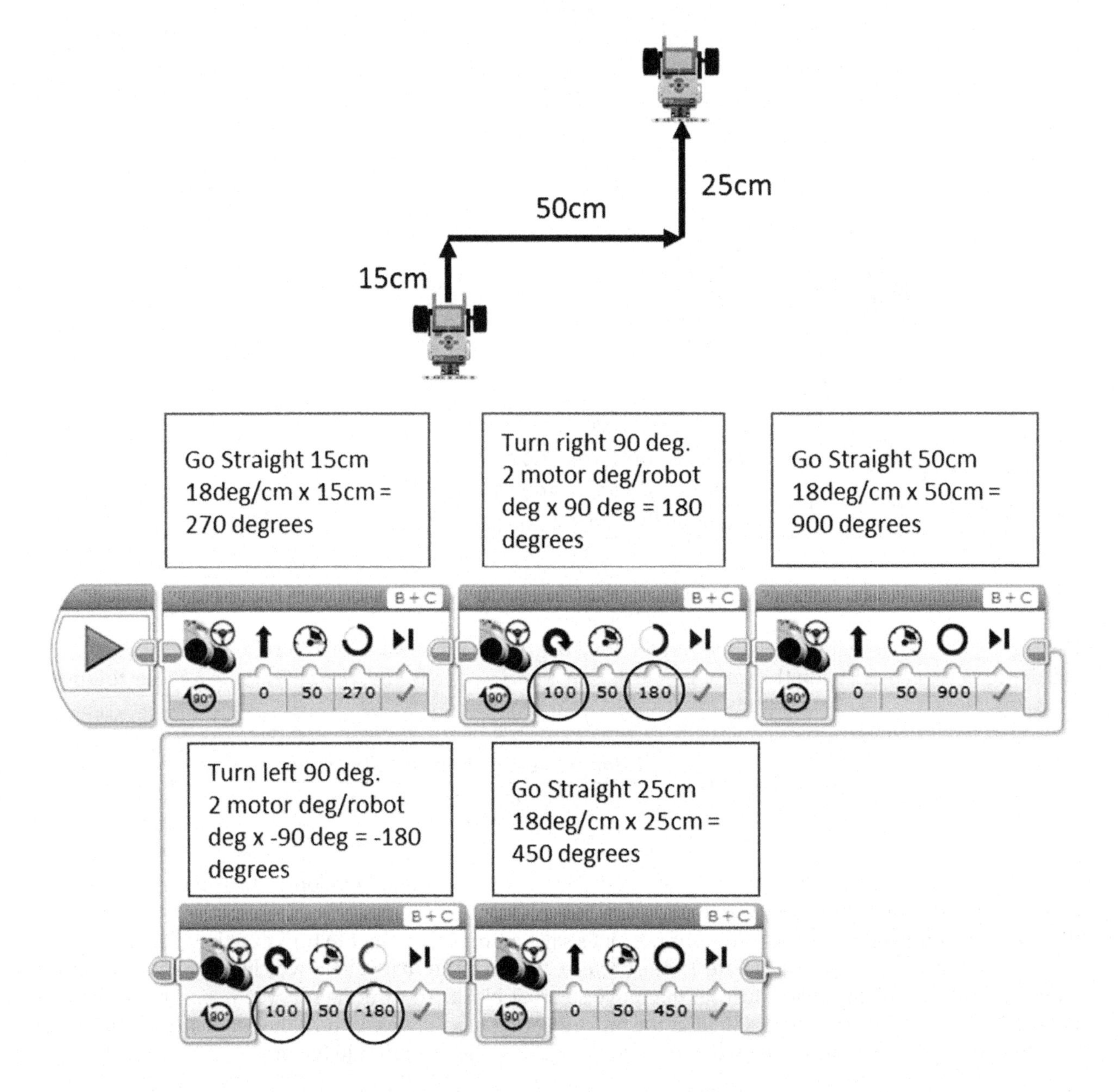

Figure 87: Simple Z curve movement exercise helps drive in the calibration concept. It is helpful to remember that when laying out a program, you face the same way as the robot, so your left and right are the same as the robot left and right as well. Pay special attention to the highlighted fields which show that the relation motion between the two wheels is set to 100 and the turn left is simply managed by setting the turn to negative, whereas the turn right is positive.

Exercises to reinforce basic robot movement

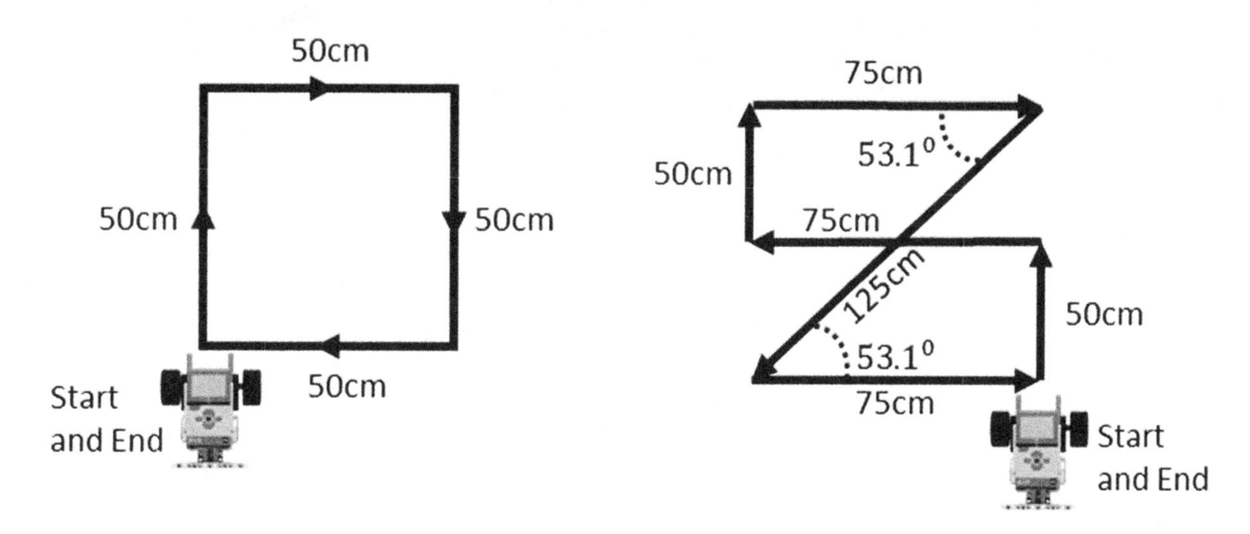

Figure 88: Robot Navigation exercises

If the robot battery is in a nearly discharged state, the robot movement is not reliable as the motors might not catch up to each other and the runs will be unreliable. We recommend having your test runs on a fully charged robot.

Many participants and teams prefer to use the Move Tank block instead of the Move Steering block since the Move Tank block allows us to control the motor power on each motor explicitly and thus makes visualizing the actual robot movement simpler. There are pros and cons to using Move Steering vs the Move Tank block and you can choose the one that you like better. In the end, for the most part, you can perform what you need with either block and use the other one where one does not meet your needs. We are showing the two blocks in equivalent configuration in Figure 89 and how it affects the robot motion.

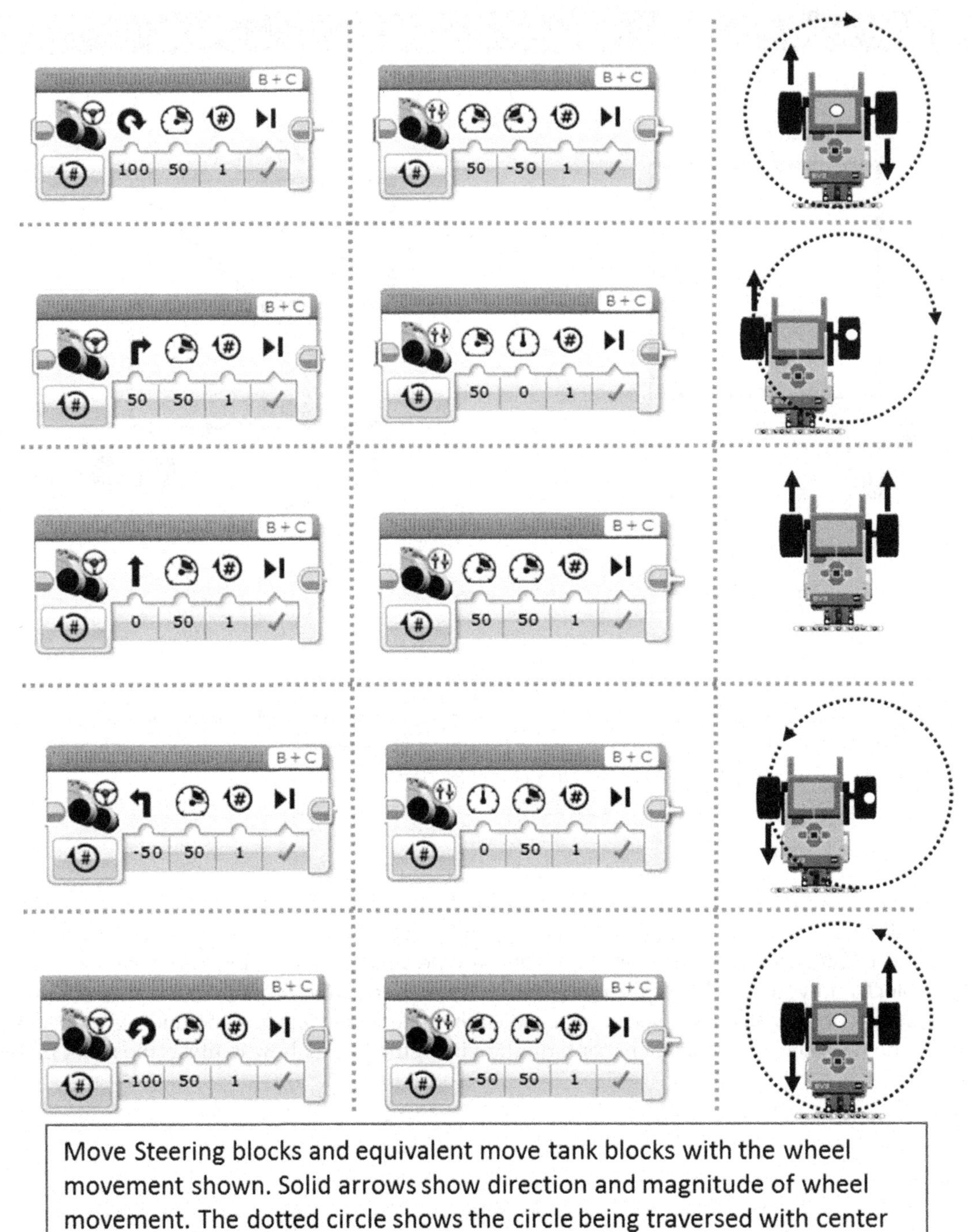

Move Steering blocks and equivalent move tank blocks with the wheel movement shown. Solid arrows show direction and magnitude of wheel movement. The dotted circle shows the circle being traversed with center and direction of circle traversal shown.

Figure 89: Move Steering and Move Tank blocks shown where they are equivalent along with the effect on the robot motion.

Summary

Basic robot movement with calibration is the first step to understanding that robots are mechanical systems that can be controlled by carefully measuring their characteristics and applying the understanding to robot path planning. Many beginner teams perform pretty much entirety of their robot path planning using nothing, but the movement based on calibration. Although advanced teams should use sensors and other methods to improve the robot movement, a team can get pretty far using just the calibration values and measurement based movement (odometry.)

Basic Robot Navigation

In Robotics competitions and especially in FLL, where most of the tasks are based on successfully interacting with a model that is placed in a fixed location, the most important factor to your success is the ability of the robot to get to its intended destination without any mistakes in a consistent manner. When teaching students and when coaching teams, we call this the *rule of 8 out of 10* i.e. when practicing a robot run, your programming practices, your robot design and your navigation paradigm MUST ensure that the robot reaches its destination 8 out of 10 times. Sometimes, human mistakes and at other times, unpredictable behavior on part of the robot can cause the robot to get waylaid and not perform the task at hand. However, anything less than 8/10 success rate will only cause the participants to get false hope that the programming and robot design work fine when in fact, some of the times, they are getting lucky.

Navigation Paradigms

Robot path planning using purely odometry is quite different from Robot Navigation using multiple markers and reference points. It is important for participants to understand this right at the beginning. In order to explain the difference between the two we use the analogy of hiking in woods vs taking a boat in the middle of the ocean.

When you are trying to find your path from a starting point to a destination in an unfamiliar environment, you can use either the simple method of navigation known as *dead reckoning.* An alternative approach is to rely on sensors, other navigation techniques and reference points on the robot competition field for the robot to locate and orient itself.
First, let's discuss the concept of *dead reckoning*. If a boater, with access to only a compass and map, decides to take a journey across a featureless ocean where the shore and islands are too far and not visible, then they must rely on the concept of dead reckoning to try and reach their destination. *Dead reckoning* is a navigation paradigm where you know the speed at which you are moving across a terrain and additionally you know the direction that you are moving in. Based on how long you have been moving, you can roughly approximate where you should have reached on the map after a certain period e.g. if you started at a known position X and rowed at a speed of 4 miles per hour in the North direction, then you should ideally be 4 miles north from your start location in one hour assuming you moved in a perfectly straight line which, is incidentally, almost never true. Given access to a map, you can figure out where you are with a reasonable amount of certainty. This paradigm was used by early sailors to chart courses through the ocean and can certainly be used by your robot.

When the robot path planning is performed using the *dead reckoning* paradigm, it is referred to as *Odometery (see* Figure 90*)*. As we explained in the previous chapter, by calibrating your robot, you can make it go any distance that you want and turn it in any direction that you want. One would think that this should be sufficient for the robot to accurately reach the destination every single time without fail. However, it is a great source of frustration for the students as well as coaches when using *Odometry* for robot path planning does not consistently work and they are left wondering why.

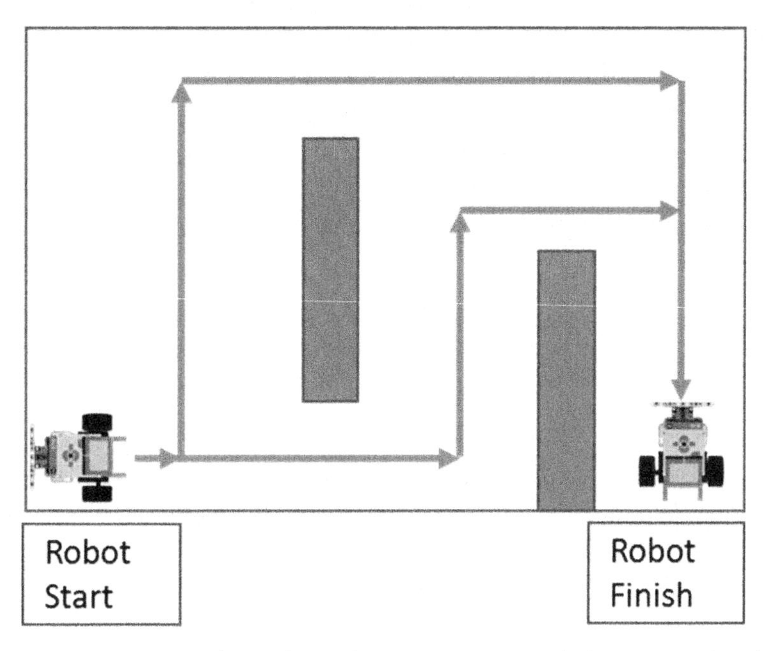

Figure 90: Notice how the Robot can use any of the two paths shown around the two obstacles to reach its destination by simply using measuring the distances and turning left and right. This is an example of using Odometry to navigate the terrain.

In popular culture, especially comics and animated movies, robots are portrayed as intelligent, incredibly powerful and precise machines that can understand language, emotions and perform a great variety of tasks completely autonomously. Although, that is the goal of robot research, the current state of technology is far from it. Robot navigation firmly plants the idea in student's minds that the programming is what makes the robots intelligent.

Although *Odometry* works well as a path planning paradigm for small distances, it suffers greatly as the distances increase over half a meter and especially when multiple turns are involved. The more number of turns in a path, the greater the error accumulated and higher the chance that the robot will not reach its intended destination. This brings us to the concept of *robot drift*.

Figure 91: Robot drift is the most common problem encountered when solely relying on Odometry for path planning. As the robot moves, it will drift from its intended path to the left or right. The longer the distance, the more the drift.

Robot drift is the amount of distance that the robot moves sideways when going to its destination in a straight line. Depending on the tires and how well constructed your robot is, the drift can vary widely. It is not a great idea to solely rely on *Odometry* for robot navigation since over distances greater than 50cm or so, the robot drift can put the robot significantly away from its intended path. If you are relying solely on Odometry, after a few straight moves and turns, the robot will be quite a bit away from its intended path since the errors in each of the segments that it travels keeps on accumulating.

The wheels that come in the EV3 education kit, in general have been very inconsistent in terms of drift and have frustrated kids and parents alike. Apart from the fact that they drift significantly, they inconsistently may drift left and right in successive runs. Due to this variability in the amount and direction of the drift, it becomes very hard to correct for them in your programming. In general, wheels that are excessively squishy should be avoided. Instead, try using wheels that are rigid and are reasonably wide so they grip the travel surface well.

To summarize, *Odometry* is best used for small distances where the robot drift will not pose a significant challenge. For larger distances, it is best to start correcting the robot drift and other robot movement errors using other path planning techniques that we will discuss next.

Before we start looking at other path planning techniques, let's define the word *Navigation* in context of robot path planning. The word Navigation is used to describe the method of using reference points or markers during path planning to ensure that the robot can follow the intended path. Odometry uses only one reference point i.e. the starting point for navigation and thus, unless supplemented with other methods, forms a weak path planning method. This is not to say that Odometry should not be used at all. Odometry is the most fundamental robot path planning method and forms the basis of all robot movement. You just need to keep in mind its limitations and work with them.

If you are having a hard time with making your robot move on its intended path purely using Odometry. Try putting a 0.1-0.5 second delay between successive move blocks using the Wait block which is explained in a later chapter. The reason behind this is that often the robot has not come to a full stop between blocks and motion from the previous block carries over into the next one causing errors.

Here are some of the key Navigation Paradigms that are used to supplement and overcome all the issues inherent in robot path planning using only odometry.

1. Squaring/Aligning to the wall - Covered in this chapter
2. Wall Following (mechanically) - Covered in this chapter
3. Wall Following using Ultrasonic Sensor - Covered in the chapter on the Ultrasonic Sensor
4. Color detection and Line Squaring using Color Sensor - Covered in the chapter on the Color Sensor
5. Line/Edge following using color sensor - Covered in the chapter on the Color Sensor

Before we begin to go over these navigation techniques, let's go over how to use *Odometry* in the optimum manner using the following concepts:

1. Aligning the robot perfectly at start
2. Taking care of the gear slack
3. Using a starting jig effectively

Aligning the robot perfectly

Aligning the robot perfectly means placing the robot in the same place when starting in a consistent manner. We cannot stress this point enough. Not aligning the robot perfectly can put you at a disadvantage in the best case and downright ruin your run in the very worst case. The first requirement to have a perfectly aligned robot is to have a flat bumper as we discussed in *Parts of the robot* section in the previous chapter. FLL competitions are in general held atop a smooth mat and the location where the robot must start usually contains a grid and other markers that you can use to align your robot perfectly. To perfectly align the robot, here are the key rules:

1. Align the robot bumper perfectly parallel to a line on the mat - Many participants make the mistake of aligning the robot against the wall. It is crucial to understand that given the variations in how the tables are setup, the mat might not be flush against the table wall (Figure 92), in fact, it might not even be parallel to the table wall!

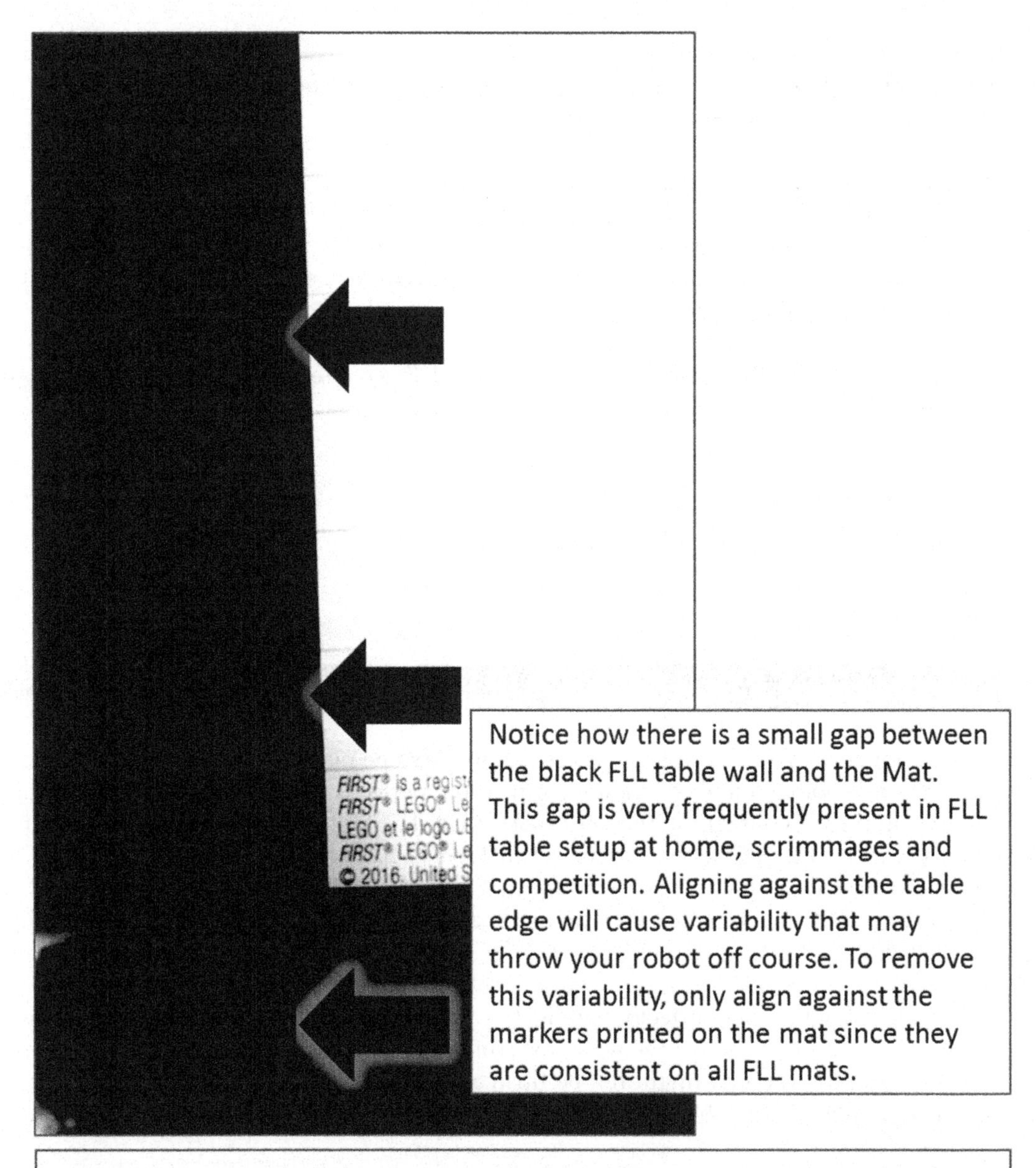

Figure 92: There is quite often a significant gap between table walls and the FLL mat edge. Thus, if you use the table wall as the starting point, depending on the gap, your robot will likely have variability in its starting point and thus variability in its mission performance on different tables.

2. *Things on the mat are precise in relation to other elements on the mat, not on the table.* Trying to start against the wall is, in general a bad idea (Figure 92.) Here is a visual explanation of how to align the robot on the mat.

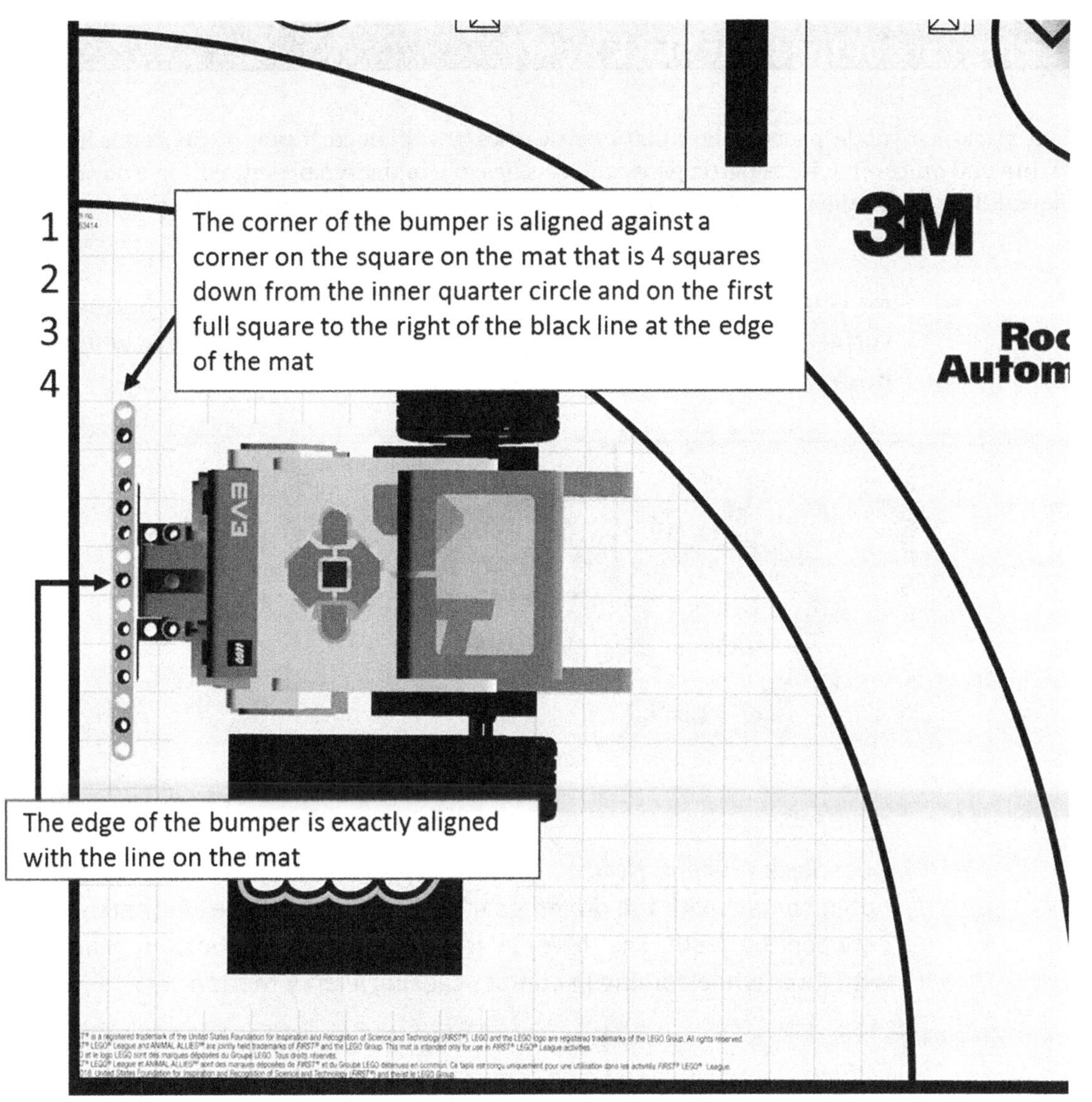

Figure 93: Aligning the robot at a fixed location on the board based on the reference markers on the mat itself. The robot's bumper is 4 down from the top most square and 1 to the right of the black line at the edge and perfectly parallel and aligned with the vertical line. Knowing this coordinate, the robot can be repositioned at the same location consistently. This mat is the one that was used in the 2016 Animal Allies FLL season.

Taking care of the Gear Slack

Gear slack is a subtle phenomenon that often causes unexplained tilting of the robot immediately at the startup and quite often leads participants and coaches to wonder what went wrong even after they perfectly aligned the robot at the initial start point. See Figure 94 for illustration of gear slack.

No Gear Slack Present: Notice how the driver gear's tooth's bottom surface is in contact with the driven gear's tooth's top surface which will push it firmly and gently

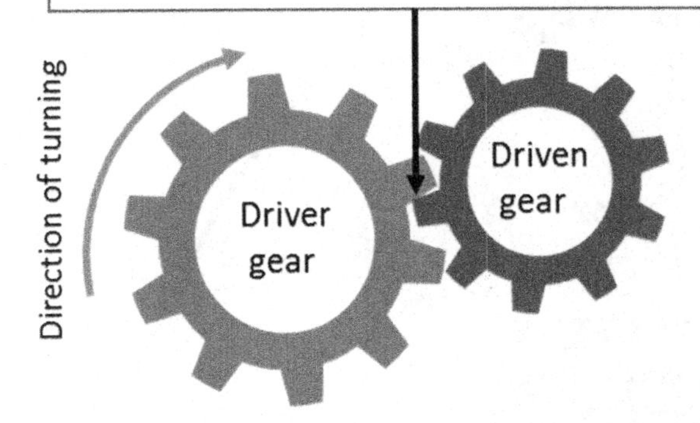

Gear Slack Present: Notice how the driver gear's tooth's bottom surface is not in contact with the driven gear's tooth's top surface i.e. there is slack between the teeth. The driver gear's tooth will hit the bottom gear's tooth with force while coming in contact causing a jerky motion.

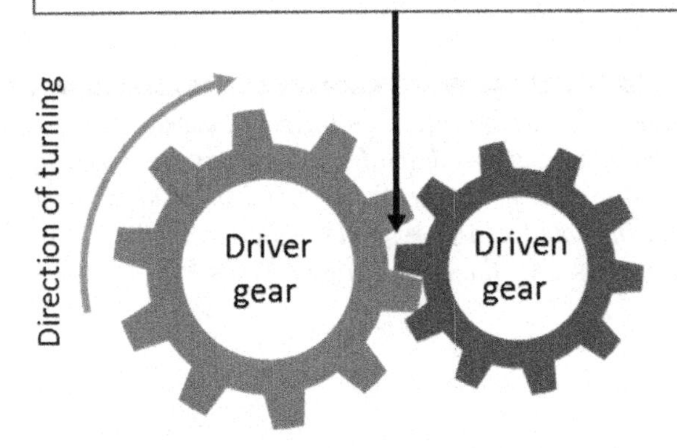

Figure 94: When there is gear slack present, the driver gear's teeth's leading or driving edge is not touching the driven gear's teeth's following edge. As the motors engage at high power, the driver gear's teeth slam into the driven gear's teeth causing a jerking motion as soon as the robot starts. The problem worsens when there is different amount of gear

slack in the two different motors. Despite careful alignment, gear slack causes the robot to assume an incorrect position as soon as it starts moving.

Essentially, whenever two gears are turning as part of a mechanism, the teeth of the driver gear are pushing the teeth of the driven gears. When you are setting up the mechanism, depending on how you setup the mechanism, the teeth might be touching each other or might have a slight separation. When the driver gear engages with the driven gear, depending on whether it was touching the driven gear before engaging, it will either smoothly push the driven gear or if it was not touching previously i.e. there was some spacing or slack between the gears, it will hit the driven gear like a hammer hitting a surface, causing a jerky motion at the startup. This will likely orient the robot in an unexpected direction at the very onset causing the robot to drift. To avoid gear slack, align the robot at the starting point and with one finger gently push the robot towards the front without applying undue pressure. Observe that there is a little bit of play where the robot can rock back and forth just the smallest bit without actually moving. Push the robot towards the front so that both the wheels completely lock and pushing further will actually start the wheels turning. This will put the driving gears in both the motors in contact with the driven gears and remove the gear slack. This will remove jerky starts and give a much more consistent startup experience.

Starting Jig

Starting Jig refers to a mechanism that is created and placed in the base and acts as an aid to positioning the robot correctly. While practicing robot runs, one might think that the starting jig is an unnecessary mechanism. However, quite often, during the heat of the competition, participants tend to misplace the robot and having a Starting Jig helps orient the robot without any errors.

2016 Animal Allies Mat

2014 World Class Mat

The actual base from two separate FLL competitions. The one on the left is from FLL Animal Allies 2016 and the one on the right is from FLL World Class 2014 season. The Animal Allies base has a nice grid in the base that can be used for alignment. In the World Class mat, there is no such thing, markers are present only on the edges, making this exceedingly difficult, making it a great candidate for a Starting Jig.

Figure 95: The Base design from two separate FLL seasons shows the variability that can make a starting Jig an absolute necessity.

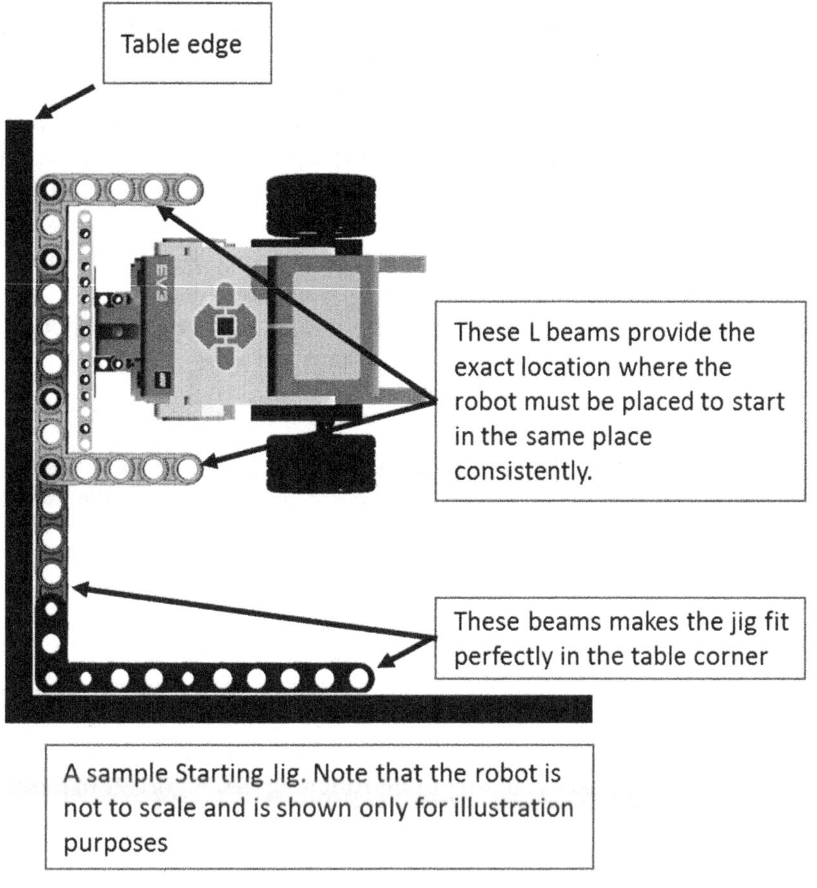

Figure 96: Starting Jig and how to use it.

Figure 97: An angular starting jig can be used to start a robot in an angular fashion. The robot is not to scale and used for illustration purposes only.

Using a starting jig is a very useful tactic for starting the robot in the same place consistently and takes out a lot of uncertainty in the alignment process. A few key points to keep in mind when using the starting jig are:

- There is quite often some space between the FLL mat and the table borders. Placing the jig all the way back against the wall will cause your robot to be at slightly different positions on different tables.
- Only the markers on the mat are on fixed positions. The various models and markers on the mat are in fixed location with respect to the mat. Thus, trying to align the starting jig right against the mat border or mat corner is a great idea.
- When pushing the robot all the way against the starting jig, remember the gear slack as discussed and counter it by gently nudging the robot forward after aligning.
- Don't make the beams surrounding the robot too long in case the robot starts hitting its stall walls. The walls are there simply as an aid in alignment, not to box the robot.

Using wall and flat surfaces for navigation

One of the easiest ways to be able to remove the drift from the robot's intended path is for your robot is to periodically align against flat surfaces using a bumper. One key fact to remember about the robot movement is that when moving straight, with any good set of wheels, it is very unlikely that the robot wheels will slip on a surface. Given that the robot wheel is unlikely to slip when moving on a FLL mat, we can use this information to our advantage. When a robot moves using the calibration values, without slipping, it will very closely match the distance that you program it to move. Unfortunately, when a robot moves straight, it does drift sideways in spite of all the care you can put into aligning. See Figure 91 to refresh your memory on robot drift.

The process of aligning against a wall simply involves the robot moving backwards and hitting a wall or other flat surface with its bumper. If the robot bumper is not perfectly flat against the wall, then as the robot keeps spinning its wheels in reverse and keeps trying to go backwards. Eventually the portion of the bumper that is away from the wall will come in contact with the wall. This does one important thing i.e. it puts the robot flat against the wall, which is a fixed reference point. Once the robot is flat against the wall and it moves a fixed distance from the wall using calibrated values, you have high confidence that the robot is in a known position on its intended path. To illustrate this better, let's look at an exercise.

We are showcasing an example in Figure 98 where the robot can move for long distance using the wall and the bumper based alignment to move in a straight line at a fixed distance from the wall.

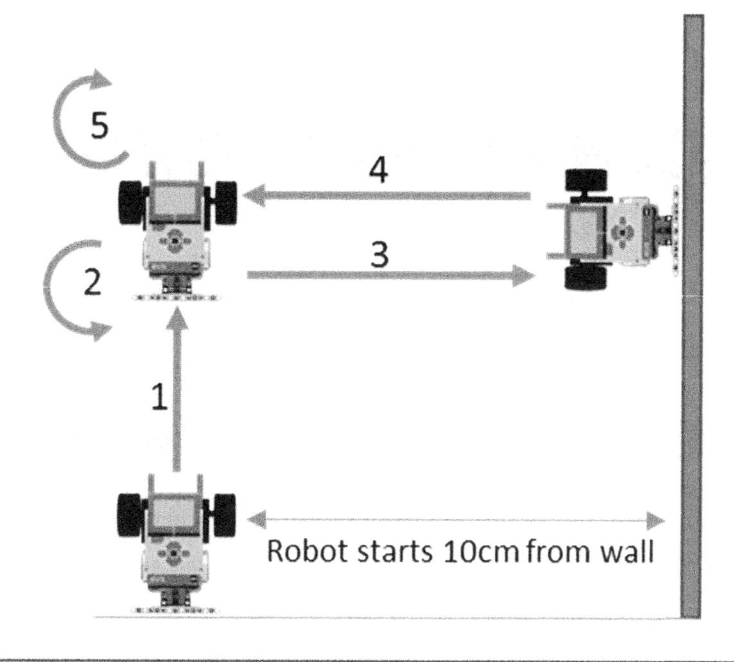

- Step 1: Robot starts 10cm from the wall and goes straight for some distance.
- Step 2: Robot turns left 90 degrees to orient its bumper facing the wall
- Step 3: Robot moves in reverse for more than 10 cm i.e. something akin to 15cm so its bumper hits the wall and the wheels on the robot keep spinning. However, because the wheels are spinning, even if the bumper hit the wall at an angle, at the end of the process, the bumper will be completely flat with respect to the wall.
- Step 4: Robot moves 10 cm from the wall.
- Step 5: Robot turns 90 degrees to the right

Repeat steps 1-5 as many times as required to travel the required straight distance. This method of moving straight while aligning to the wall with the bumper will keep the robot at nearly 10cm from the wall throughout the movement by correcting the robot drift with every wall alignment operation.

Figure 98: Wall alignment with bumper to move straight and counter drift

Exercises with basic alignment

Here are two exercises for you to cement the idea of using wall alignment to make your robot traverse the intended path over long distances by supplementing Odometry with wall alignment.

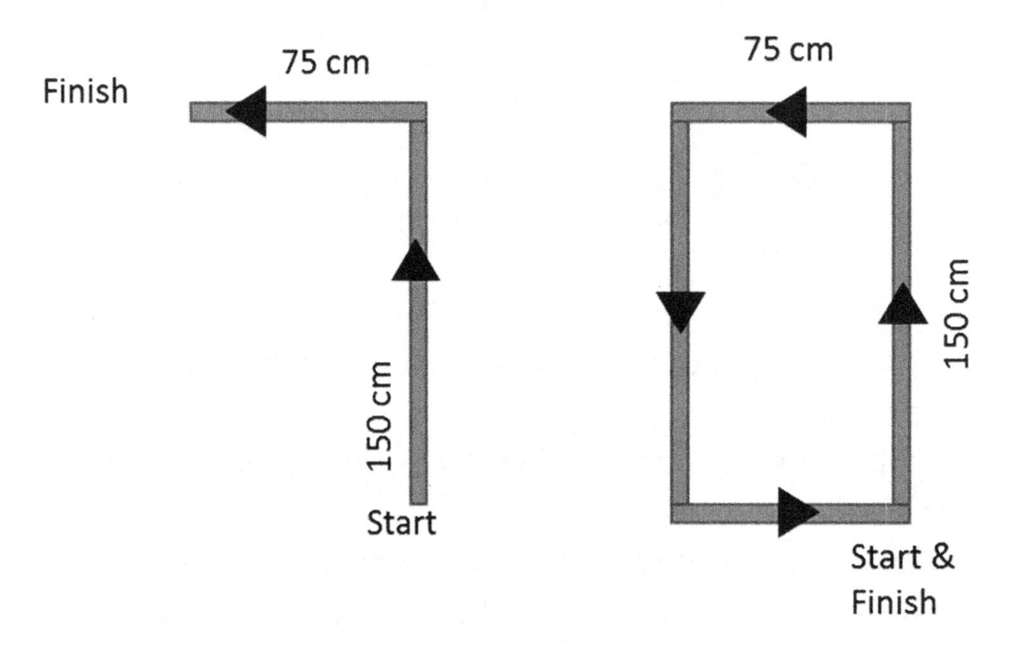

Figure 99: Two exercises, traversing an L shape and a rectangle using bumper based alignment. Arrows indicate direction of movement.

Wall Following using wheels

Wall following is another fundamental technique used in FLL when you have a situation where you need to move close to a wall for long distances. This technique works best when the robot starts from the base next to the wall and must move parallel to the wall.

The technique is best explained with a visual. See Figure 100 for an illustration:

Two wheels that spin freely, attached on one side of the robot allow the robot to follow the wall smoothly. To ensure that the robot *follows* the wall without any hiccups, the motor away from the wall is operated at a very slightly higher power than the one closer to the wall. This practice keeps pushing the robot into the wall causing the *wall following wheels* to continue making good contact with the wall.

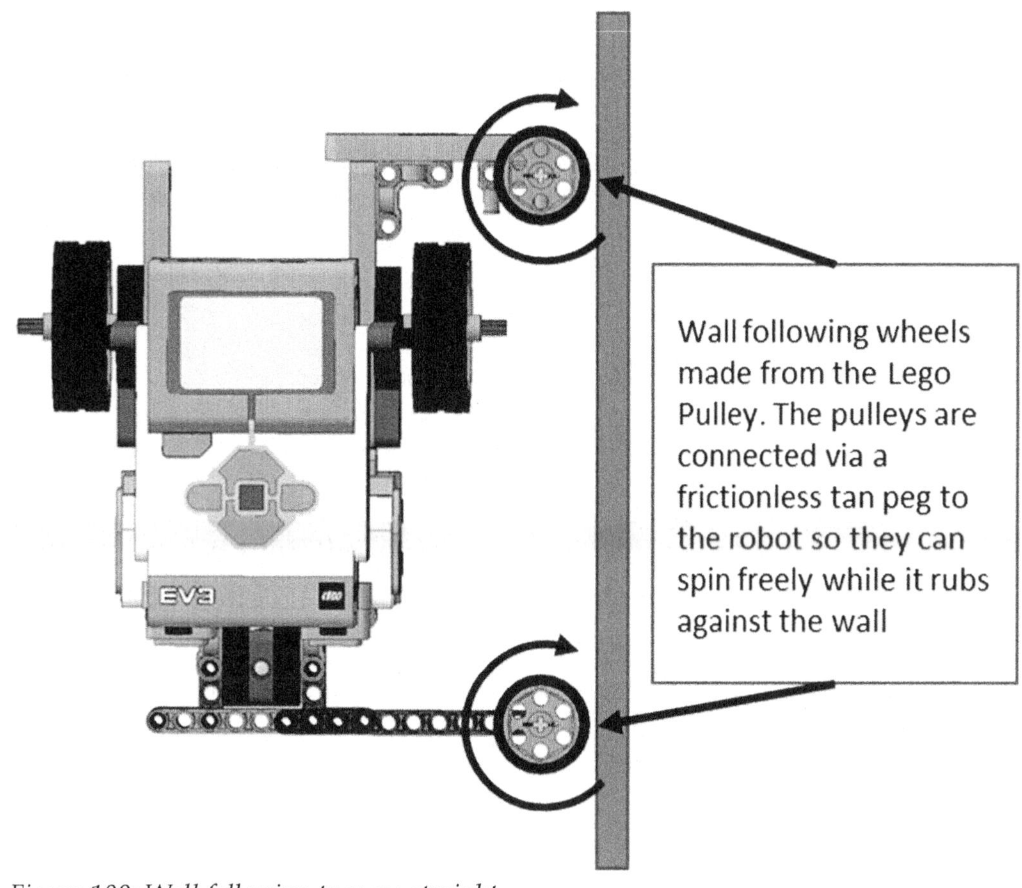

Figure 100: Wall following to move straight.

The wall following wheels must be completely outside the body of the robot. Additionally, no part of the robot except the wall following wheels should be touching the wall.

For sake of clarity, below we are showing the pieces required to make the forward portion of the wall following assembly in the above robot. Additionally, Figure 102 shows a program that will follow the wall with the wall following wheels attached on the right side and the wall to the right of the robot.

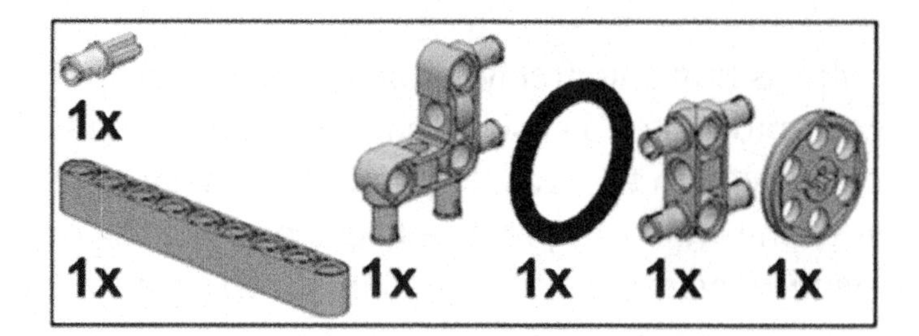

The top wall following assembly and its constituent parts.

Figure 101: A wall following assembly expanded for clarity. The key part to note is the tan peg which is a frictionless peg and spins freely when inserted in a beam.

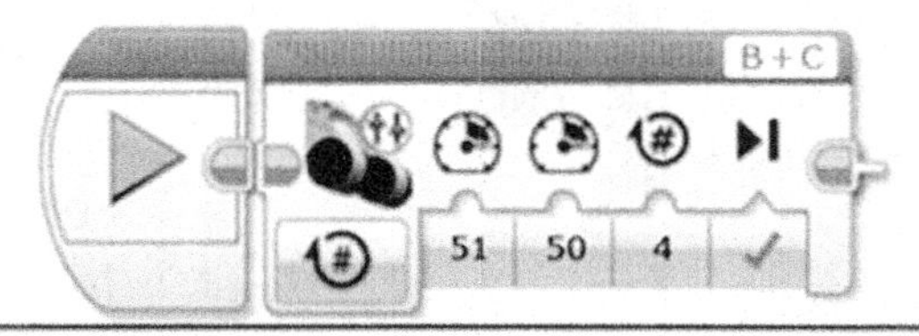

Using the move tank block, we can supply different powers to the different motors. In this case, the B motor is at 51 power and C motor is at 50 power causing the robot to keep pushing slightly into the wall while moving ahead. This is what causes wall following with wheels to work.

Figure 102: Wall following program using the move tank block. One of the few places move tank is clearer to use than move steering

Sometimes you may find that you cannot be right next to the wall from the base itself. In such situations, in order to approach and follow the wall using the wheels, you need to approach the wall at a slight angle. This works as long as the angle is a small angle. Larger angles are more likely to make your robot get stuck to the wall when both the wheels don't make contact with the table wall. Additionally, when you are trying to leave the wall i.e. peel off the wall, you need to use unequal power on the two wheels to try and get off. This is again fraught with errors and although the robot will eventually leave the wall, often it may not be exactly in the position or at the angle you desire. To fix this, you will have to use other navigation markers. In general, we recommend using the wheel based wall following when you can move forward and backwards right from the base without any obstacles on the way.

Summary

In this chapter, we went over some of the most fundamental issues required for you to be able to move your robot in a precise and consistent manner. The techniques presented in this chapter such as the Starting Jig, aligning consistently and taking care of the gear slack will make any rookie team be well prepared for the FLL competition. Techniques such as Wall Alignment and wall following will set a team apart and provide greater certainty in FLL Robot Games.

MyBlocks

Although it is great performing the calculation for moving and turning to reinforce the concepts, eventually, performing the calculation and setting all the inputs on the Move Steering or Move Tank blocks does get tiresome and error prone. To remove chances of an error and to encapsulate the logic of calculating the degrees based on the robot calibration values, we wrap all this logic in entity known as Myblocks. Myblocks are user created blocks that you can create and use in your programs. Myblocks allow us to create custom logic in an easy and portable manner.

Pre-requisite to Myblocks

Before we can use MyBlocks, we need to gain a bit of understanding in using variables as well as mathematical operations in EV3. Traditionally, variables and math blocks are considered advanced topics and are usually covered much later. However, we find that once participants are reasonably familiar with the calibration and how to use mathematical principles for accurate straight motion and turning, it is a lot easier for them to wrap all that functionality in a MyBlock to make future progress a lot smoother.

Variables

Variables in EV3 programming are same as covered in a Math course in an elementary school. In a much simpler way of speaking, a variable is like a container in which you can store things and later retrieve them. So, the most apt analogy that we drive for students is to equate it with a suitcase. You put things in a suitcase and then you can take things out of a suitcase later (See Figure 103). The analogy provides a good starting point if you are completely new to programming but for anyone familiar with a programming language such as C, Java, JavaScript, Python or any other language, using variables in the EV3 programming is a bit unintuitive.

Variables are like suitcases that can contain values. Here, a suitcase represents a variable. The name of the variable is *counter and it* contains a value 10.

Figure 103: Creating a variable

Like a real suitcase, you will either be (a) putting something in the suitcase or (b) taking something out of a suitcase. In EV3 terminology, putting something in the suitcase is called *writing* to the variable and taking something out of the suitcase is called as *reading* from the variable. The part that gets confusing sometimes for kids is that a variable is a concept that describes a place to store a value and is different from its visual representation in the program. Its visual representation in EV3, which incidentally is also a block with a picture of the suitcase needs to be placed every time you need to *write to* or *read from* the variable. This does not keep creating new copies of the variable. Rather all those blocks are referring to the exact same storage location. Not only that, you must specify in the EV3 variable block that it is in either a read or write mode as shown in Figure 104.

All variable blocks with the same name in an EV3 program whether they be in read or write mode refer to the same value storage location. For some reason, this takes a bit of time to sink in kids minds even when they are comfortable with variables through math class.

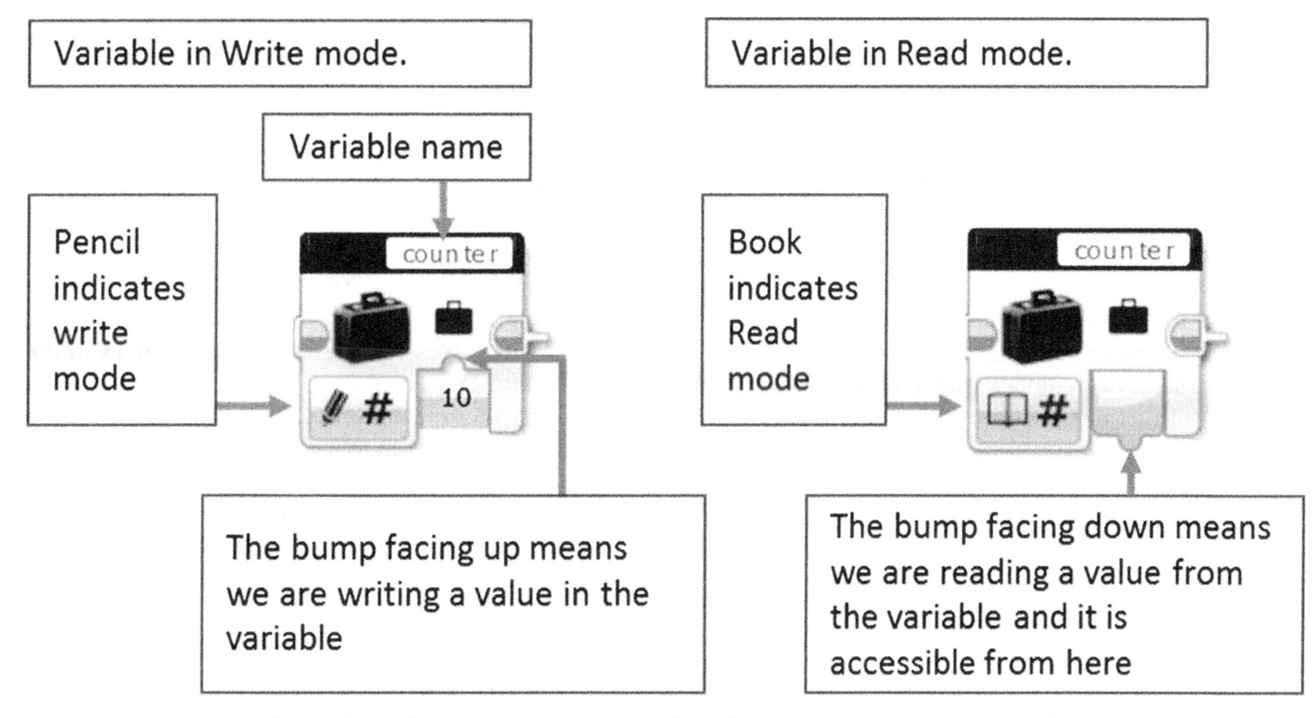

Figure 104: EV3 variable in write and read mode. Even when two variable blocks are placed on the EV3 program, as long as they have the same name, they refer to the same storage location.

Although EV3 variables can be of any of the Text, Numeric, Logic, Numeric and Logic Array types (see Figure 105 for all the options) in competitions, we have rarely, if ever needed anything beyond the Numeric variable type. In the variable block, numeric type is represented by the # symbol. For sake of clarity, Numeric type variables can only hold number values. They will produce an error if you try to put other kind of values in them.

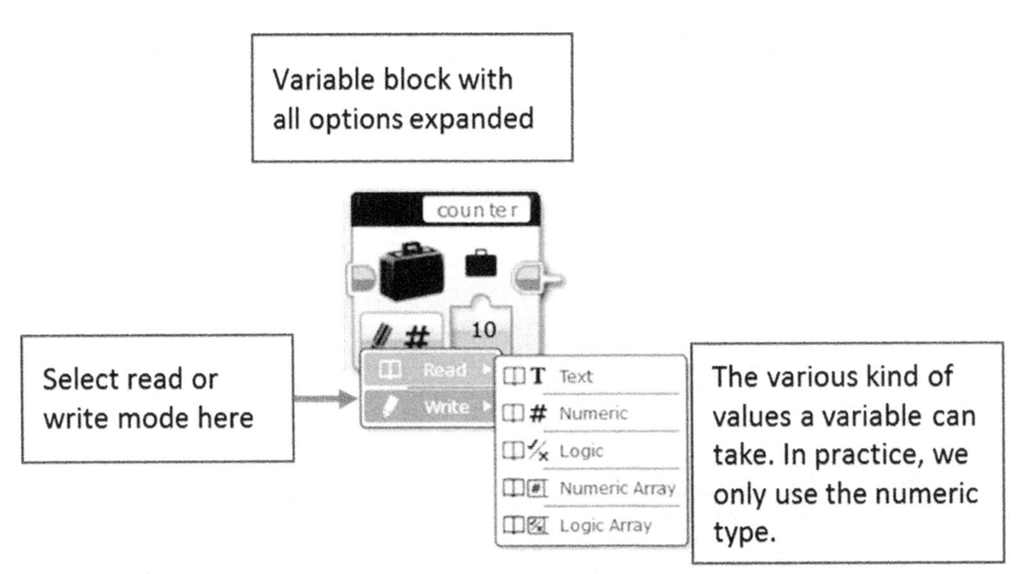

Figure 105: The various options in EV3 variables

A variable gets created as soon as you drag a variable block onto the EV3 editor and assign it a name by clicking the name field and hitting *Add Variable (see* Figure 106 for illustration). You may create a variable by using the variable block in either *read* or *write* mode. The variable will get created the first time you place a block with a unique variable name. It is important to remember that unless you use the variable block in the *write* mode to create a variable as well as provide it an initial value, the variable by default will have a zero value. In general, when programming, it is not a good idea to rely on default values of variables since different languages have different ways of assigning initial values to the variables. C language will often not initialize variables whereas a language such as C# will report an error if you attempt to read the value of a variable before assigning it a value. Thus, depending on language, not assigning a value to a variable before using it can often lead to unexpected results. For sake of consistency and clarity, it is always a good idea to create a variable using the *write* mode and assign it an initial value, even if you assign it the default value of 0 - if the variable is a numeric variable.

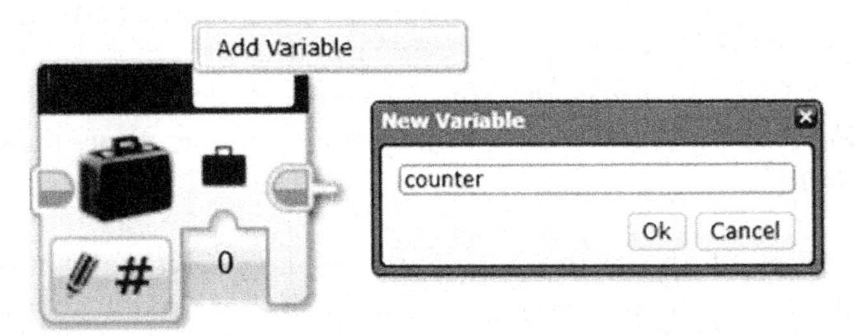

Figure 106: Creating a new variable named Counter. Clicking at the top right section and selecting "Add Variable" brings in the "New Variable" dialog where you can supply the name.

If you look in the red palette of the EV3 tool, you will also note another variable block that has the picture of a suitcase lock on top of it. This block refers to variables whose value does not change after initial assignment. Such variables are called constants. Although we are in favor of using constants in other programming languages, we do not recommend using these blocks in EV3 programming since these variables only have a value and cannot be assigned a name. This will make it hard to remember at a later point of time what they are representing. Regular variables contain all the features of this block and allow you to name them and hence are preferred.

Data Wires

Usually, when we are using variables, we assign them fixed values only in the beginning or the first time we create them. In later parts of the programs, we want to read their values and use it as input to other blocks as well as update their values dynamically. The way this is done is by connecting the output of a block or variable to the input of another block or variable. The connection of values from one block to another is performed via a concept known as *data wire* represented by a colored line that starts from under one block and goes to the next one as shown in Figure 107.

Before we start using *Data Wires* to connect blocks, there is a fundamental tenet of EV3 programming that we need to review. The EV3 program is a *linear flow* of blocks i.e. blocks that come earlier in the program are executed earlier whereas blocks that come later in the program are executed later (Figure 107). Thus, if a block depends on the value of a variable and wants to read its value, then the variable should have had the required value written into it some place earlier in the program. This sounds obvious, but we have seen kids make mistakes on this topic and thus bringing it to your attention.

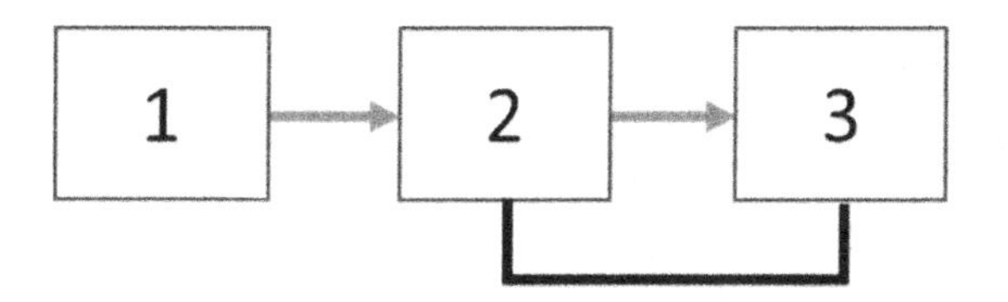

Figure 107: EV3 programs flow linearly and thus even though there is no direction specified, the red line, which denotes a data wire which transfers values from one block to another, is transferring values from block 2 to block 3.

How to connect two blocks using data wire is best explained visually, so refer to Figure 108 for the explanation.

In traditional programming languages, variable writes i.e. assignments are done from right side to left side e.g. the line "distance = counter", which implies that the value from the variable on the right side, "counter" is placed in the variable on the left side i.e. "distance". In EV3, this is inverted as shown in Figure 108 because of the linear flow of the blocks from left to right. This is quite often counter-intuitive to both kids and adults.

---

1. counter

Spool Icon

A variable in read mode. When you hover your mouse over the output section, the mouse pointer converts into a thread spool indicating that this output can be connected to another place.

2. counter distance 0

Two variables, *counter* and *distance*. We want to connect the output of *counter* to *distance*, so we left click on the output of *counter* and while holding the left mouse button, drag the mouse towards the input of *distance* variable. The yellow line that shows the connection is called the *data wire*.

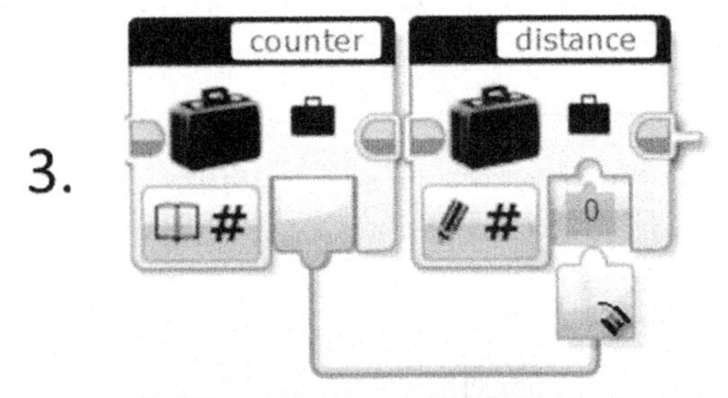

As the data wire from the output of *counter* reaches the input of *distance*, you see the input show a blue fill indicating, that a connection can be made.

Once the *data wire* has reached the input of *distance*, you can let go of the mouse button and a solid gold connection is formed indicating that the value of *counter* will be transferred to *distance*.

Figure 108: How to connect two blocks using a data wire. Keep in mind that in majority of cases, the type of the output should match the type of the input e.g. numeric output should ideally be connected to numeric input unless you are trying to do something out of the ordinary such as displaying the value of a variable on screen.

Even though we have mentioned earlier that you should always make sure that the type of the output of a block should be the same as the input of another block if you are planning to connect them using data wires, there is one exception that comes up quite often. The exception is when you want to display the value of a variable on the screen. To display anything on screen, you need to use the *Display Block.*

Display Block

The *display block,* as the name suggests, displays text on the EV3 screen and hence requires a text input. If you are trying to display the value of a *numeric* variable on the screen, simply connect the output of the *numeric* variable to the *text* input of the display block and the display block will automatically convert the numeric value to text and display it appropriately. The *display block* is present in the green (action) tab. Below (Figure 109) we explain the display block visually.

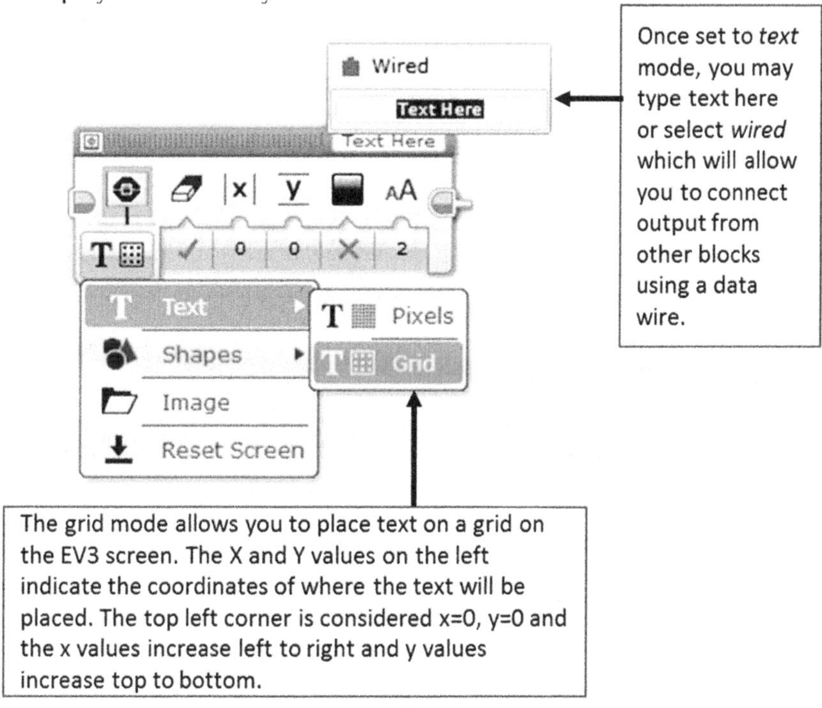

Figure 109: The display block with the text. Note that depending on the font size used, as shown in the last input to the block, you may be able to fit different amount of text on the block.

Figure 110 shows the input of the display block connected with data wires to the output of a variable to display it on the EV3 screen.

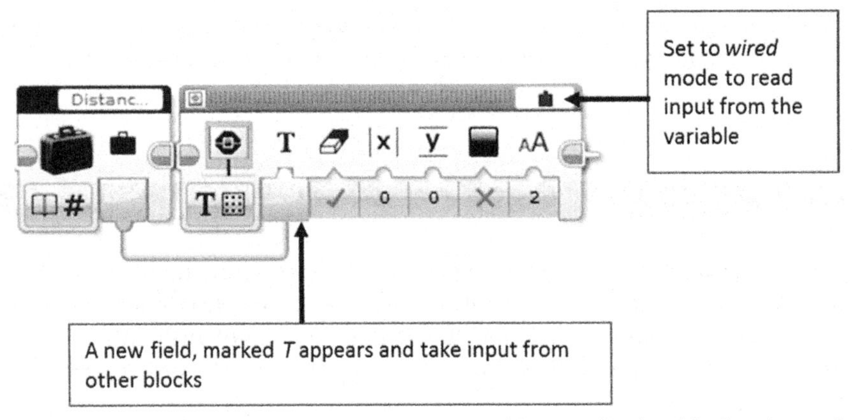

Figure 110: Connecting a variable to a display block using a data wire

---

Even though the above shows a variable connected to a display block, if you have a program with just the above two blocks along with a start block, nothing will display on the screen because the EV3 brick will finish the program in a fraction of a second. You will need to use a Wait block at the end to see the output on the screen. The reason for this is explained in the next chapter.

The Math Block

The *Math Block*, as the name suggests, is a block that can perform math on numbers or numeric variables. It is available in the red palette (Data Operations.) See Figure 111 for details.

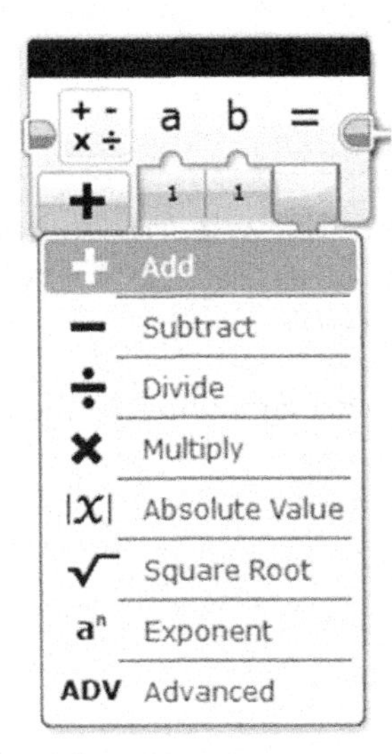

The Math block usually has two inputs and one output. You can either type a number directly in math block or connect a variable to its input using a data wire. The result is available through the output section. The Absolute Value and Square root need only one input and hence have just one input section.

Figure 111: The math block with the options expanded.

The one key thing to note in the options for the *Math Block* is that it has an advanced mode. The advance mode allows user to accept up to 4 inputs, supports a lot more operations than what are shown in the UI. Additionally, this block allows one to write complex math formulas as shown in Figure 112. In particular, it supports Modulo, Exponent, Negate, Floor, Ceil, Log base 10, Log base 2, Sine, Cosine, Tangent, Arc Sine, Arc Cosine and Arc Tangent.

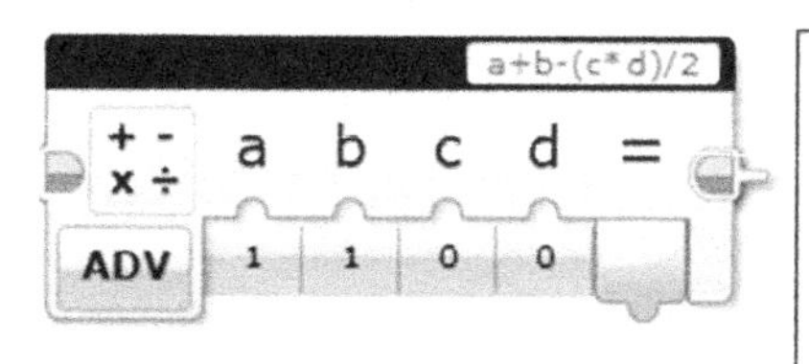

The Advanced section in the math block allows you to type the equation that you want at the top right. It supports up to 4 variables. If you don't want to use a variable, simply ignore it in the equation at the top corner and it will not be used.

Figure 112: The Advanced selection in the Math Block can simplify writing complex equations and reduce the number of blocks in your program.

Exercises

Here are a few exercises for you to get familiar with variables. It is recommended; however, to familiarize yourself with the Wait block in the next chapter to before attempting them. Programming, in general, requires multiple concepts almost simultaneously to be able to perform a lot of common tasks. This is just the nature of it. We try avoiding jumping between topics, but unfortunately, this one is a bit unavoidable. You may skip the exercises and come back to them when you are done with the Wait blocks and have covered the Wait block with the brick button.

Exercise 1

- Create a numeric variable in write mode and name it TestVariable. Give it an initial value 100
- Display the variable's value on screen
- Wait for EV3 Center brick button press
- Read the value of the variable, increment it by 2
- Display the value on screen
- Wait for EV3 Center brick button press

Exercise 2

- Create four variables - call them RedVar, BlueVar, GreenVar and ResultVar
- Give them initial values of 30 for Redvar, 50 for Bluevar) and 90 for GreenVar
- Calculate the equation GreenVar - (RedVar * BlueVar)/2 and put the result in ResultVar.
- Display the Value of ResultVar on screen
- Wait for EV3 centre brick button press
- Repeat the exercise with and without using the Advanced mode in the Math Block.

Exercise 3

- create a variable called RobotDegrees
- Make the robot go forward 100 degrees using move steering block
- Using the motor sensor block, get the degrees and put it inside RobotDegrees
- Wait for Center brick button press
- Using the RobotDegrees variable make the robot go back the amount RobotDegrees

Exercise 4

- Create a variable called RobotDegrees
- The variable can have negative or positive degrees
- Always run the robot *forward* by RobotDegrees no matter whether the number is positive or negative. Test by setting RobotDegrees as negative.

MyBlocks for Moving Straight and turning

Armed with the information on how to create variables, how to connect the output and input of two blocks using data wires and how to use Math Blocks, we are finally ready to create a myblock for moving straight and turning. First, let's create the myblock for moving straight.

To create a myblock, it is necessary to first write a program using variables where the distance is provided as a variable (let's call it distance) and the *Moving straight* ratio (let's call it GoStrRat) is also provided as a variable. Recall that in the previous chapter, we calculated a Moving straight ratio of 18 using a sample robot. Using these variables, and the sample Moving straight ratio we can now write our program for the robot to travel 50 cm as:

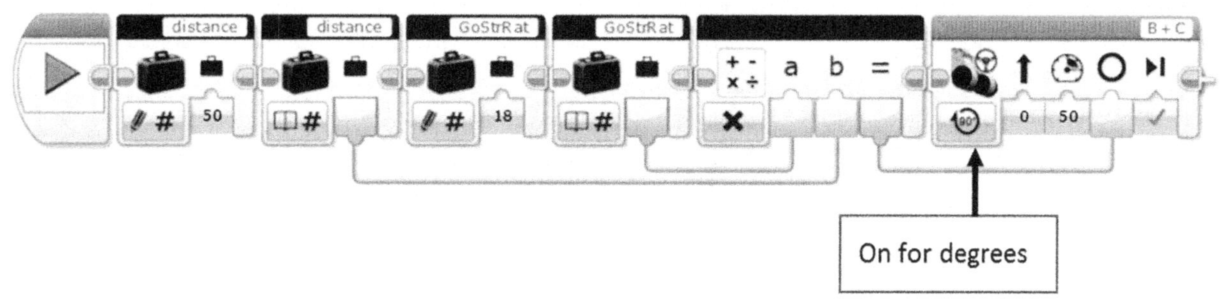

Figure 113: A program to make the robot travel 50cm using variables and math block. Here we calculate the number of degrees required in the move steering block by calculating the value of distance x GoStrRatio and feeding it as the degrees in the Move steering block.

To convert this into a myblock with the distance being fed as an input, select all the blocks except the distance variable blocks and the start block. Note that once a block is selected, a slight, light blue halo surrounds the block. You may select multiple blocks by left clicking and holding the mouse button and dragging the mouse over the blocks you wish to select. Alternately, on Windows you may also keep pressing Ctrl button and clicking the blocks you wish to select. This is how it will look like when you are done.

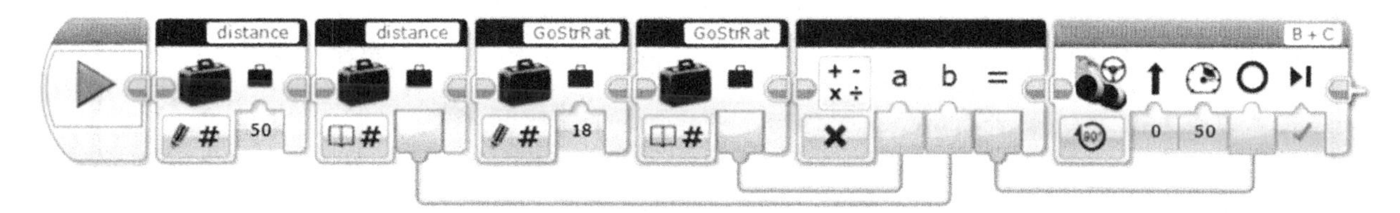

Figure 114: Step 1 in creating a myblock. Select all the blocks except the variable blocks that you will be supplying as input. Here, only the distance variable will be supplied as input since the GoStrRat is fixed and will not change. Notice the blue halo around the blocks that are selected.

Next, from the menu options in EV3, select Tools -> "My Block Builder" which will show the following window:

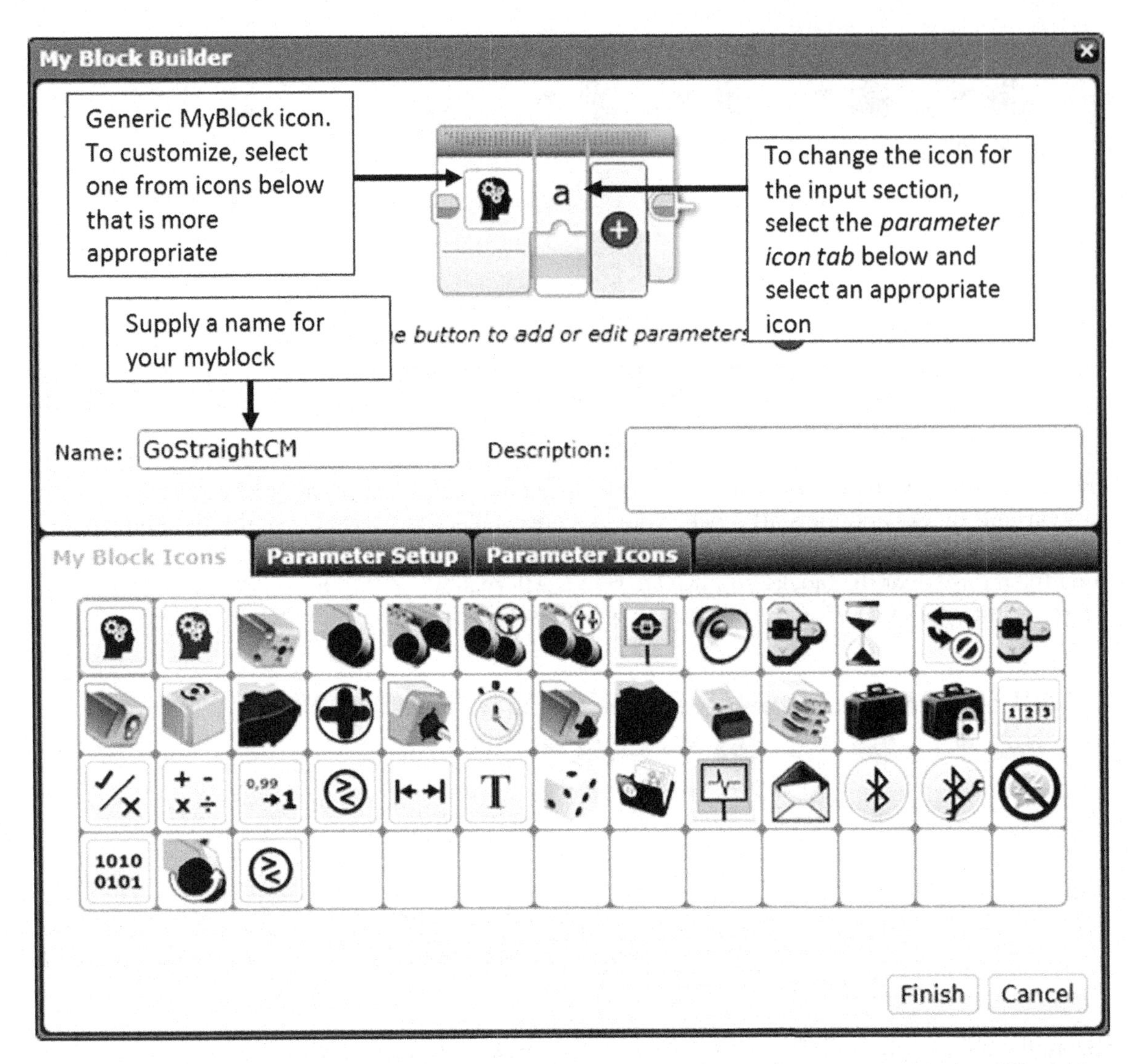

Figure 115: The My Block Builder Window. It shows how your myblock will look like. Currently, it shows that this myblock will have one input. You know that the input the block is going to be the "distance" variable because we left it out when selecting the blocks for inclusion within the myblock. You should definitely customize the myblock icon as well as the parameter icon to make their meaning clearer at a glance.

Here we show the My Block Builder with appropriate icons selected to represent that the myblock is a robot movement for some distance block and that the input is in centimeters.

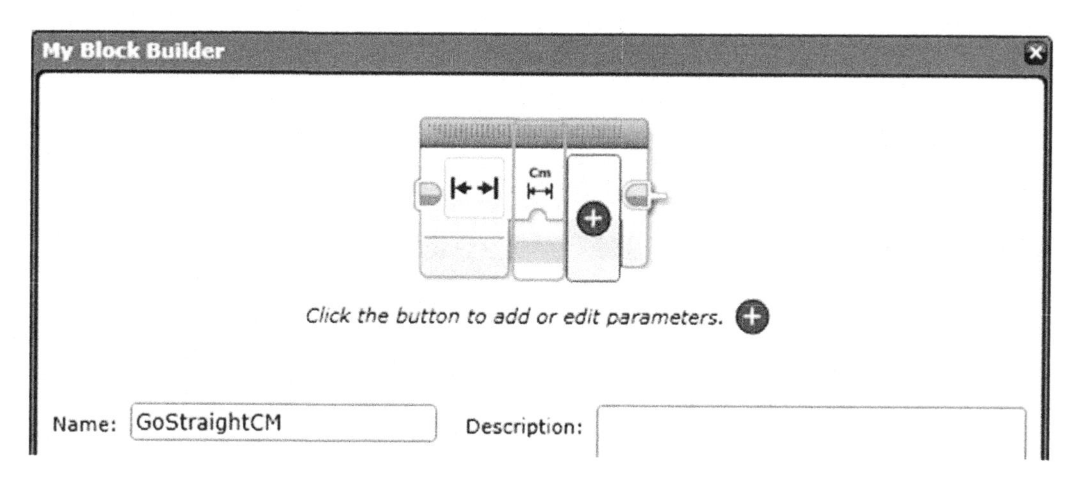

Figure 116: My Block Builder with appropriate icons selected for both the block as well as the input

Finally, hit the *Finish* button and your program will suddenly look like the following (Figure 117):

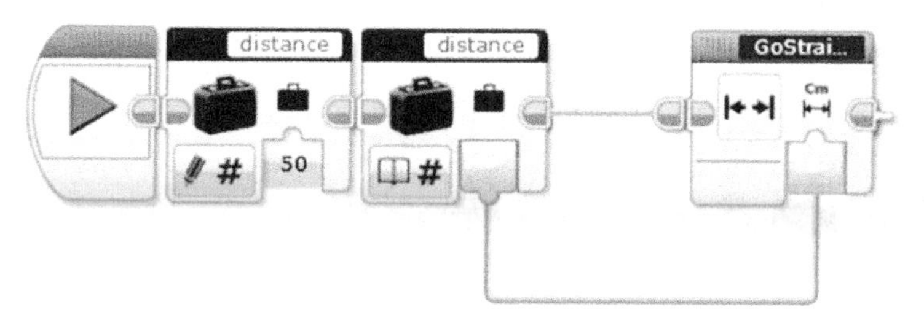

Figure 117: The program after the Myblock has been created.

At this point, you have created the myblock for Moving straight and you can use it just like any other block in the EV3 palette. This block will be now available in the light blue tab at the bottom. Also note that, at this point, the *distance* variable blocks are superfluous since the GoStraightCM block can accept input by simply typing the distance in its input field as shown below.

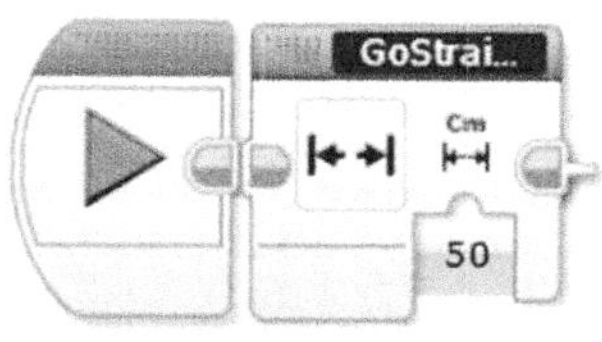

Figure 118: A myblock can accept input directly in its input field and have other blocks connect to it using data wires just like the original EV3 blocks. Here, we can type the distance directly in the input of the myblock.

The above Moving straight block works both for going forward as well as going backwards. To move forward, supply the degrees in positive and to go backwards, supply negative degrees.

Below, we are showing the program for Turning using the sample turning ratio of 2.0.

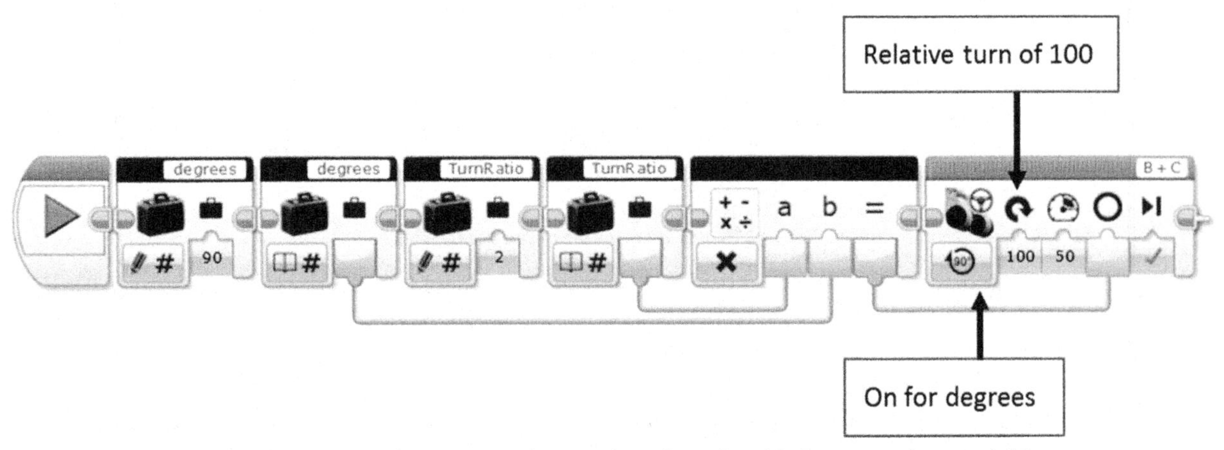

Figure 119: The program for turning the robot 90 degrees using variables.

Just like the Moving straight MyBlock, to turn the above into a myblock, select everything except the degrees variable blocks and the Start block and use My Block Builder to create a myblock. The Myblock we create will look like the following:

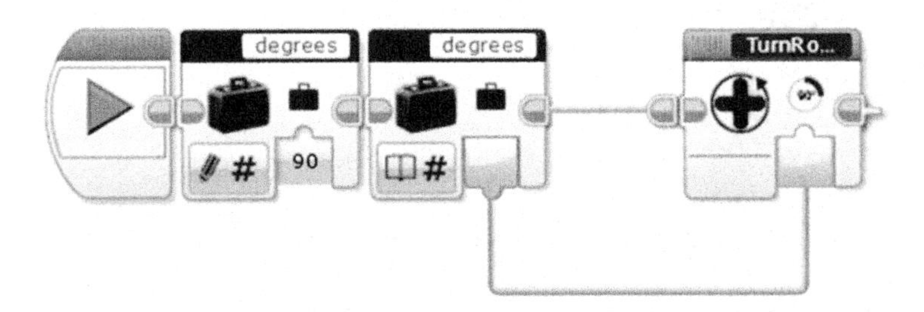

Figure 120: The TurnRobot myblock after creating a myblock from the above program.

The above myblock works both for turning right and turning left. To turn right, supply positive degrees and to turn left, supply negative degrees as input to the myblock.

If you make a mistake when creating the myblock and want to correct it, simply double click the myblock and it will open up in a new tab. You can fix the mistakes and save it. The changes will automatically reflect in all the program within the *same project.* An open myblock is shown in Figure 121.

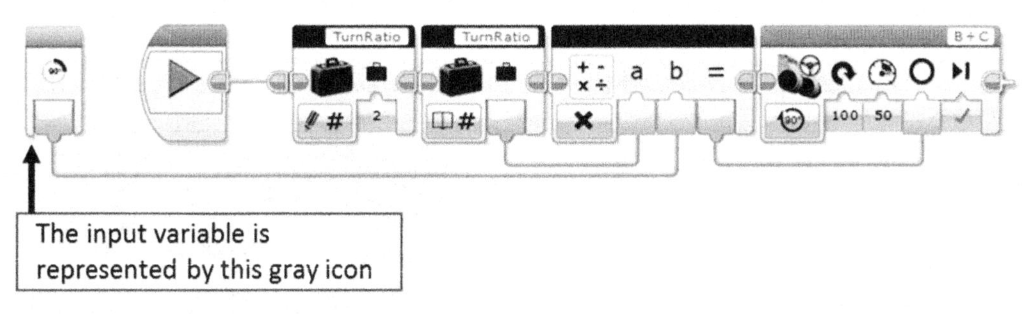

Figure 121: An open myblock. You can double click a myblock in a program to open it and correct any mistakes.

Editing a myblock will update all the programs that use the myblock in the current project. It WILL NOT have an impact outside the project however.

EV3 blocks are available only within a project and do not automatically transfer from one project to another. Once you have created a myblock, it is a great idea to export it and save it in a safe location so that you can find it and include it in other projects.

To export myblocks from a project and then later import them in another project, you should utilize the wrench icon to the left of the EV3 window and then use the My Blocks tab and use the import and export buttons. See Figure 122 for illustration.

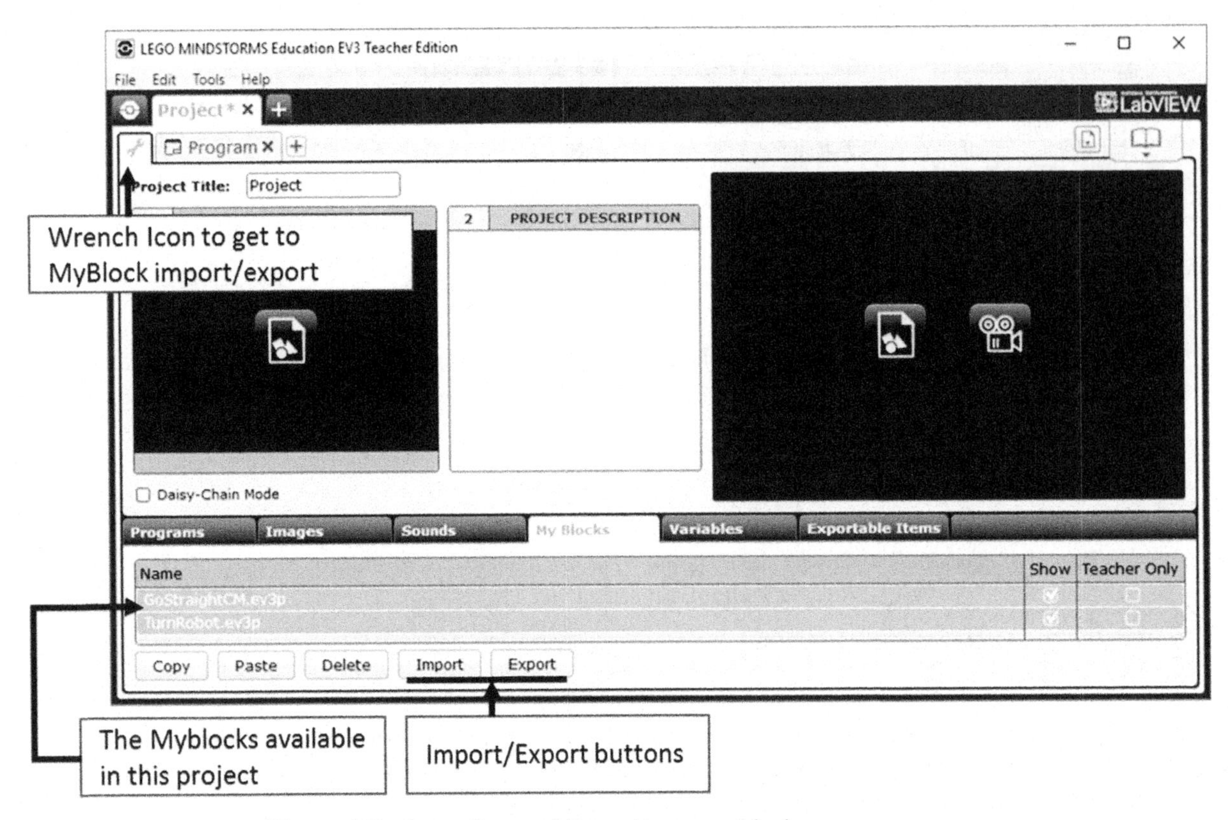

Figure 122: Importing and Exporting a myblock

Finally, as we finish this chapter, it will be a great idea to repeat some of the exercises in the robot movement chapter using the myblocks and see how MyBlocks simplify programming a great deal.

Summary

MyBlocks are the staple of FLL programming. Right from the move and turn MyBlocks, you will be encapsulating a lot of reusable functionality in myblocks. As you get more experienced, you will be using more and more myblock to streamline and optimize your programs.

Basic programming blocks

In this chapter, we are going to look at some of the essential blocks that are part of any FLL program. We will primarily focus on the flow control blocks, but we are also going to briefly explain the single motor block since it is heavily used for driving attachments. We have already explained the Move Tank and Move Steering blocks in earlier chapters but had omitted the single motor blocks, which is explained below.

The Single motor block

The single motor control block comes in two flavors. One for controlling the large motor and the other for the medium motor as shown in Figure 123. Both these blocks are very similar to the move steering or move tank blocks except that they control only one motor at a time. Normally, the single motor block is used for driving attachments and the number of degrees or even rotations that it needs to turn are experimentally verified based on the range of motion that is needed for the purpose at hand.

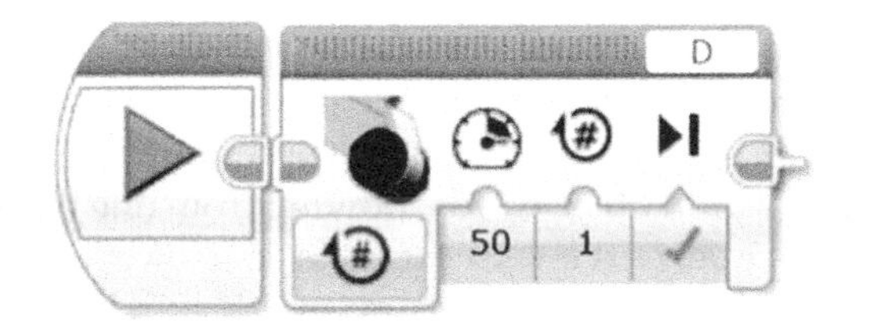

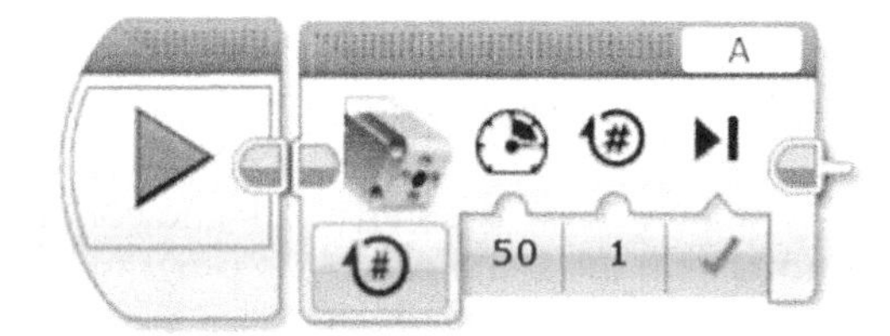

Figure 123: Two programs showing the single motor block. The left program turns a large motor connected to port D and the right one does the same for a medium motor on port A.

Flow Control Blocks

Although a majority of the FLL programs are written using nothing more than linear placement of blocks without using any conditional statements or repeating instructions, both conditional statements and repetition of instructions is used inside MyBlocks that are then directly used in the robot path planning. In this chapter, we will be explaining about the flow control blocks which, allow you to create conditional statements and to repeat a set of instructions. The flow control blocks are shown in Figure 124.

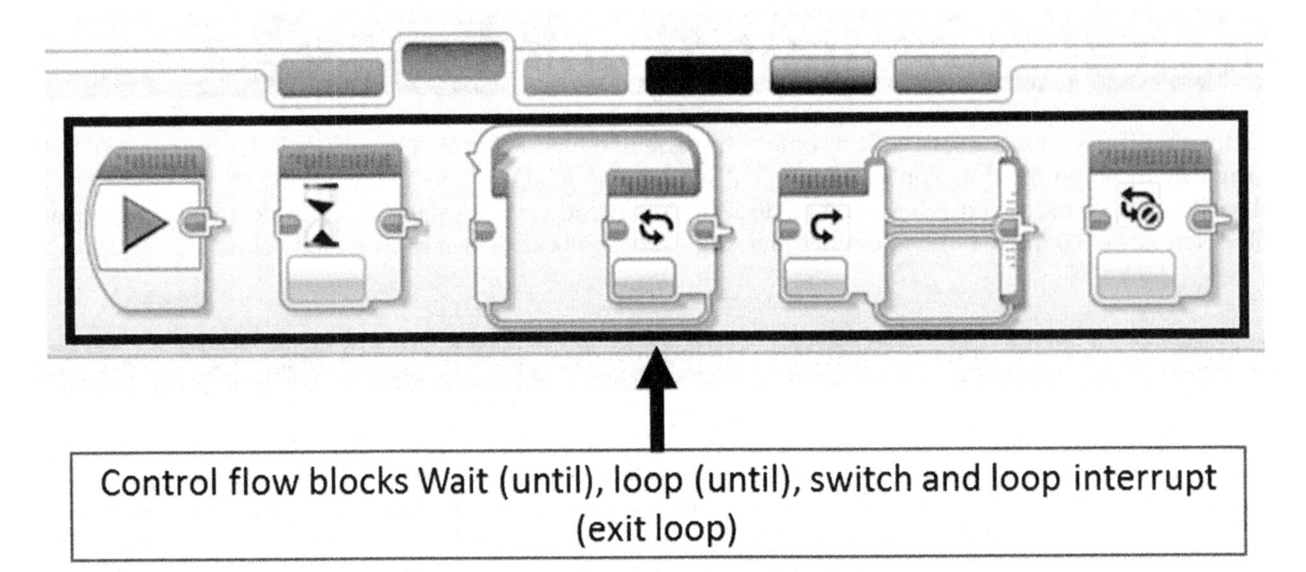

Figure 124: Flow control blocks

Flow Control is a programming terminology which refers to blocks that change the way your program executes. Since a program is a series of instructions, it is logical to assume that the program will start from the first block i.e. the start block and continue executing - or in other words, flowing from one block to another till the last block in the program is executed in sequence. Flow Control blocks, however, change this linear flow. Instead of flowing from one block to another without interruptions, the flow blocks can:

- Make the program wait until a certain period of time has elapsed or until a certain condition is fulfilled e.g. wait until the robot has reached the end of the wall and is pressing against the wall.
- Make the program repeat a set of blocks for a certain period of time, a certain number of times or until a certain condition is fulfilled e.g. go straight 50cms and turn left - repeated 4 times to make the robot finish traversing a square.
- Make the program take one among a few different options depending on a condition e.g. if the robot's color sensor sees red color, it turns left versus if it sees green, the robot turns right.

In other words, the *flow control* blocks make your program behave slightly differently from the usual block after block linear execution. Instead, like rivers, they can fork the block execution stream to go along different paths and then merge them later at a later point of time or make them repeat a set of blocks or even make them wait at a certain point in your program. In the following sections, we go over the flow control blocks in detail and explain where and how they are used as well as some of the pitfalls that you can incur when using them. One key thing to keep in mind is that while we introduce these flow control blocks, we will be using some of the EV3 sensors to illustrate the *flow control* blocks although we have not yet covered the sensor blocks. The reason is that flow control blocks are very heavily reliant on the sensors, using them in conditional statements and as means to terminate a repetition and blocks. It is a bit of a chicken and egg problem trying to introduce the flow blocks first with the help of sensor blocks or the sensor blocks with the help of control flow blocks. However, we believe that to be able to write meaningful and non-trivial sensor based programs, you first need to understand the flow control blocks. If the material in this chapter does not make sense at first, plough through it and continue onto the sensor blocks in the following chapters. We promise that things will make sense a second time around once you have a bit of context from all around.

In later chapters, you are going to see Sensor blocks that are stand alone and can be used to get the state of the sensor without needing the flow control blocks. However, in a majority of your programs, you will be using the Sensor Blocks as part of the Flow Control blocks. Keep this in mind as you go through Flow Control and later Sensor Blocks

---

Flow Control: The Wait block

We like to call the wait block as the *Wait Until* block since that better represents its actual function. The wait block is an interesting block in that it allows the robot to wait for an event to occur. E.g. in its simplest incarnation, the wait block can wait for a certain amount of time. This simple version has limited usefulness. However, the wait block can also wait for other conditions, especially, waiting for events triggered by sensors such as *Wait until the touch sensor detects a touch* or *Wait until the color sensor detects a certain color.* A full listing of all the useful Wait block options is shown in Figure 125. *Wait* block is one of the few blocks where the program stops and waits for the event specified by it to take place before it moves on to the next step. One key fact to keep in mind about the wait block is that if you had switched on a motor before the wait block, the wait block will not stop the motor; instead the motor will keep on rotating until the wait condition is fulfilled or program is interrupted manually, or the program concludes. This is quite commonly used in robotic challenges where sensor use is intended.

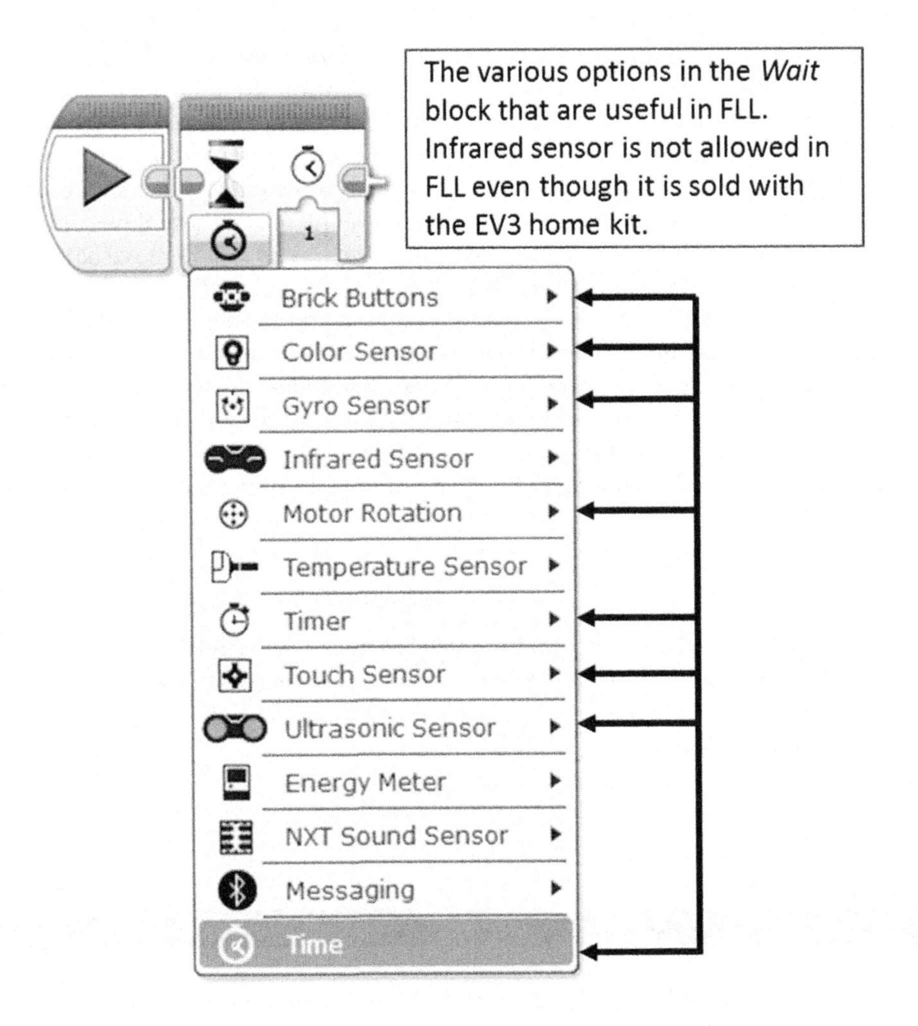

Figure 125: The various options in the Wait or Wait Until block. Although the highlighted options are usable in FLL, practically, only the Brick Button, Color Sensor, Gyro Sensor, Motor Rotation Sensor, Touch Sensor and Ultrasonic Sensor are the ones commonly used.

Note that the wait block does not stop motors that are already running. They will keep on running until the condition specified by the wait block is met.

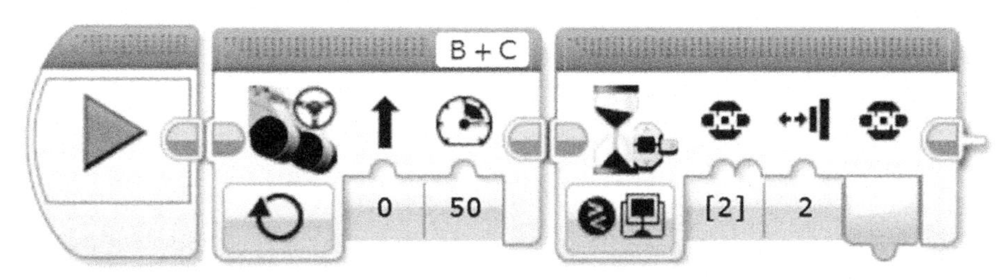

Figure 126: The Wait Until block with the wait condition set as Wait for the center brick button to be pressed. The robot will continue to move until this happens.

When using a wait until block, it is extremely important to specify the pre-wait until condition properly. For example, with wait-until is set for pressing touch sensor and the wheel rotation is set to number of rotations (*Figure 127* (a), (c)), as precondition, then the robot will attempt to complete a rotation even if the touch sensor may have been pressed. This will cause the robot to lock at the contact point of the wall Figure 128. Additionally, between each one full rotations the robot will turn on and off causing a jerky motion. Instead the condition should be set to on (*Figure 127* (b), (d)).

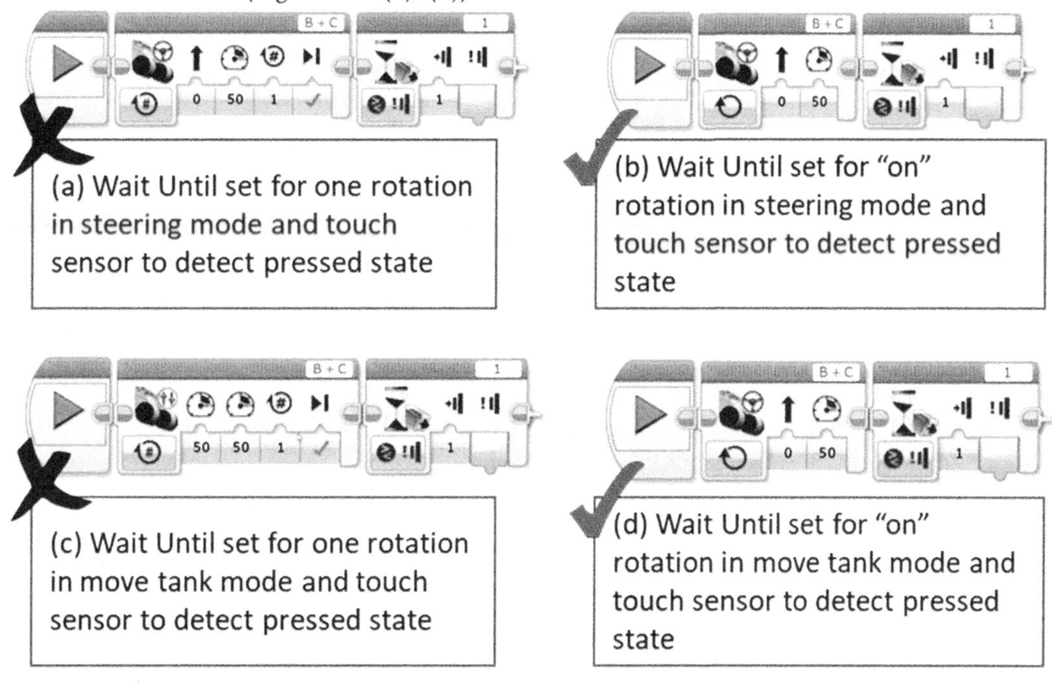

Figure 127 : The wait block with wait condition set as touch sensor to be pressed.

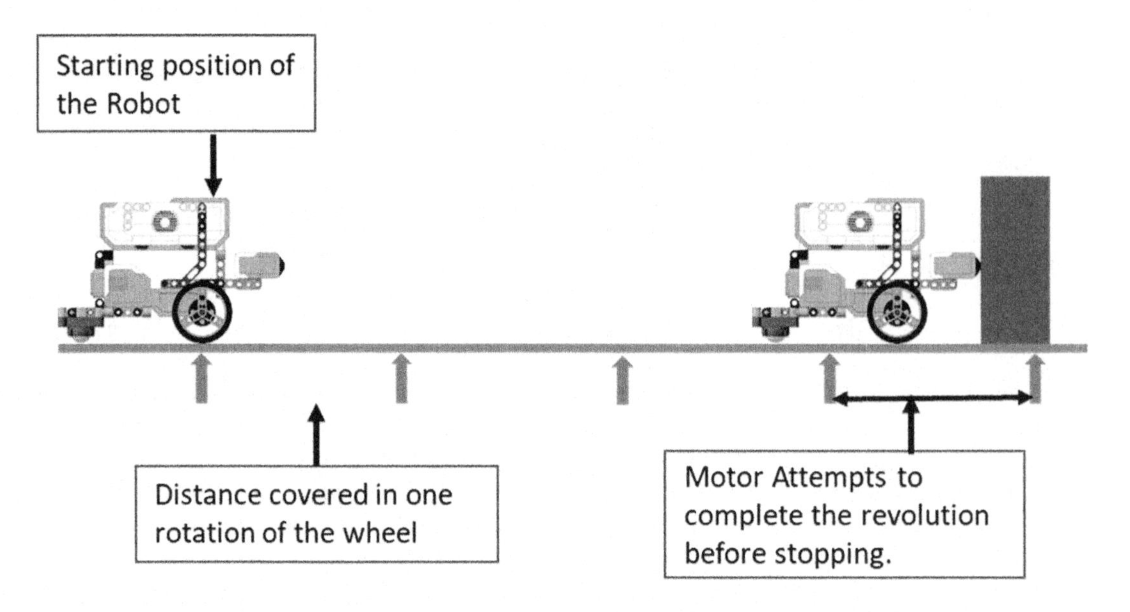

Figure 128: When Pre-Wait rotational condition is set to full rotation instead of on, then Robot will attempt to spin the motor full rotation and hence will lock the motor (Figure 127 (a),(c)).

As we mentioned above, Wait block is one of the rare blocks that actually makes the program stop until the condition specified by it is met. Some other blocks that make the robot stop until the operation that the block is performing is fully executed are: move steering/tank blocks with fixed number of rotations or degrees, single motor blocks and sound block. The motor blocks are of special importance to keep in mind because if a motor gets stuck for some reason, the program will wait and wait for the motor to finish making the amount of movement programmed. This will cause the entire program and thus the entire robot to stall.

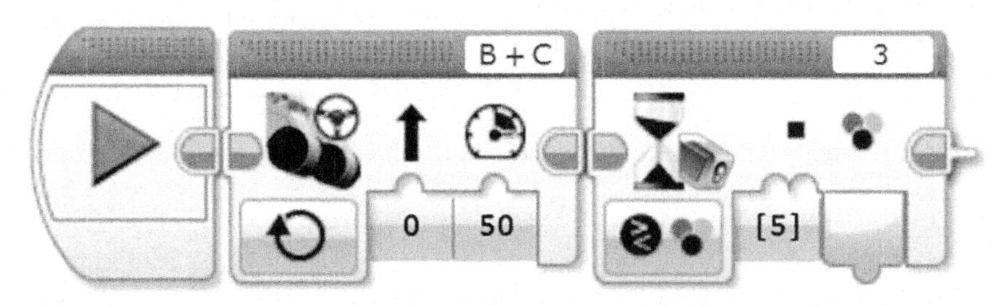

Figure 129: The Wait Until block with the wait condition set as Wait for the color sensor on port 3 to detect red color. The robot will continue to move until this happens.

Flow Control: The Switch block

The switch block is a conditional block that chooses one among multiple sets of blocks and executes one set depending on whether a condition is fulfilled or not. The switch block is primarily used to test whether a sensor is in a certain desired state.

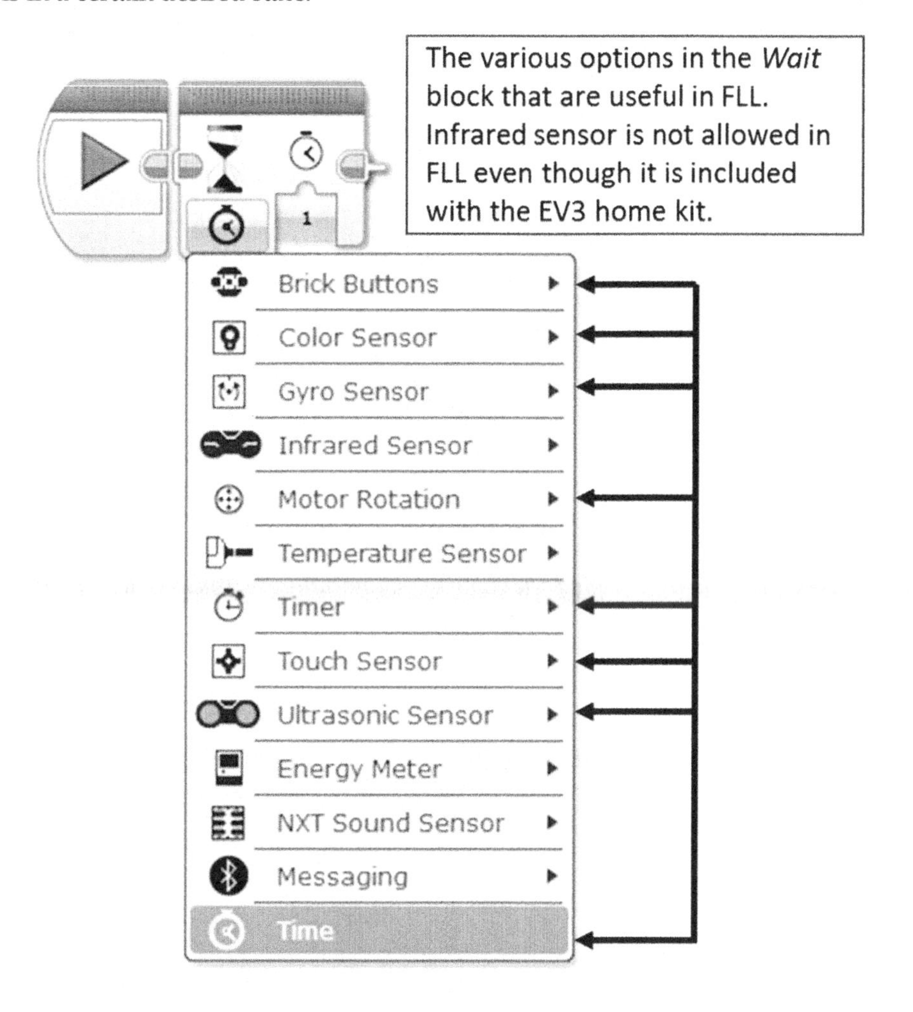

Figure 130: The switch block with all the expanded options.

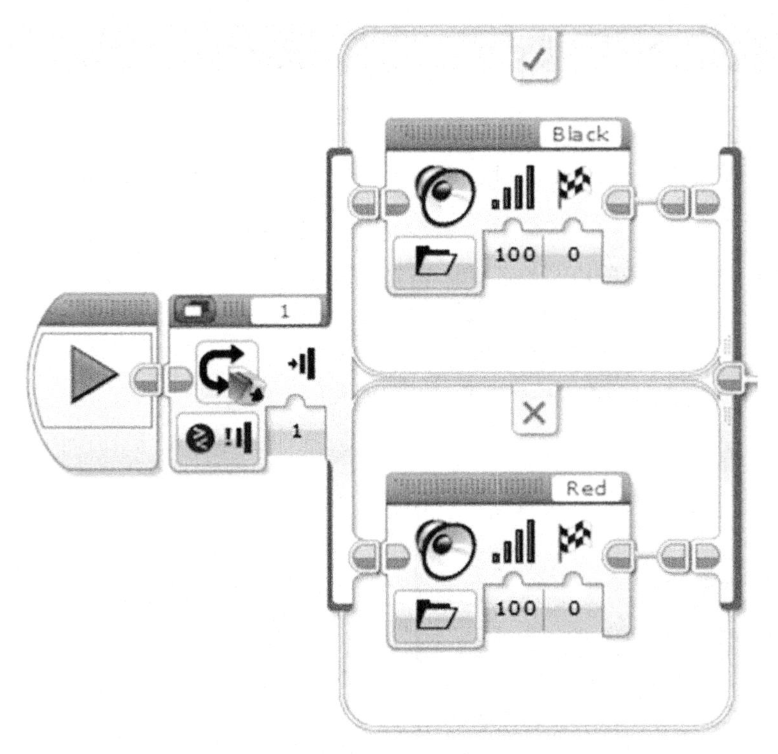

Figure 131: A simple switch block that will test the touch sensor for whether it is pressed or not and if pressed, it will play the "black" sound and otherwise it will play the "red" sound.

The switch statement has a tabbed view as well apart from the expanded view. In large programs, it is convenient to use the tabbed mode to save space and to be able to navigate over the program a bit easier.

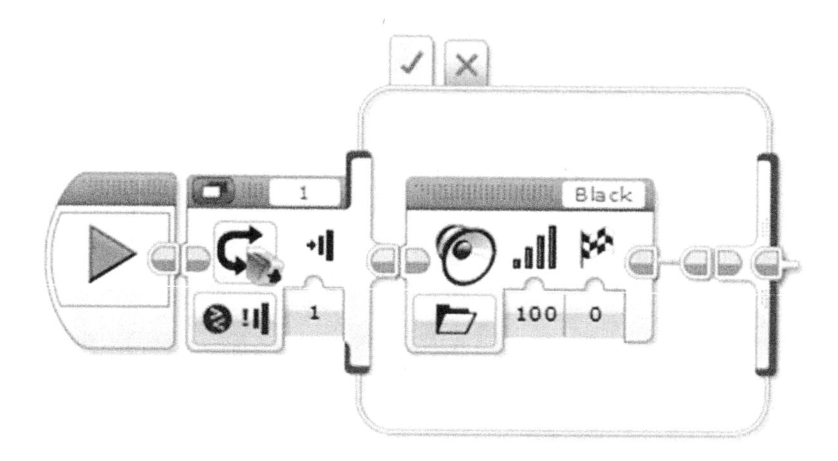

Figure 132: The tabbed view of the switch block. For the most part, we find it a bit more confusing than the expanded version since there is very little difference in terms of shading between the two tabs and it often confuses students and coaches alike.

Even though the default state of the switch block, when placed in a program shows only a True (check) or False (cross) indicating that the switch block can test a condition and tell you whether it is True or False, the switch block is not limited to just these two options. For cases where multiple responses are possible to a conditional statement, the switch block supplies the ability to create multiple options for running blocks depending on which one of multiple responses is the one that matched. This is one of the hidden in plain sight secrets with the switch blocks and students quite often are caught unaware of this possibility.

We are showcasing the switch block's ability to have multiple cases in Figure 133 where the color sensor can measure the color it is seeing and takes different actions for red and black colors. For all other colors, including if the color is not recognized by the color sensor, the switch condition matches the third option known as the *default* option. When using a multiple-state switch block be especially careful with the *default* option since this option is the one that will match if none of the other options match. In many cases, especially when a switch is embedded inside a loop, it makes sense to leave the default case as empty.

It is perfectly fine to leave any of the options in the switch block empty. All it means is that in case the condition specified by that option becomes true, you do not want to perform any action. This is something that catches beginners by surprise as they feel that every option must contain something.

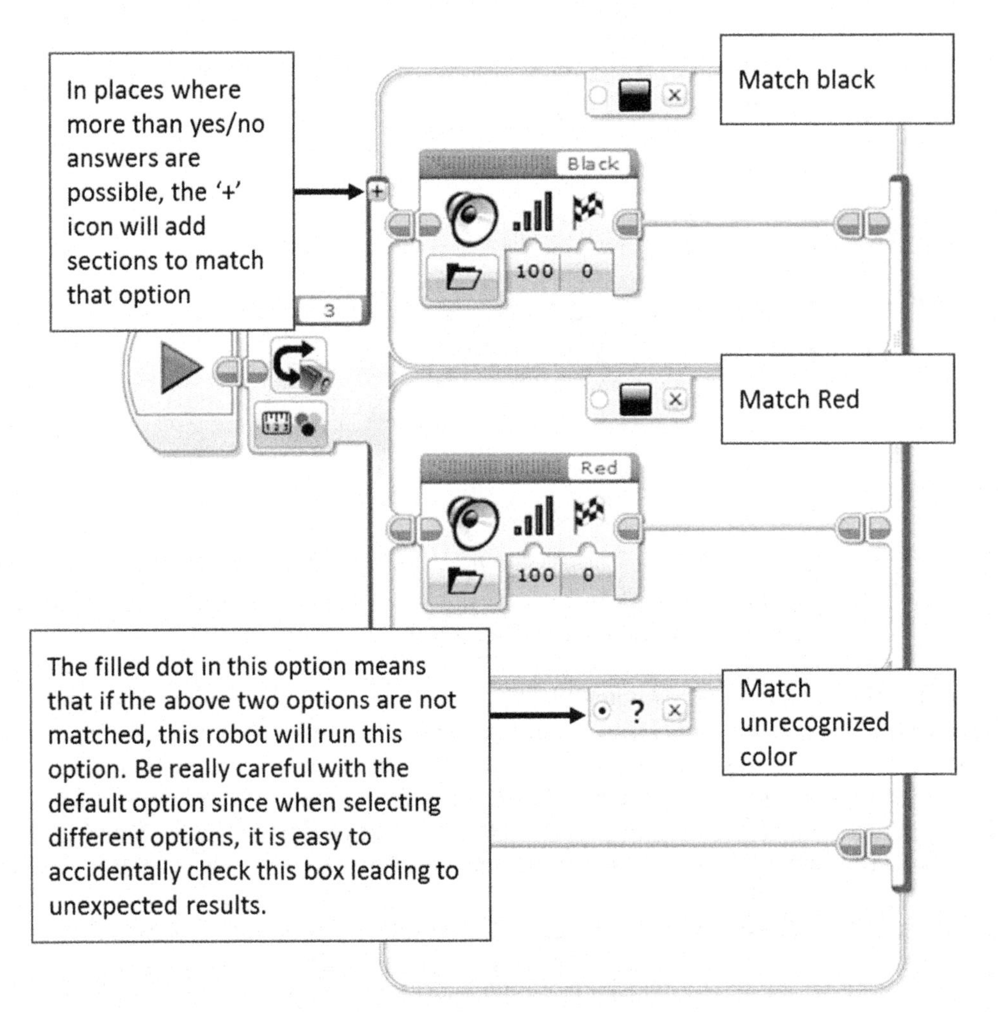

Figure 133: In the above switch block, the EV3 brick will produce "black" sound for black color and "red" sound for red color when the color sensor detects these colors. In case the color sensor sees something other than these two colors or a color that it does not recognize, it will not do anything. The unrecognized color is set as the default option i.e. if the sensor is not seeing either black or red, the switch statement will execute the blocks in the unrecognized color option.

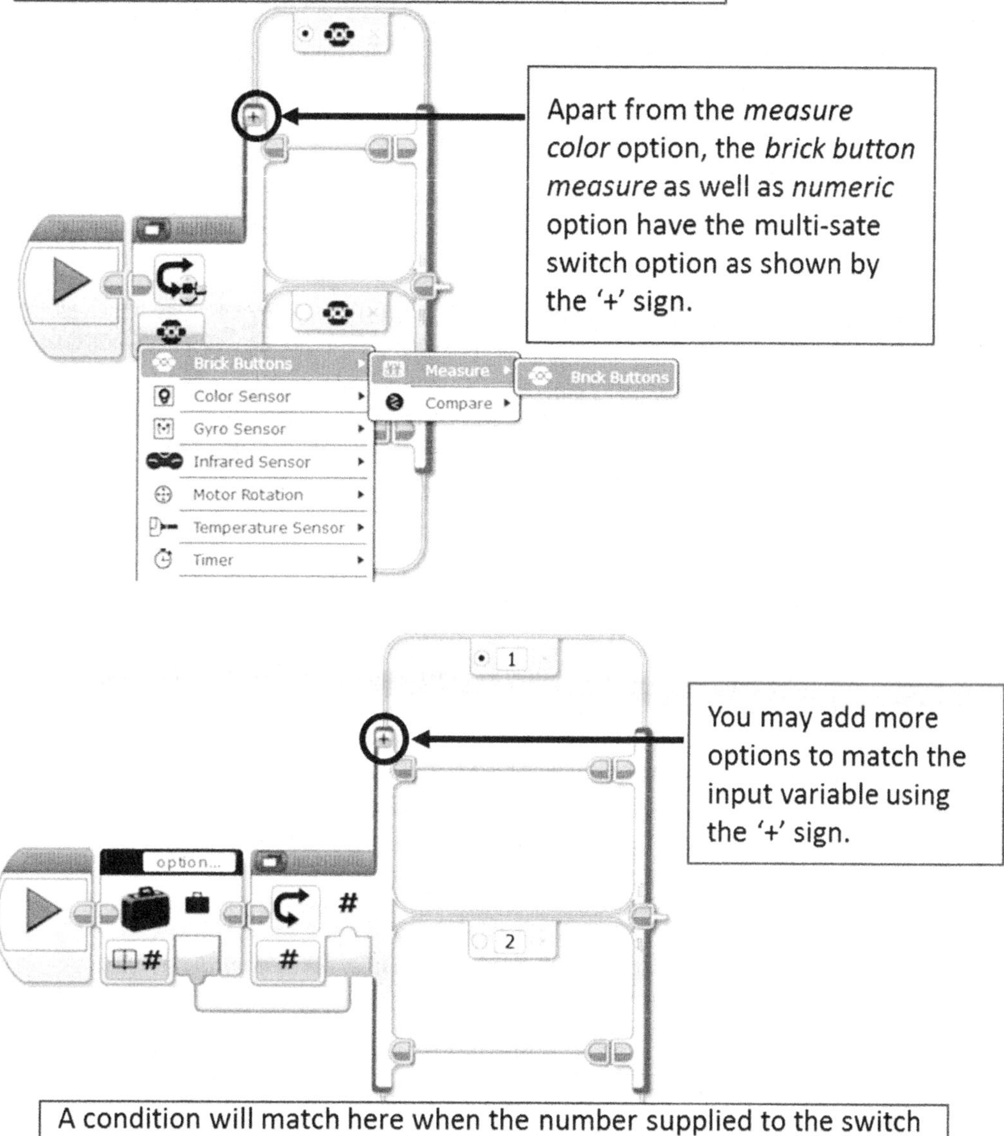

Figure 134: The Measure Color, Measure Brick Button and the numeric options have the ability to create a switch with multiple states.

The switch statement can have multiple cases in response to inputs which do not simply allow the true/false answer. In such case, remember that there is always a default case (identified by the filled in dot to the option left) and you should be aware of which case you have selected as the default case. The default case matches when none of the other cases have matched and the switch will execute this option when nothing else matches.

Flow Control: Programming mistakes with the Switch block

One key fact to remember about the switch block is that unlike the wait block, it does not wait for the condition to become true or false. Instead, when the switch block is executed, it evaluates whether the condition specified in the switch block is matching *at that instant of time* and executes the appropriate section. This is quite often a case for confusion among students when they write programs like the one shown in Figure 135 which will not produce any robot motion.

The reason for this is the switch statement behavior which performs an instant evaluation of the touch condition. To better explain, the first block switches the motor on. Immediately after switching on, the robot checks whether the touch sensor is pressed in the switch block's condition. If a person was holding the touch sensor in its pressed state at the very onset, the robot will immediately stop both the motors. On the other hand, if the person was not pressing the touch sensor, the robot will enter the false condition and exit the program since there are no more blocks after the switch. Most kids feel that there is something wrong with the robot instead of being able to understand this nuance of the switch block. If the above behavior is desired, it is better to use the Wait Until block with the condition set as the press of the touch sensor.

The switch block is a non-waiting block and reads the value of the sensor the instant the switch condition is checked. If you need to wait for a sensor to get in a desired state, you must use the Wait block or put the switch in a loop block covered in the next section

Note that if you do not fully understand the program below since we haven't covered the touch sensor so far, revisit this section after going over the Touch Sensor chapter.

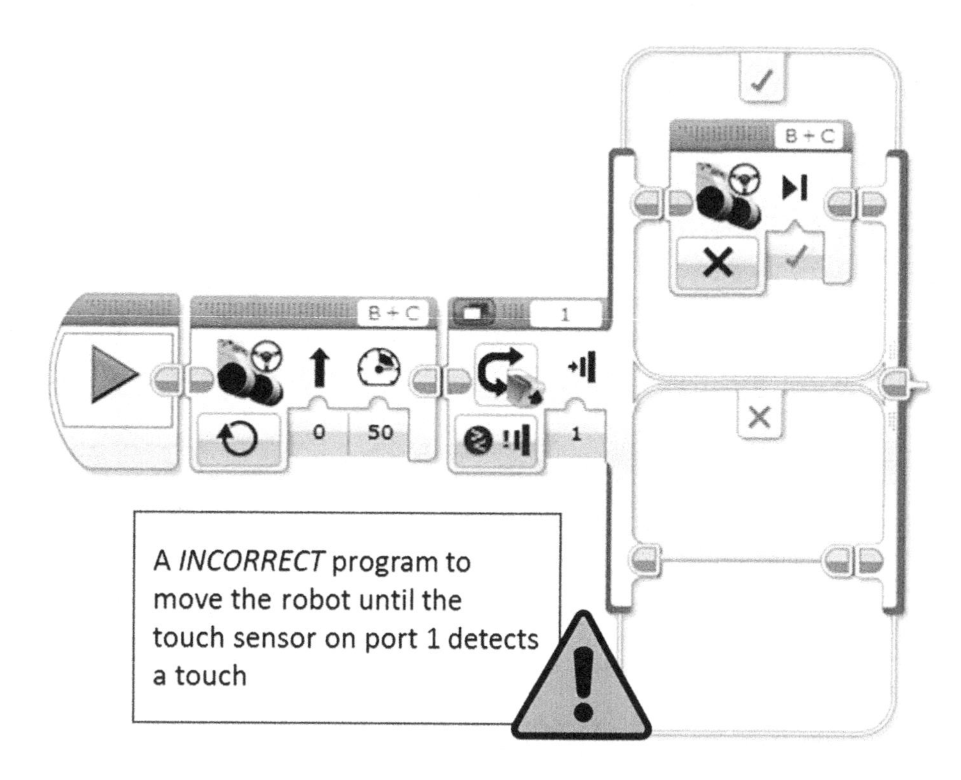

Figure 135: Many kids write programs like the above expecting that given that they switched the driving motors on and are then are using a switch block and switching the motors off when the touch sensor detects a touch, the robot will keep moving until the touch sensor detects a touch. Unfortunately, the above program will not do anything at all!

Flow Control: The Loop block

Loop block is a block that can hold other blocks inside it and run them repeatedly until a condition is fulfilled. This condition is called as the *loop termination condition.* Before we start explaining the loop block let us look at the loop block described in Figure 136 with options expanded to showcase its capabilities.

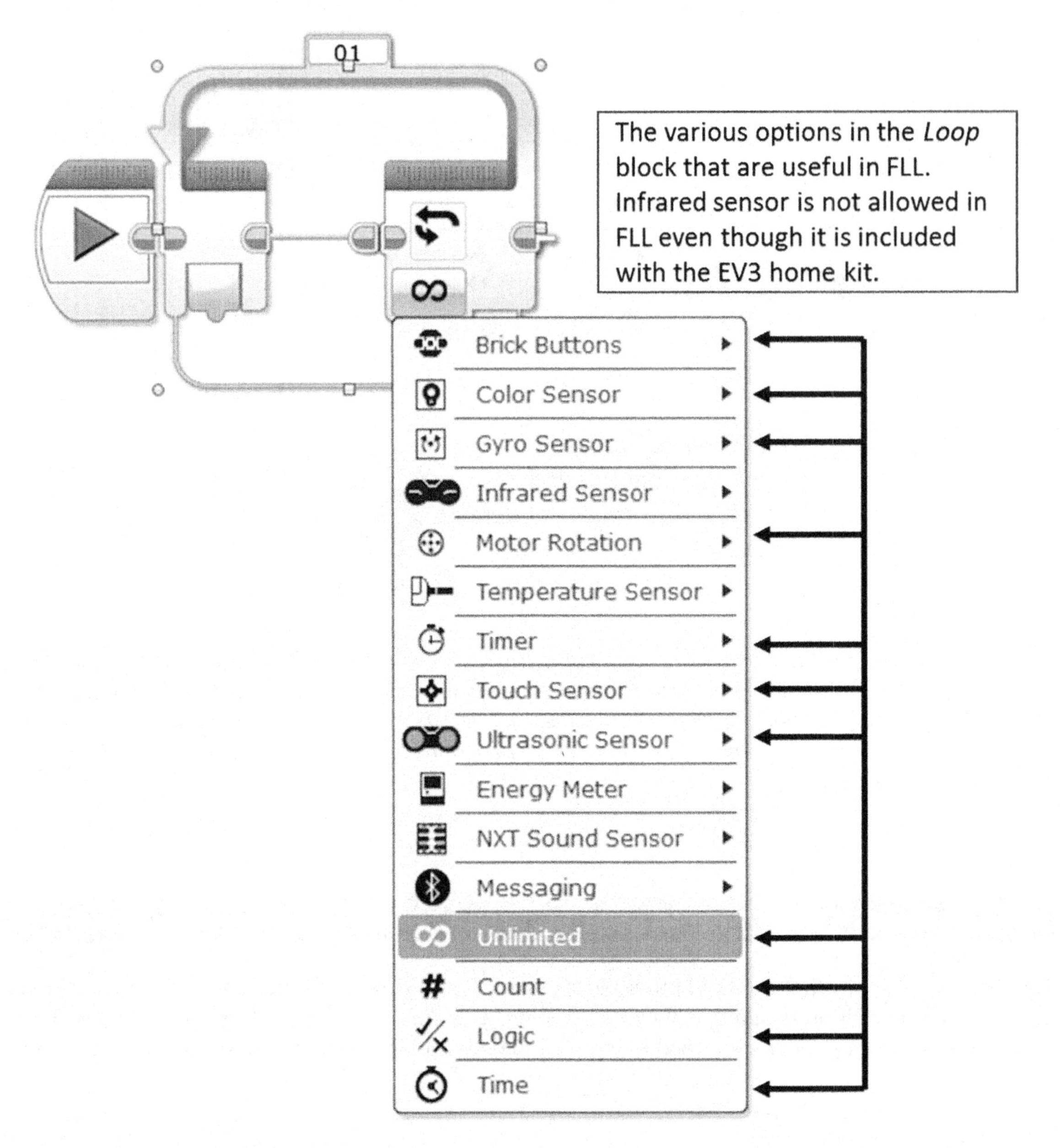

Figure 136: The various options in the loop block that are commonly used.

As showcased in Figure 137 , the loop terminates when the ultrasonic sensor detects a distance less than 10cm, you can select the loop to terminate on one of multiple conditions based on sensors. Examples of such conditions are: sensor getting in a desired state, after a certain period of time, a count or even when a logical condition becomes true.

If you are familiar with programming using other programming languages, the loop block's exit condition in EV3 software might seem a bit unusual to you. The reason for this is that unlike most programming languages where you exit the loop when the loop termination condition turns *false*, in the EV3 loop block, you exit when the loop condition turns *true*.

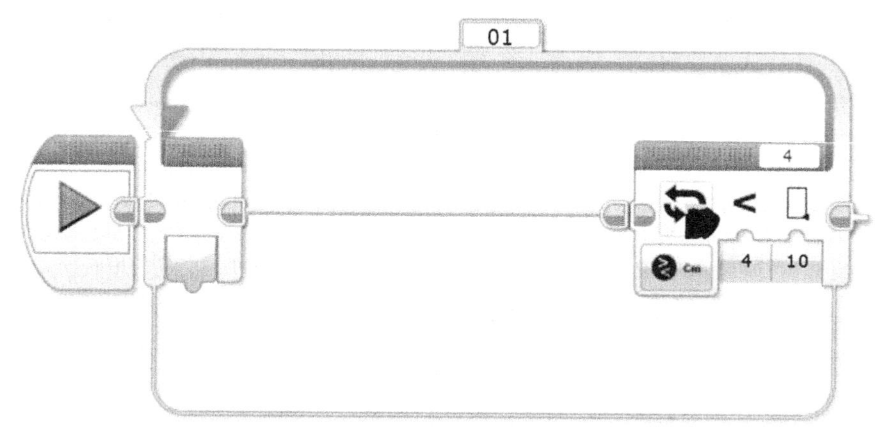

Figure 137: The above loop will exit when the ultrasonic detects the distance to be less than 10cm.

Additionally, we can use the *Move Steering* or other motor movement blocks in an interesting way within a loop i.e. placing a motor block within the loop with simply the *on* mode. Counter to what you would think i.e. you would expect that the motor would switch on and off repeatedly, the EV3 is smart enough to realize that you are switching the motor on again and again and simply keep the motors on continuously until the loop condition is met (Figure 138.)

Although a bit counterintuitive, this behavior of the motors in loops makes some key scenarios such as PID control (an advanced smooth control based on mathematical operations) of motors possible. PID control is later covered in chapter *PID algorithm for wall following using Ultrasonic and line* following using Color Sensors.

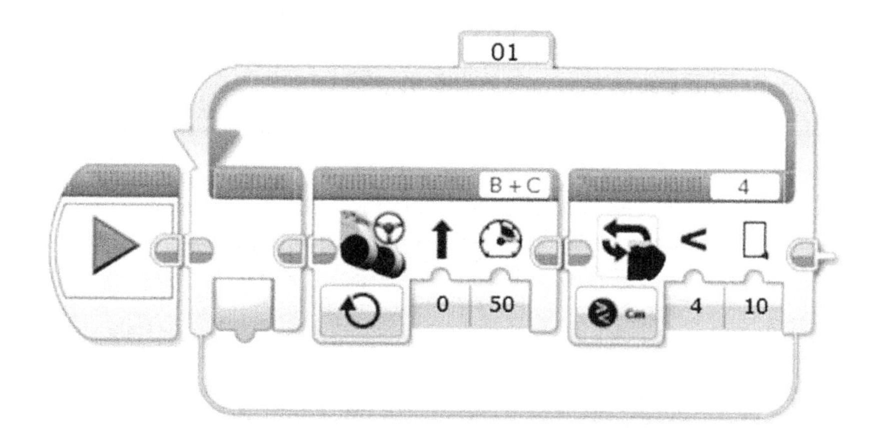

Figure 138: The above loop keeps the robot moving until the ultrasonic sensor on port 4 is less than 10cm from an obstacle. Note that the loop repeatedly turns the B and C motors on using the move steering block. One would expect that this would lead to the robot

moving in a jerky fashion as the robot switches on and off. However, the robot behavior in this case is simply to keep the motors on until the loop condition is met. This curious paradigm which allows extremely quick motor control is frequently in line following and wall following to great success. These topics are covered in later chapters.

Note that, if instead of using the On mode, if you use the on for degrees mode in move steering or move tank block along with the loop block as shown in Figure 138, the motors will indeed switch on and off over and over which cause jerky motion. The reason is that when move steering or move tank blocks are used in On for Degrees/rotation mode with supplied degrees/rotation, the motor runs and waits for the number of degrees/rotation to finish and then switches the motors off before moving to the next block. Unlike the On mode, which keeps the motors on continuously in a loop, this causes the on-off jerky behavior. Additionally, similar to the condition shown in Figure 128, the Robot will overshoot to complete the condition even when the exit condition is met.

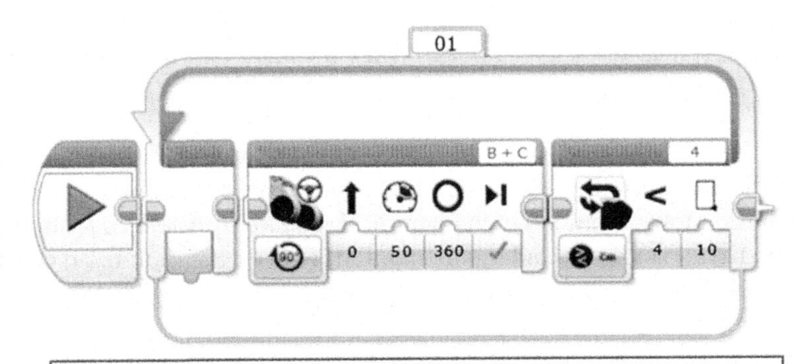

An incorrect way to keep the motors on while waiting for the robot to get within 10cm of an obstacle. Since the move steering block is using *On for Degrees* mode, the robot will move one turn of the driving motors, stop and repeat the behavior over and over causing a jerky motion.

Figure 139: An incorrect way to keep the motors on until the robot reaches within 10cm of an obstacle. This program will move the robot in a jerky fashion and will likely approach the obstacle much lesser than 10cm before exiting the loop.

The Loop Interrupt

Quite often, when using loops, apart from exiting the loop when the loop termination condition is met, we might need to quit the loop based on other conditions inside the loop. Example of this are conditions which might be based on a switch block present inside a loop. For example, consider the case where the robot is programmed to move until a color sensor mounted on the robot sees red color on the floor. In this scenario, the color sensor would be facing down, looking at the floor. However, it is possible that on the path to the red color, the robot might hit an obstacle and might need to stop. To account for this condition, we must mount a touch sensor on the front of the robot that can detect this collision. If the robot collides with the obstacle and the touch sensor triggers the pressed state, we need to exit the loop which is checking for the presence of the red color and keeping the motors on till that period of time. This exit of the loop requires us to use the switch block inside the loop to get out of the loop. This is where the *loop interrupt* block comes in as it allows us to exit the loop immediately from anywhere within the loop. Take a look at Figure 140 to better understand the above example.

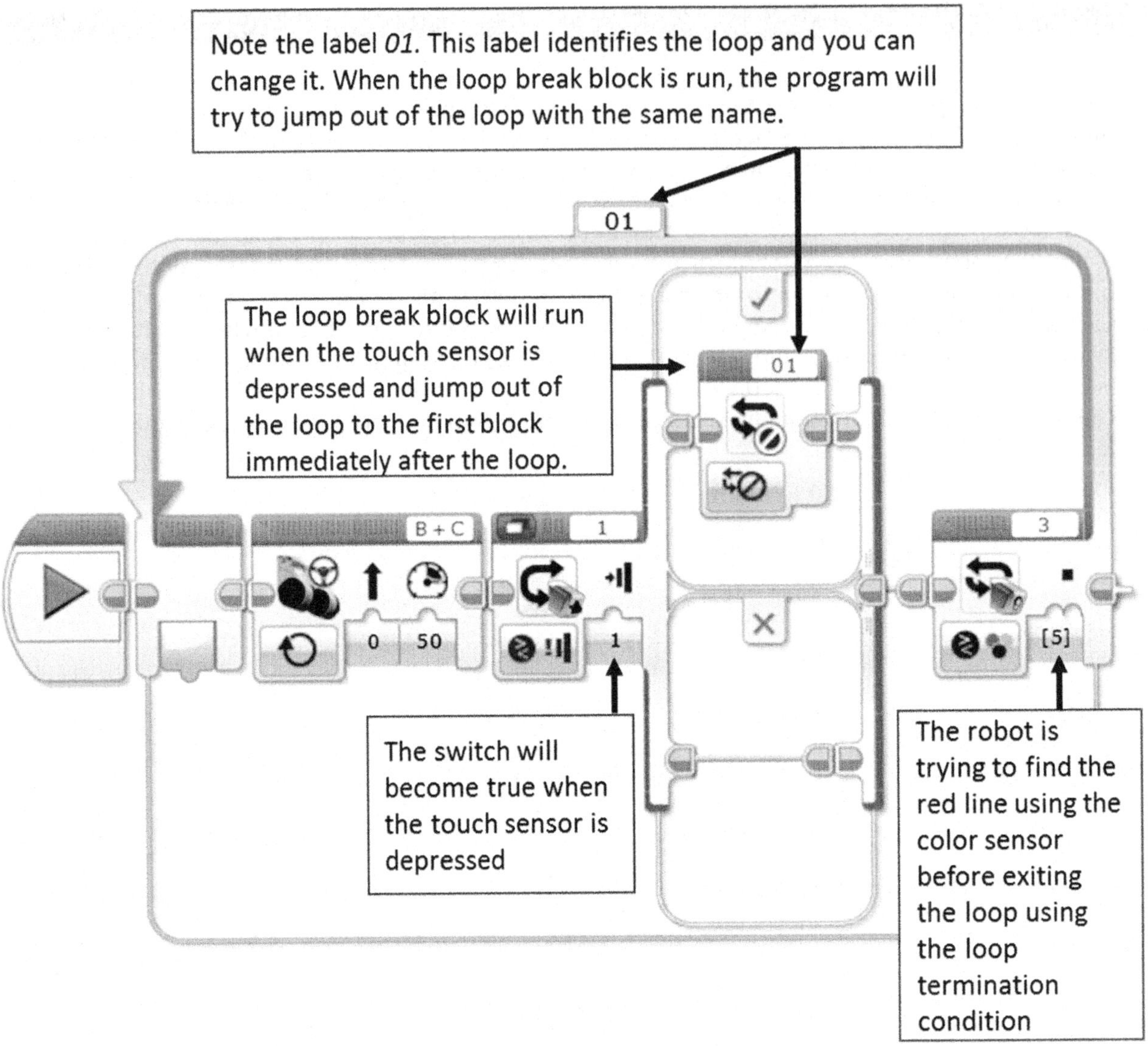

Figure 140: The loop interrupt

Flow Control: Programming Mistakes with the loop block

In general, the loop block works quite well, but when used in conjunction with the switch block based on a sensor value, and a count based loop and blocks that take very little time to execute, it may lead to unintuitive results. For example, the program shown in Figure 141 which, students quite often write to replicate the behavior of the program in Figure 138 will likely not work and will lead students and coaches confused alike.

The reason the loop will not work as expected are the following (1) The Ultrasonic Sensor detection of the distances is a sub millisecond operation and (2) Switching the motors on is again a sub-millisecond operation.

Both operations in conjunction are going to take an extremely short, unpredictable amount of time. Thus, 200 iterations of the loop are likely going to last less than a fraction of second, causing the robot to turn the motors on and then almost immediately switch them off making it feel like the program did not do anything. This is in general true of any loop where all the operations inside the loop take combined take very little time. Such loops are not suited for a counter based operation.

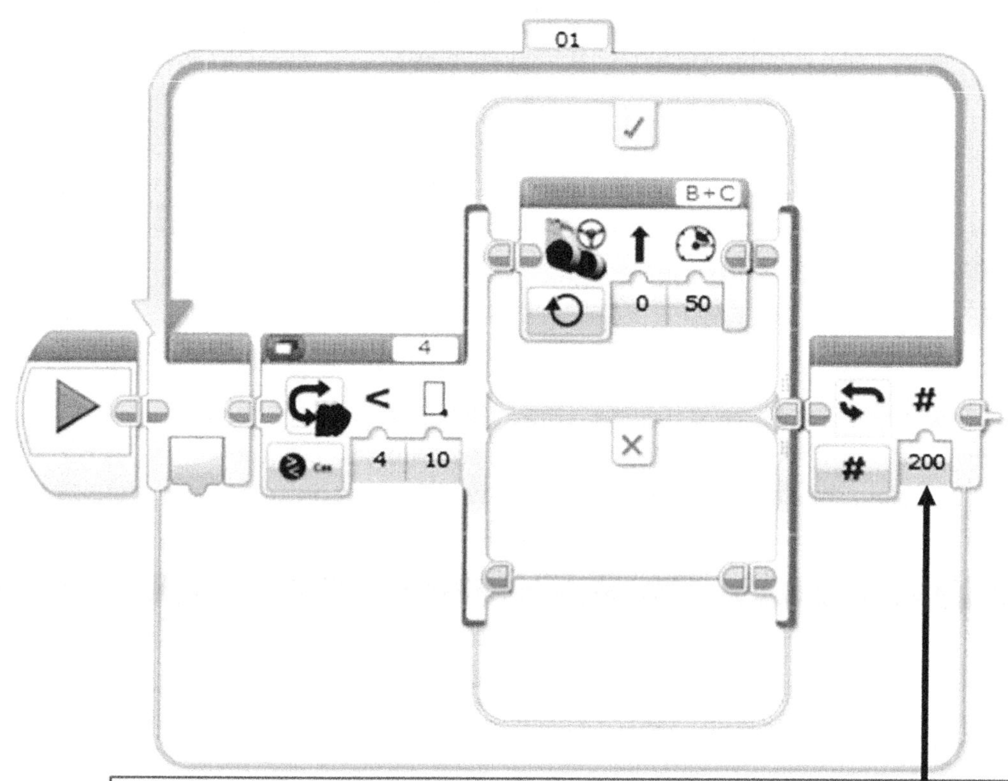

Figure 141: The above loop is going to produce very unpredictable results and cause a lot of confusion.

Any loop where all the operations inside the loop take combined take very little time is not suited for a counter based operation when you are trying to perform a time-based task.

Another common beginner mistake is when all the loops and loop interrupts in the program have the same name (Figure 142). The EV3 software really does not help here and provides every loop and loop interrupt block the same name of *01*. The loop interrupt jumps out of a loop which has the same name as the loop interrupt. If multiple loops and loop interrupts have the exact same name, the operations become unpredictable since you will not be sure which loop a loop interrupt is trying to jump out of. It is a good idea to name every loop uniquely and ensure that the loop interrupt blocks inside the loop refer to the loop that they are enclosed in. Additionally, note that the loop names don't need to be numeric. You can provide descriptive names that easily make you understand what the loop does.

Make sure to name each loop and the loop interrupts contained in it uniquely. Additionally, it is a great idea to name loops for what they do instead of using the numbering scheme that the EV3 software uses by default.

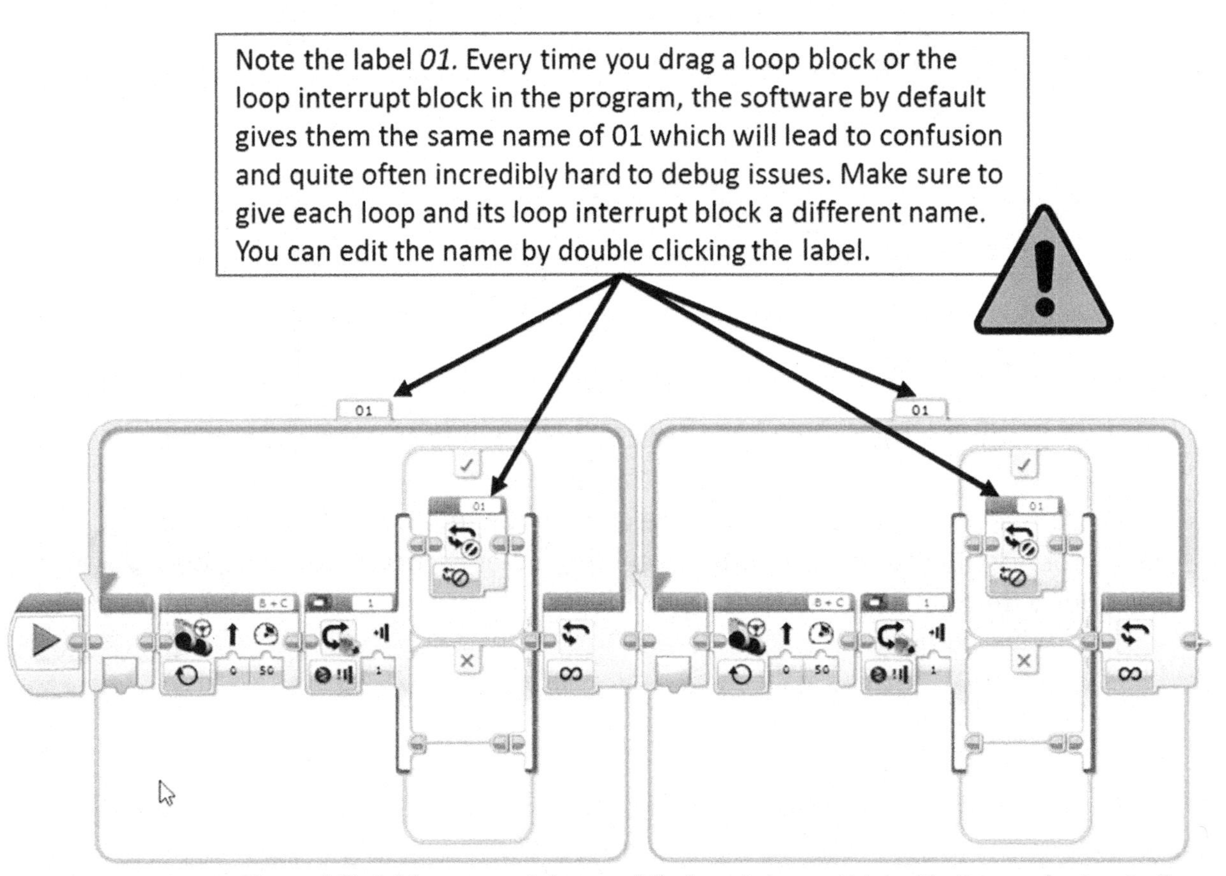

Figure 142: Make sure each loop and the loop interrupt(s) inside it are referring to the same name to avoid incorrect behavior.

Summary

Flow Control blocks are some of the most important blocks because they allow you to make conditional decisions, wait for events and to create repeating conditions. The Flow control blocks form the backbone for robot logic and intelligence and without them, you could only do very simple exercises. Given that flow control blocks and sensors go hand in hand, it would be a great idea to revisit this chapter after you go over the sensors in later chapters.

Touch Sensor and the Brick buttons

The touch sensor in the LEGO Mindstorms is a sensor that detects when a protrusion or switch in its front portion hits a surface and is depressed. The touch sensor essentially is a spring loaded on-off electrical switch and functionally equivalent to a switch in the home. It is normally off but when the protrusion in the front is pressed, the switch turns on, indicating a press. The touch sensor is shown in Figure 143.

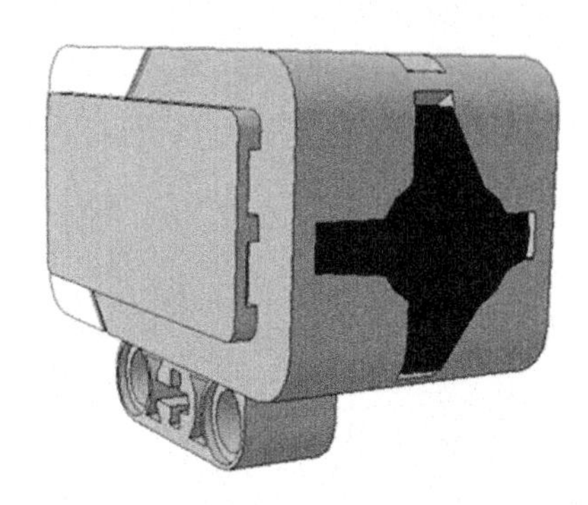

Figure 143: The EV3 touch sensor. The red plus shape in the front is the portion that gets depressed when hitting against an object and registers the touch.

Although the touch sensor can be used to detect when the red switch is depressed, if the object in front of it is not perfectly flat, sometimes, the switch may not make contact or more likely, not make enough of a contact. This will cause a problem since in this case, the touch sensor might not register a touch. To solve this problem, quite often, it is customary to attach an axle to its switch and attach a large flat object such as a gear to the other end of the axle. This expansion of the switch's effective surface area that can be pressed increases the chances of registering a touch significantly. This design with an axle and a gear is shown in Figure 144.

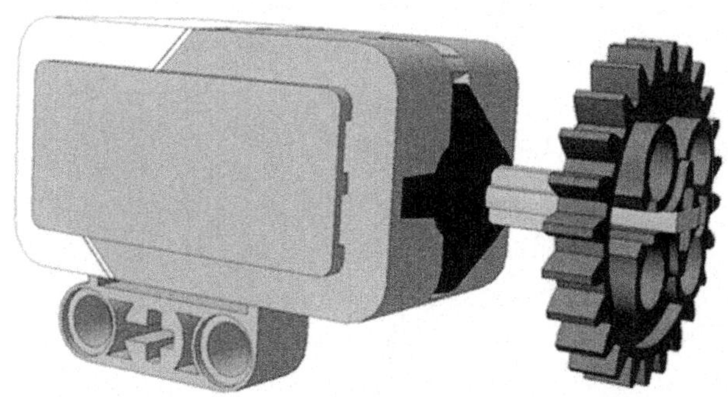

Figure 144: The touch sensor has an axle hole in the front and it is quite common to insert a small axle in it and attach a gear or pulley wheel in the front to increase the surface area that triggers the touch to increase successful chances of detecting a touch.

Programming with the touch sensor

The touch sensor is a straightforward sensor to use and program and we start with the states the touch sensor can detect. The EV3 touch sensor can detect three different states:

- When the switch gets depressed. This event occurs as soon as a non-depressed switch is depressed and will continue to occur while the switch remains pressed. In one single press of the switch, if you check the state of the switch in a repeated fashion, it will register true as long as the switch remains pressed.
- When a switch is not in the depressed state. This event occurs and continues occurring while the switch is in the non-depressed state which, incidentally is the default state of the sensor. If you are checking for non-depressed state, it will keep registering true until the switch gets pressed.
- When a switch is depressed and then let go. This event happens exactly once during a switch depress and then its immediately following letting go of the switch.

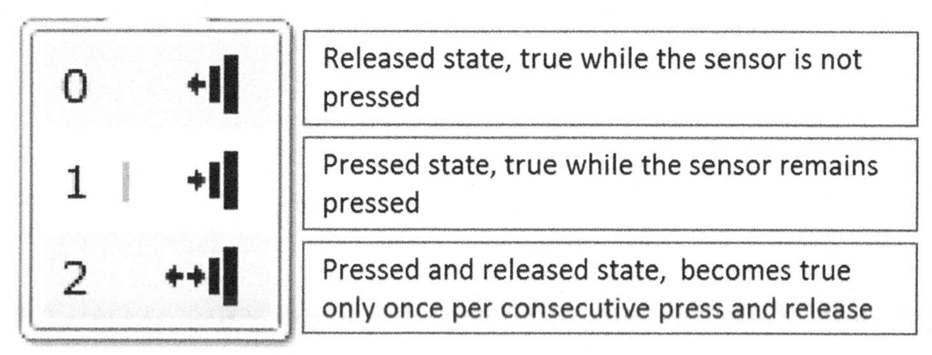

Figure 145: The three states of the touch sensor.

Even though these options seem fairly obvious, we are listing them out separately because many students and coaches have been bit by this nuance and end up with a program whose behavior they cannot explain. Let us explain with the aid of a program in Figure 146 which tries to increment a variable by 1 every time the touch sensor is pressed and display its value on the EV3 screen. The program will behave as expected only when using the *pressed and released* state and will increment the variable multiple times if you rely on just the pressed state or just the released state.

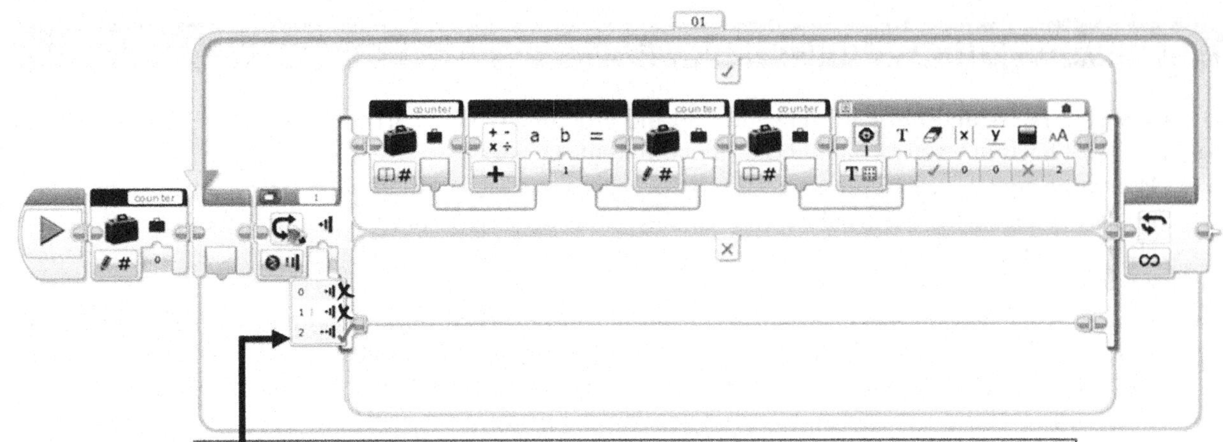

Only the pressed and released state i.e.state #2 is going to perform as expected and increment the counter once per press. In the pressed state the counter will go up by a large amount each time the touch sensor is pressed and in the released state, the counter will keep on increasing without you having to do anything at all! The reason is that all the blocks in the program take very little time and will execute many times in the tiny amount of time you press or release the button because of the loop. Each time the loop executes, the switch block will check whether the touch sensor is pressed or released right at that instant of time and if so, increment *counter* by 1.

Figure 146: The effect of using the pressed, released or pressed and released states of the touch sensor on your programs. In this program, we are trying to increase the value of the variable "counter" by 1 every time the touch sensor is pressed. Only the pressed and released state works in the expected manner. The other two states will lead to incorrect results.

Another example, which explains the same concept using a wait block is shown in Figure 147.

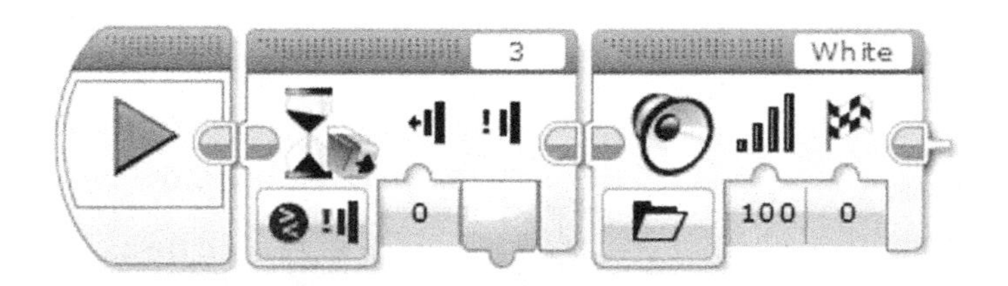

Waiting for a touch sensor release. This program, unless you are holding the touch sensor pressed before the program starts running, will not wait at all and instead will speak the word *white*. The reason is that the *released* state is held when the sensor is not in pressed state.

Figure 147: Waiting for the touch sensor release. This program will not behave as expected.

Mounting the touch sensor

It is a good idea to mount the touch sensor in such a way that when it hits the obstacle, the side the touch sensor is mounted on can become perfectly perpendicular to the obstacle. In the sample robot we use in this book, we would mount the touch sensor as shown in Figure 148.

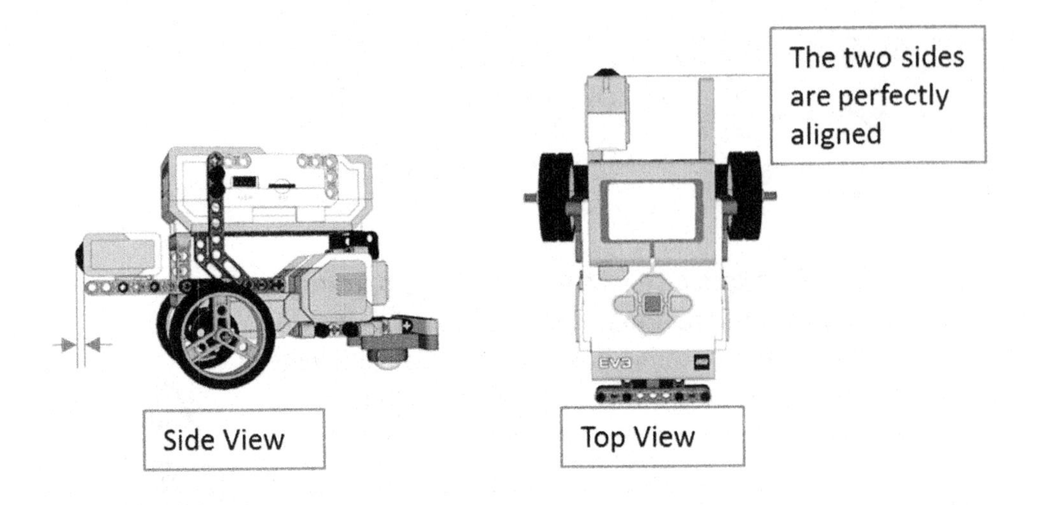

Mount the touch sensor so that only its red sensing portion is extending beyond the robot. It is a great idea to ensure that the side the sensor is mounted on, apart from the red sensing portion, is completely flat to a wall as shown.

Figure 148: Mounting the touch sensor

With a touch sensor mounted in the front, as shown in Figure 148, we are providing an exercise in Figure 149 that you can attempt to cement the understanding of the touch sensor, aligning, loop block and switch block.

The robot is trying to find the door on the left in the alley based only on the touch sensor input.

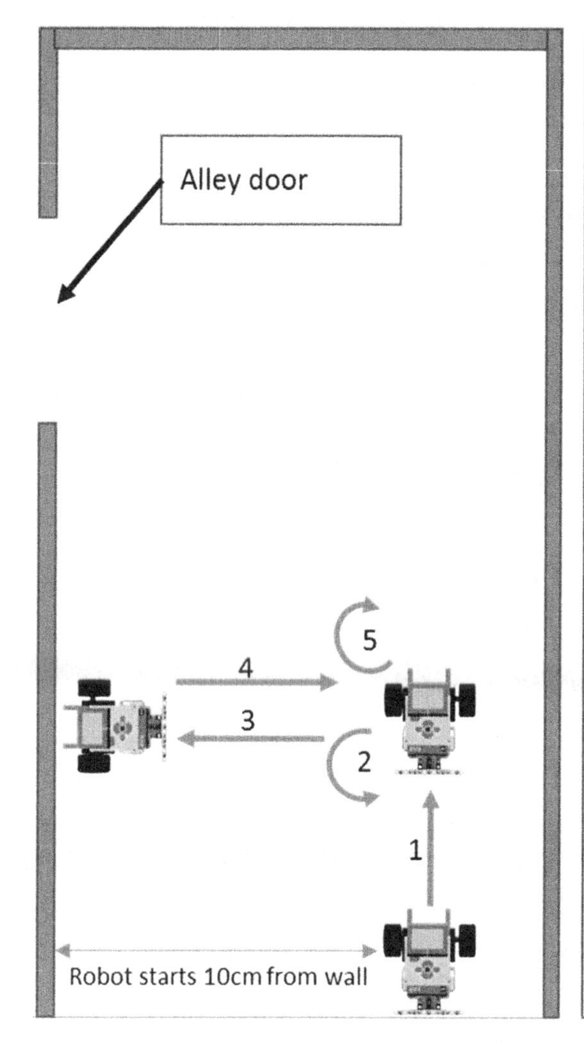

- Step 1: Robot starts 10cm from the wall and goes straight for some distance.
- Step 2: Robot turns left 90 degrees to orient its front to the alley wall with the door.
- Step 3: Robot moves forward for more than 10 cm i.e. something akin to 15cm so its touch sensor hits the wall if the wall is there. If the wall is there, since the robot front is perfectly even, the robot will become perpendicular to the wall and sensor will be in pressed state. If so, move to the next step. If the wall is not there, the touch sensor will not be in the pressed state at the end of the 15 cm movement and you can declare success and stop the robot.
- Step 4: If door not found, Robot moves back, in reverse, 10 cm from the wall.
- Step 5: Robot turns 90 degrees to the right

Repeat steps 1-5 as many times as required to find the door.

Figure 149: Find the door in the alley wall using just the touch sensor.

Wall aligning with a touch sensor

We learned about aligning to a wall by hitting the bumper against a wall and spinning the wheels in the Basic Robot Navigation chapter. In that technique, the bumper would keep pushing into the wall to straighten itself with respect to the wall.

Using the wall to align a robot is perfectly reasonable strategy and is one of the most commonly used aligning strategies in FLL. However, if you have the space for two touch sensors and your design affords mounting them in a manner so that both the sensors are perfectly parallel and in line with each other, you could use two touch sensors to become perfectly straight to a wall in a more intelligent manner as shown in Figure 150.

Once sensor 1 has hit the wall and enters the pressed state, keep turning the motor close to sensor 2 until sensor 2 is in the pressed state as well. This will align the robot to the wall in a more intelligent manner.

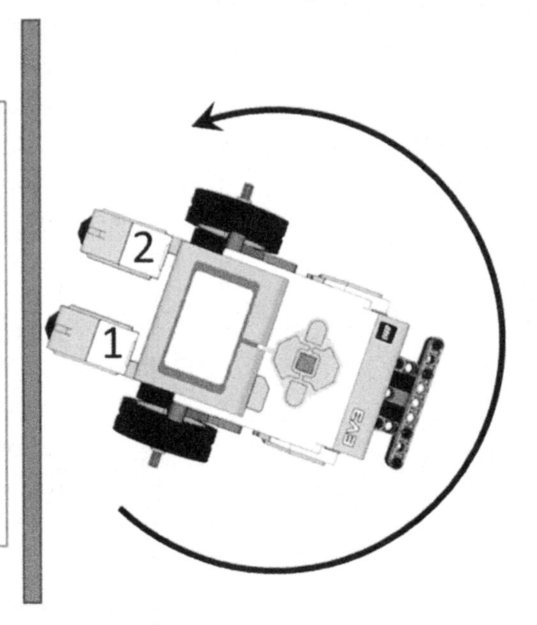

Figure 150: Intelligent wall aligning using two touch sensors

Wall Following with a touch sensor

The EV3 Education Core kit comes with two touch sensors and the sensors can be used to great effect in a sensor based wall following using an attachment model shown in Figure 151. In fact, this attachment can be used to solve the *find the door in the alley* exercise (Figure 149) discussed earlier in this chapter in a more predictable manner.

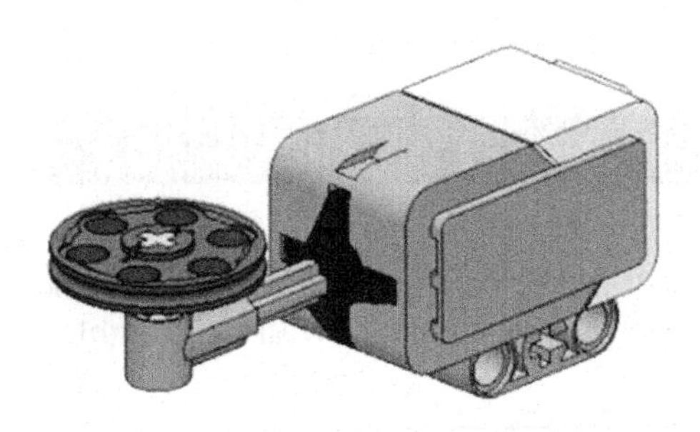

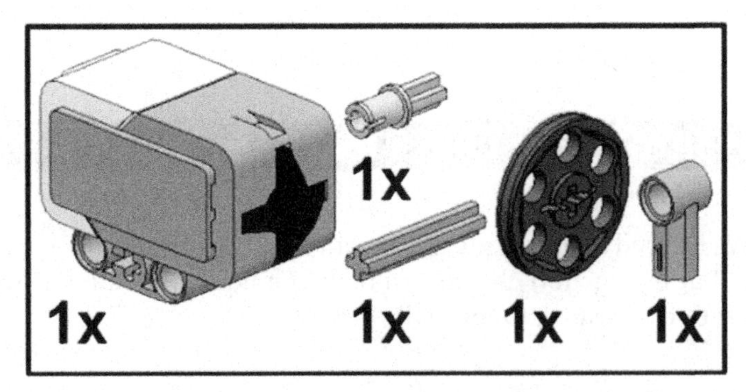

Figure 151: A touch sensor assembly for wall following shown along with the constituent components. The wheel can rub against a wall and as long as the touch sensor keeps

detecting the press, you can assume that the wall is continuous with no gaps. For better accuracy, you may want to use two touch sensors to ensure that if one touch sensor accidentally bounces, you do not stop wall following.

Figure 151 describes an assembly for touch sensor that can be used for wall following. In Figure 152, we show a robot equipped with the touch sensor wall following assembly. By pressing against the wall and ensuring that both the touch sensors remain in a depressed state, one can easily figure out when a gap in the wall is found. This will work no matter whether the robot is moving forward or backwards. If you already know the direction in which the wall opening is, you may be able to make do with only one touch sensor while using a free spinning wheel in the back side of the robot.

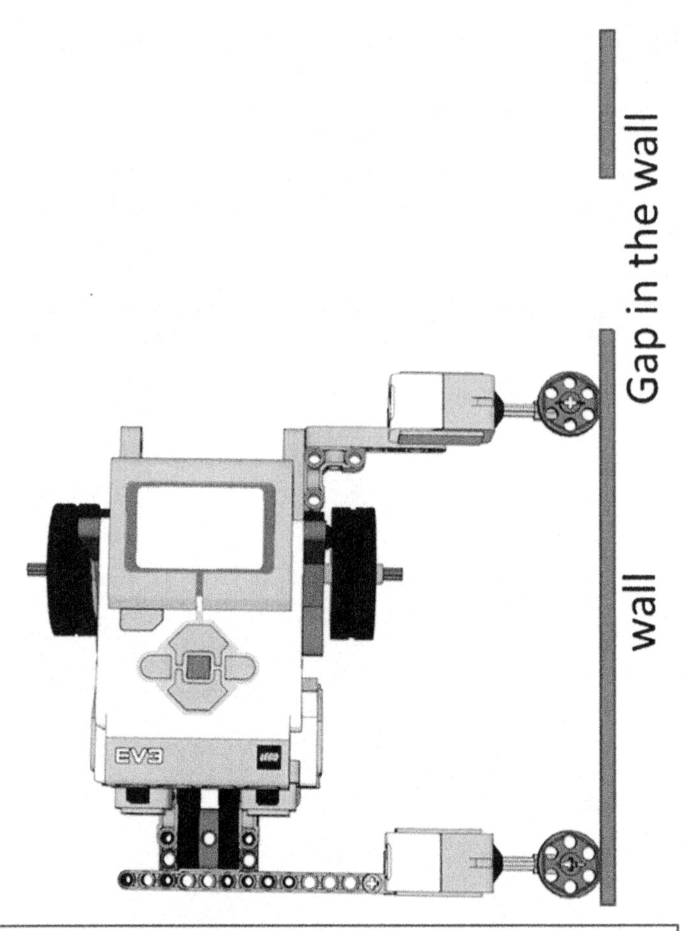

With two touch sensors mounted on the robot and connected to two free spinning pulley wheels, the robot can find a gap in the wall in a sensor based, intelligent fashion. To do this, the robot should ensure that both the touch sensors are in the depressed state. If any of sensors goes to the released state, the door is detected. Note that this robot can detect gaps in walls while going forward as well as backwards because of the use of two touch sensors.

Figure 152: Wall following using two touch sensors connected to free spinning pulley wheels

Exercises with the Touch sensor

Here are a few exercises with the touch sensor along with their associated programs. The blue box represents the FLL table with wooden walls made out of 2x4 wooden studs.

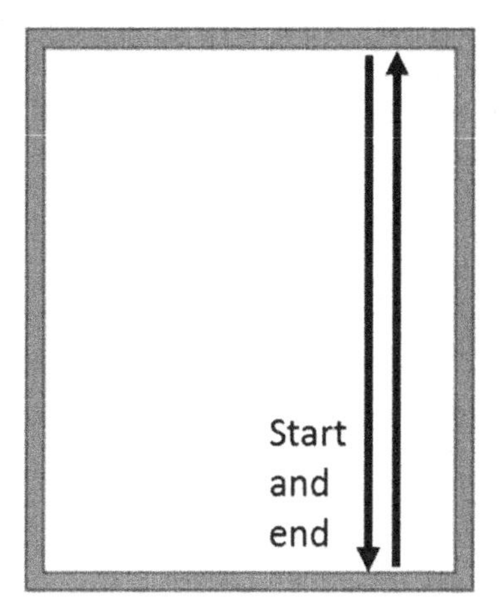

Exercise 1: Robot starts at the bottom and goes straight until it hits the wall. Then it backs up a little bit, executes a u-turn and comes back to the start

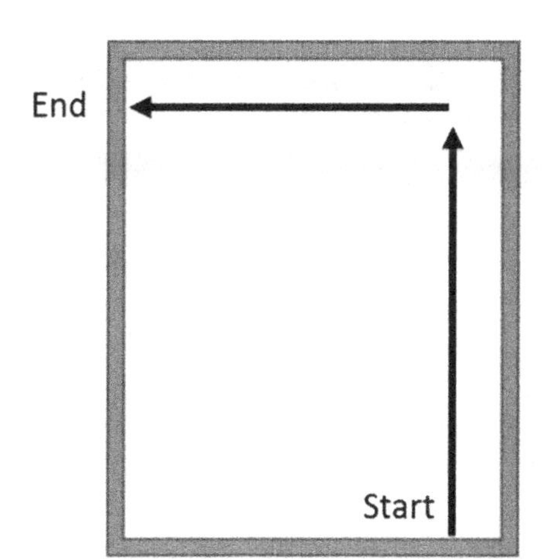

Exercise 2: Robot starts at the bottom and goes straight until it hits the wall. Then it backs up a little bit, executes a 90 degree left-turn and continues till it again hits the end wall.

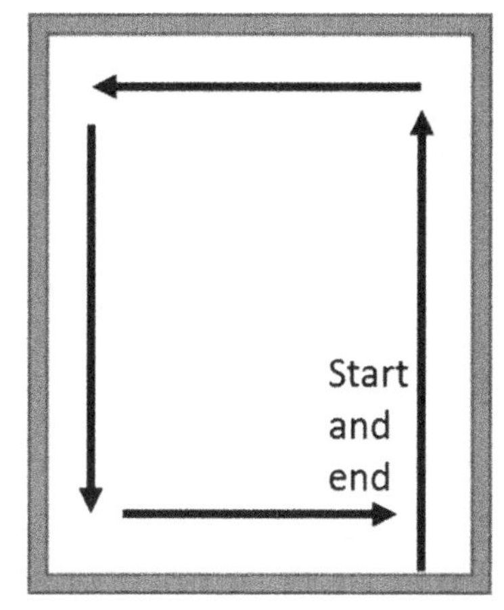

Exercise 3: Robot starts at the bottom and goes straight until it hits the wall. Then it backs up a little bit, executes a 90 degree left-turn. Repeat above three times using a loop to come back to start point.

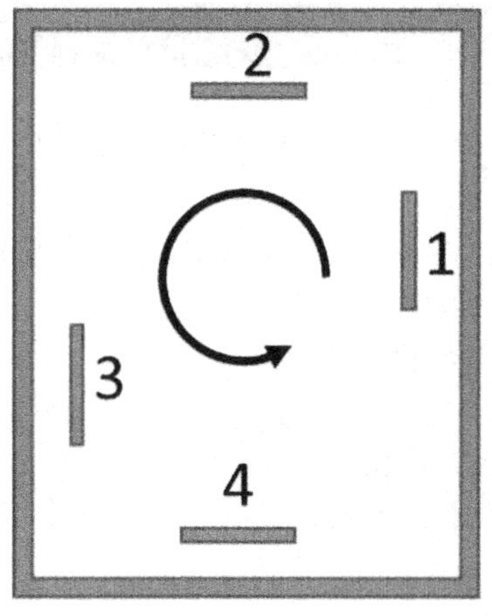

Exercise 4: Robot starts at the bottom and goes around the table in a clockwise fashion. There is a 30cm block of wood representing a electrical charging station, 25cm from the one of the walls. Positions 1, 2, 3, 4 are marked for sake of illustration to show potential positions of the charging station. The robot must locate the charging station using only the touch sensor.

Figure 153: Touch sensor exercises

Brick Button as a sensor

The brick buttons on the EV3 are normally used for navigating the menus on the EV3 brick itself. However, these buttons are programmable and can perform different actions while your program is running. Thus, the Brick buttons on the EV3 can be used as sensors by giving them special functions when your program is running on the robot. We use the brick button as the controlling sensor in the chapter on Master Program later in the book. In practice, the brick buttons, when used as a sensor behaves just like the touch sensor as each of the brick buttons can detect the *pressed, released* and the *pressed & released* states. There is a small difference between the touch sensor and the brick buttons since apart from registering button presses, the brick buttons, can also be programmed to emit different kind of light (Figure 154).

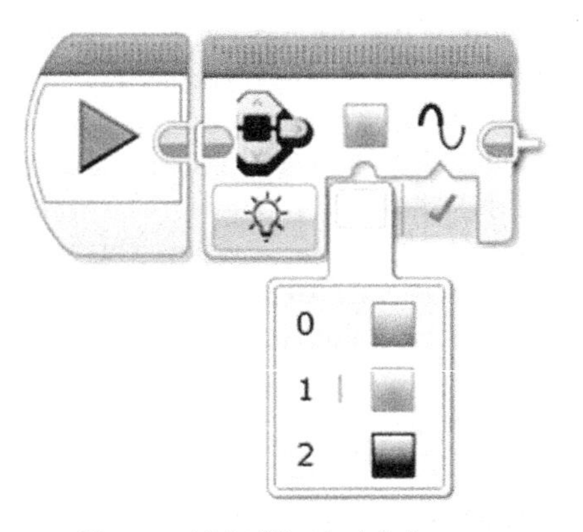

The brick button block can be programmed to emit one of the three colors, green, orange and red. Additionally, using the last input, it can be made to blink or remain steady. This can be used as a visual aid to remind you what your program is doing.

Figure 154: The brick button can emit green, orange or red light and can blink or remain steady.

Since the brick button is mounted on top of the brick and usually is used for selecting and running the programs, this sensor is not really used for sensing the robot environment. Instead, the brick buttons' touch functionality and the ability to programmatically control the brick button lights is primarily used for debugging complex programs.

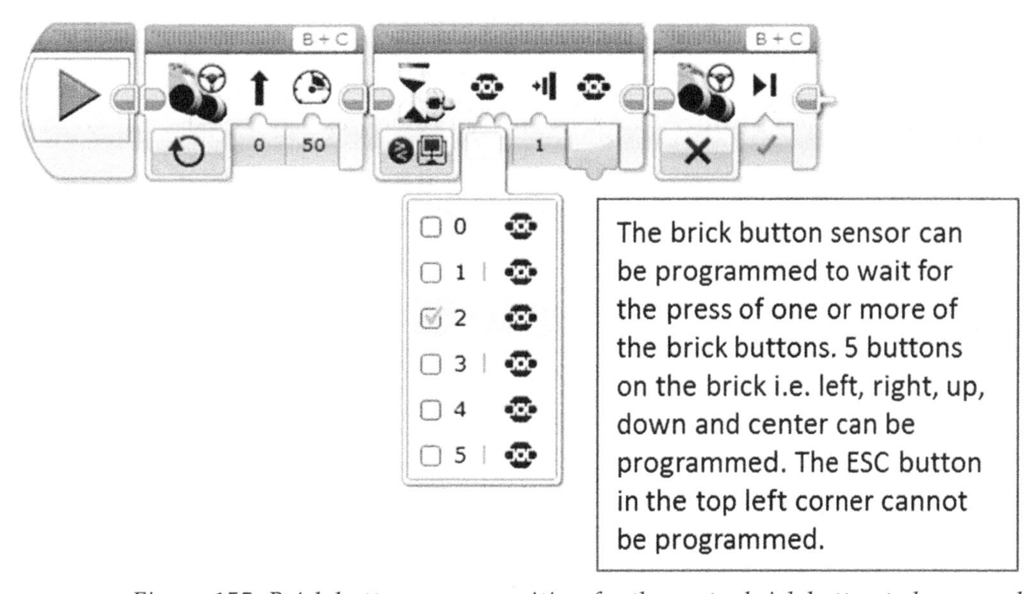

Figure 155: Brick button sensor, waiting for the center brick button to be pressed.

Since we are talking about debugging, let's quickly define it and additionally discuss a few different techniques for debugging programs. *Debugging* is the process of taking an existing program and trying to figure out why it is not operating the way it should. It involves the process of breaking down the program into its constituent components and going over them in a methodical manner to determine which step is not behaving as expected. Debugging is one of the most fundamental parts of writing program as issues in programs, also referred to as bugs are inevitable.

Usually, in EV3 programs, the errors occur during navigation or during mechanical contraption movement. In either case, usually, the robot is performing multiple actions in a quick successive fashion and often it becomes impossible to figure out what has gone wrong when things are happening quickly. Thus, it becomes necessary to be able to stop the robot at certain steps to ensure that the robot is performing the intermediate steps in a desirable fashion. After you have made sure that the robot did indeed perform the action it needed to, you need to resume the robot from that position or configuration. That is where the brick button are very useful. In its simplest incarnation, you can program the robot to wait for a brick button to be pressed after performing a step that you want to verify. Figure 155 describes one such scenario. In the example, the robot continues straight motion until the brick button is pressed.

Summary

The touch sensor and brick buttons work in similar ways and are some of the fundamental EV3 sensors. The Touch sensor can be used to detect obstacles and to align in an intelligent manner. The touch sensor lends itself to some amazing exercises that we provided earlier in the chapter. Students that go through these exercises and finish them by themselves cement the idea of the robot navigation blocks i.e. move and turn as well as the flow control blocks i.e. switch, loop, loop interrupt. They additionally understand and can use bumper based wall alignment and that navigation is key to robot path planning. Finally, the brick button is not a sensor that can be used by the robot to sense the environment. However, waiting for the brick button press is an invaluable debugging tool and has saved kids countless hours when they could insert it in their programs and figure out issues where the robot was not performing as they expected.

Ultrasonic Sensor

The Ultrasonic Sensor in the EV3 kit looks like two robot eyes that you could mount on top of a head. Students that we teach usually draw a similarity to the eyes of the Wall-E robot from the Disney movie of the same name. When powered on, the sensor lights up with red color which is an easy way to determine if the sensor is connected correctly or not. It is shown in Figure 156. The Ultrasonic sensor senses distance to an obstacle in front of it and provides the distance in your program.

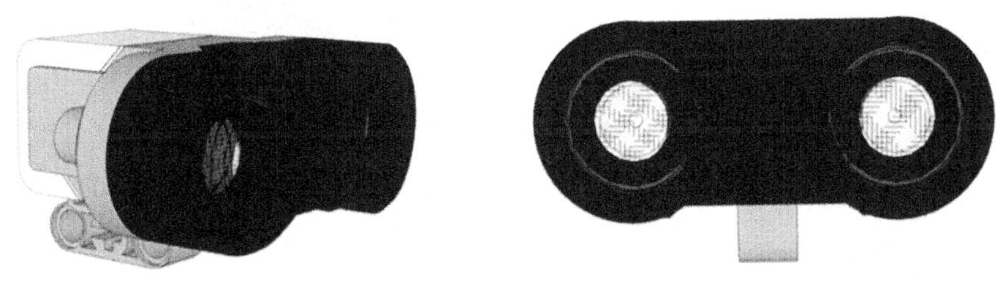

Figure 156: Ultrasonic sensor, side view and front view

The Ultrasonic sensor is based on an industrial sensor that looks exactly like the way the Lego sensor looks like. This is shown in Figure 157.

Figure 157: The industrial ultrasonic sensor

The Ultrasonic sensor works on the same principle as used by bats in nature. Bats in nature use *echolocation* to determine their current location and the location of their prey. *Echolocation* works on the basis of sound reflection. When sound is emanated by a source and eventually reflected back from an obstacle in front, depending on the time it took for the sound to travel to the obstacle and come back, one can calculate the distance to the obstacle. The reason this is possible is because the speed of sound in air is a fixed and known quantity. Although the specification for the Ultrasonic sensor says that it can detect distances from 0-255cm, in practice, it cannot detect distances less than 3cm and for most robot navigation and path planning exercises, you will likely not be using it for measuring distances significantly more than 100cm.

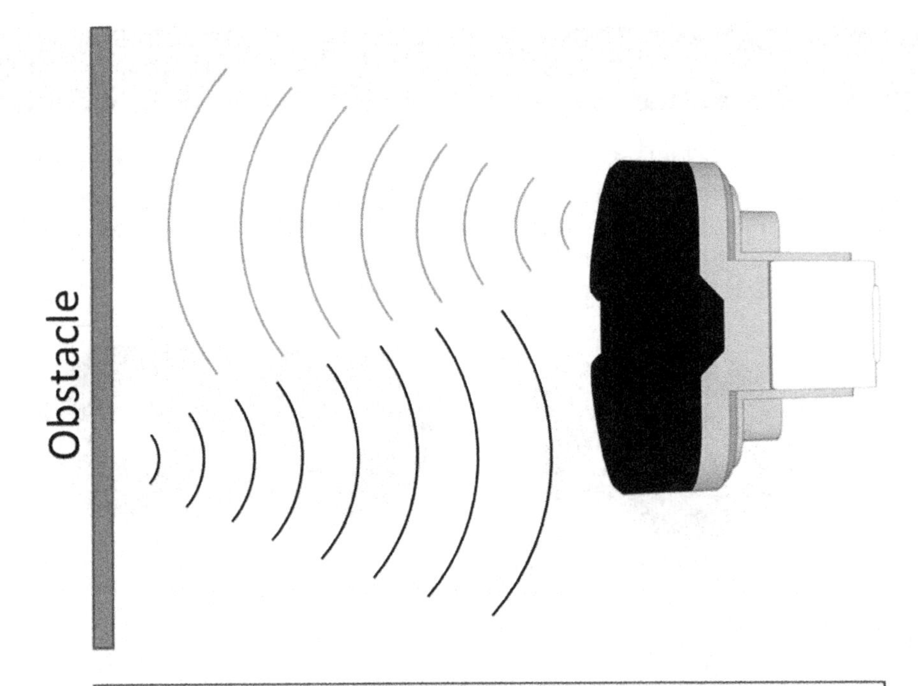

Sound waves are sent out by the ultrasonic sensor and reflect back from obstacles in front. This is called echolocation and the Ultrasonic sensor detects the reflected sound waves and based on the time taken for the echo to come back, it calculates the distance to the obstacle.

Figure 158: The ultrasonic sensor works based on echolocation.

One of the eyes of the Ultrasonic sensor creates and sends the sound whereas the other one detects the reflected sound. It is key to remember that the Ultrasonic sensor cannot detect distances less than 3 cm - or about 1.5 inches reliably.

The Ultrasonic sensor block is available in the *wait, switch and loop blocks* and is also available as a standalone measuring block when you simply need to measure the distance from obstacles without any conditions. The standalone block is shown in Figure 159 to illustrate the various options available. Although the presence/listen option is available, it has limited use in FLL and is rarely, if ever used.

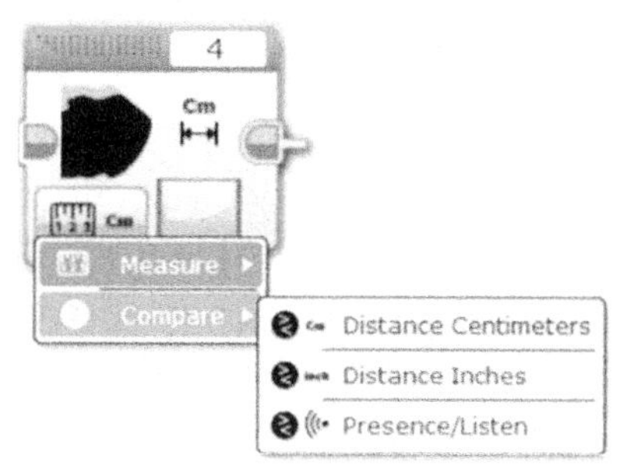

Figure 159: The Ultrasonic sensor options expanded. The Ultrasonic sensor can provide the measured distance in cm or inches.

The ultrasonic sensor works the best when the obstacle in front of it is a flat wall and when the sensor is positioned perpendicular to the flat wall as demonstrated in Figure 158. However, many beginners mistakenly assume that Ultrasonic sensor cannot be used on any obstacle that is not exactly flat. This is a fallacy, and in fact, the Ultrasonic Sensor can be used for some extremely unlikely objects that it can detect with reasonable success. Here are some Lego shapes that the Ultrasonic sensor can easily detect.

The Ultrasonic sensor works on many kinds of objects including weird shapes and objects with holes in them. Before you discount the Ultrasonic sensor, use the port view (Figure 64) on the EV3 brick to check whether the Ultrasonic sensor is detecting the object in front of it.

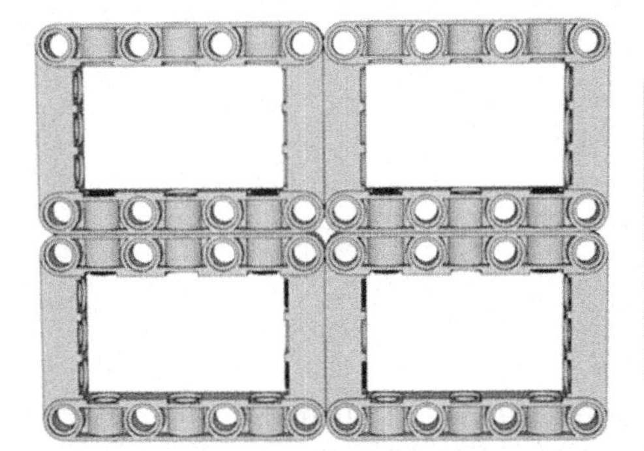

A structure built using Lego frames is detectable using Ultrasonic sensor at 5 cm distance most of the times when the sensor is reasonably well oriented.

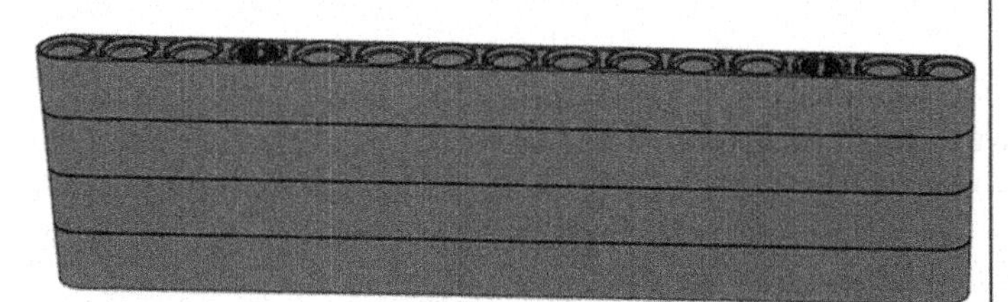

A low height wall based on Lego beams is detectable when Ultrasonic sensor is mounted closer to ground. Even when this wall is vertical, with somewhat careful positioning of the Ultrasonic sensor, you can detect this wall.

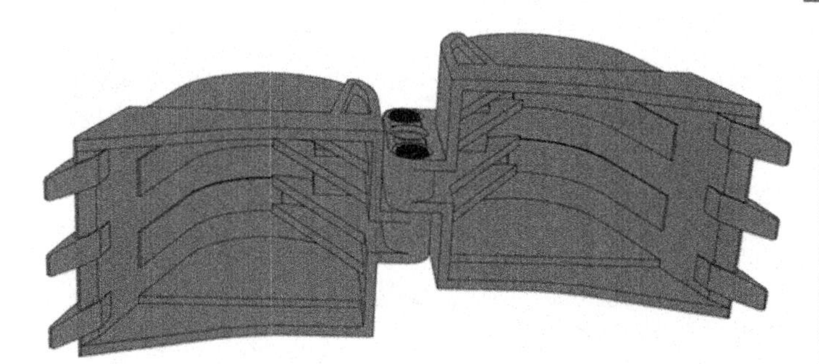

An uneven surface created using buckets (Lego design #24120) from the Lego bucket wheel excavator set can be detected fairly well by the Ultrasonic sensor.

Figure 160: Ultrasonic sensor can easily detect objects of many types, including ones with uneven surfaces and even with obstacles that have holes.

Mounting the Ultrasonic sensor

The Ultrasonic sensor can be mounted in practically any orientation on the robot without impacting its functionality. However, the most common and effective way of mounting the ultrasonic sensor is to keep both the sensing elements i.e. eyes parallel to the ground. Quite often, teams will use the ultrasonic sensor mounted upside down to make it closer to the ground. You can mount the Ultrasonic sensor almost 1-2 FLU (Fundamental Lego Units - the space between the centers of two holes in a Lego beam) from the ground and still have it work well.

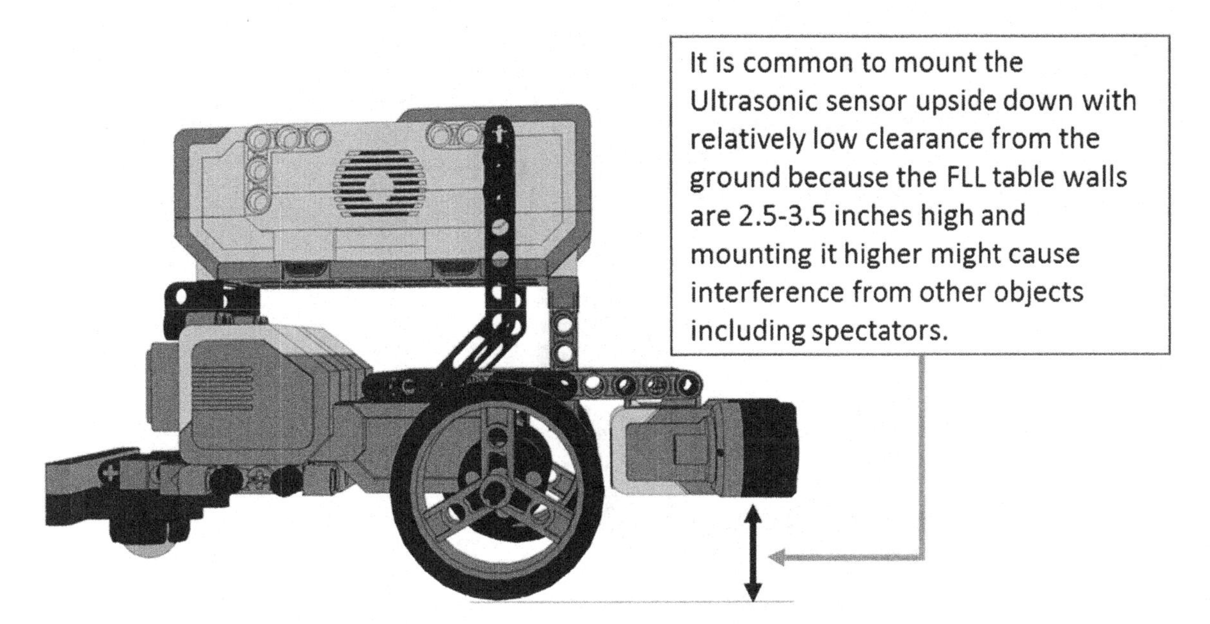

Figure 161: The Ultrasonic sensor is quite commonly mounted upside down with relatively low clearance from the ground

Examples of the Ultrasonic sensor use

In one of the simplest programs utilizing the Ultrasonic sensor, using the wait block along with the Ultrasonic sensor, we can move the robot until it reaches 5cm from a wall and stops (see Figure 162).

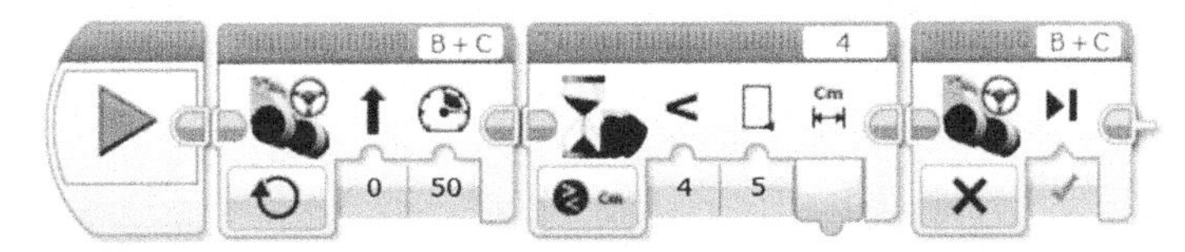

Figure 162: Moving the robot until the Ultrasonic sensor is within 5cm from the wall

Wall following with the Ultrasonic Sensor

Apart from mounting the Ultrasonic sensor in the front of the robot, if we mount the sensor on the side, we can use it to detect objects on the side and if the object on the side is a wall, we can move parallel to the wall at a distance. A robot design with the ultrasonic sensor mounted on the side is shown in Figure 163.

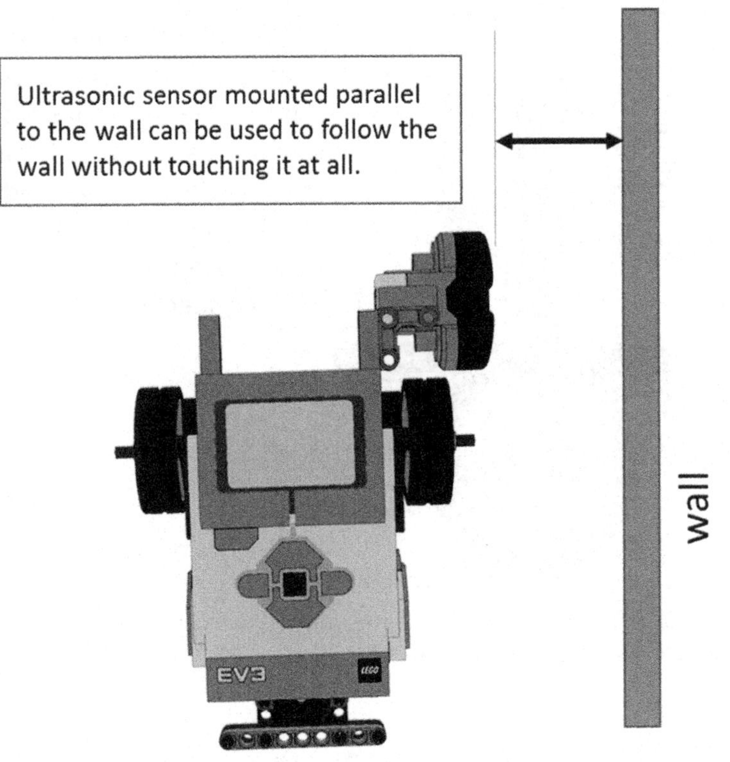

Figure 163: Mounting the Ultrasonic sensor on the side, parallel to a wall on the side is very useful for line following when you may not be able to touch the wall. The program for this is shown in Figure 164

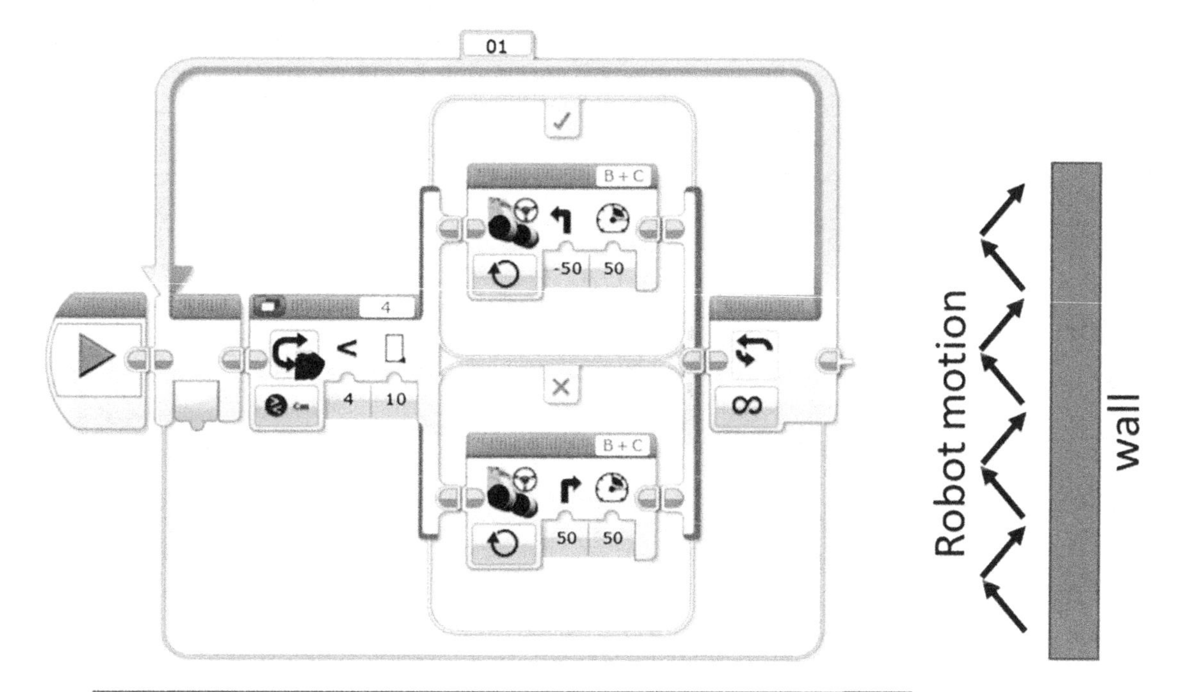

A program that uses the Ultrasonic sensor to move a robot parallel to a wall i.e. follow a wall on the right side. The Ultrasonic sensor is mounted on the right side of the robot with the sensing elements parallel to the wall. The robot moves in a zig-zag motion by only moving one wheel at a time. As the sensor gets close to the wall, the wheel close to the wall turns, moving the robot away from the wall and when the sensor gets too far from the wall, the wheel away from the wall turns moving the robot closer to the wall.

Figure 164: Ultrasonic wall following program

Exercises

In some ways, you may use the Ultrasonic sensor as a replacement for the touch sensor since touch sensor can be thought of as an Ultrasonic sensor that can only detect objects at zero distance. Thus, many of the exercises that are applicable to the touch sensor are applicable to the Ultrasonic sensor as well, so you should attempt all the exercises listed in the touch sensor chapter with the Ultrasonic sensor. We are listing a few new exercises here.

Exercise 1 below (Figure 165) is a weaving exercise new car drivers have to perform for the driver's license test. Perform this exercise with one Ultrasonic sensor even though it will be easier with two Ultrasonic sensors.

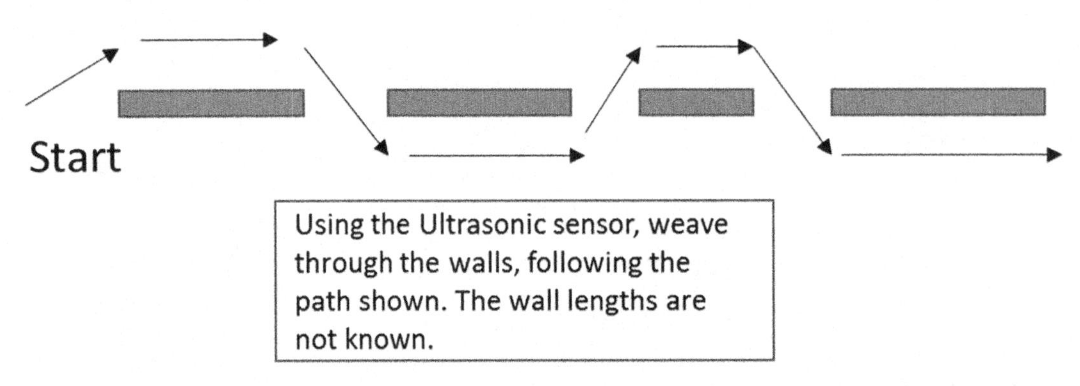

Figure 165: Weaving exercise using the Ultrasonic sensor

Exercise 2 below is based on the behaviors of young puppies. Young puppies have the tendency to follow owners all around the house and in their eagerness quite often they hit the owner from behind. The idea is to write a program that makes your robot behave in the manner of a puppy as shown in Figure 166. You should write this program in two attempts:

- Write a program where the robot puppy follows the owner, if further than 10 cm from the owner, attempts to come closer. If it is closer than 10cm to the owner, it tries to go back. You will notice that this robot dog is a bit over energetic as it never comes to rest. Rather, it keeps moving back and forth continuously. Obviously, this would tire the puppy out. That leads to the second program.
- Modify the above program so that there is a stable zone between 10cm-17cm away from the owner. If the puppy is in this stable zone, it stays put rather than moving. This will cause the puppy to behave in a saner way.

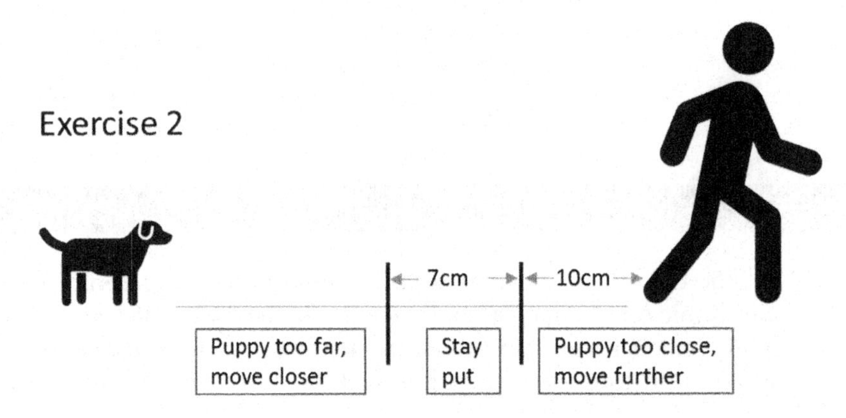

Figure 166: The Puppy follower program

Exercise 3 is an exercise in caution when approaching objects at high speed and has practical applications in FLL. When your robot has to go from one place to another, speed is of essence and it makes sense for you to make your robot travel at the fastest speed it can. However, as the robot starts approaching walls or other obstacles, it should slow down till it has successfully avoided them. This exercise is about creating a program that starts at a high motor power i.e. 100 power and as it approaches a wall, the robot drops the power down to 75, 50 and 25 at 30cm, 20cm and 10cm from the wall respectively. Finally, a variation on this exercise is to additionally use the touch sensor mounted in front of the robot to finally stop the robot when it hits the wall. The above program can be converted to a MyBlock and used in FLL to approach objects in a cautious way while at the same time saving time when the robot is far away from those objects.

Summary

The Ultrasonic sensor is an often-misunderstood sensor that can be extremely useful in FLL. It can detect obstacles from quite far and can be used to approach models in a much more controlled fashion by moving fast in the beginning and slowing down as the robot gets closer to the obstacle. It alleviates the need for wall following using wheels or touch sensor based assemblies and thus should be at the top of your mind when considering path planning in the FLL competition.

Color Sensor

The color sensor in the EV3 kit is one of the most useful sensors for FLL although its usability has been decreasing as the FLL mats in the past few years are now printed with tons of pictures and graphics with many hues and thus decreasing the usability of the color sensor.

The Color sensor in the EV3 kit should more accurately be named the color and light sensor since it not only detects a limited 7 colors (Figure 167) but additionally, it can detect the amount of light reflected from the surface in front of it. This might not seem very useful, but once you understand how the color sensor works, this becomes an extremely useful property.

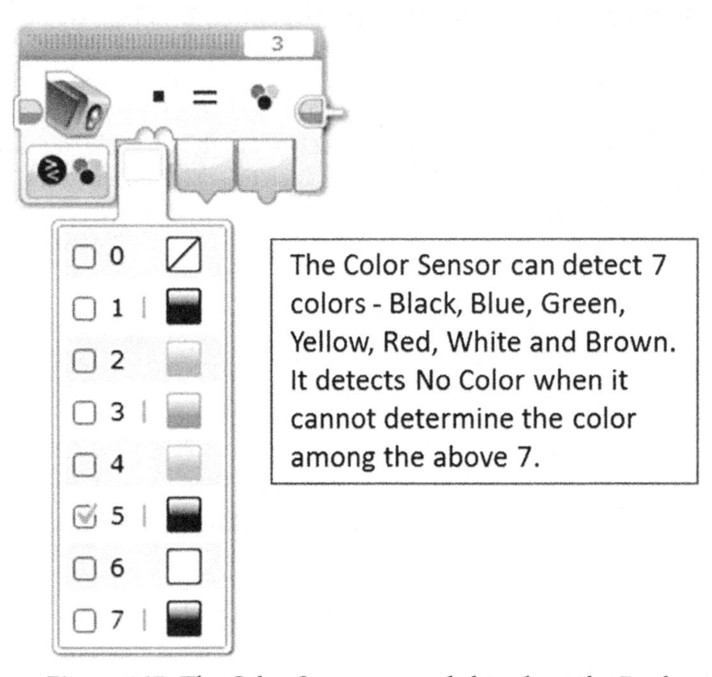

Figure 167: The Color Sensor expanded to show the 7 colors that it can detect.

One key property of the color sensor is that it contains two parts when physically viewed from the side the sensing elements are placed (Figure 168.) One of the two parts is a component with LEDs that light up the surface whose color the other component of the sensor is trying to detect. Given that the light emitted by the LED component is sufficient to illuminate the surface and provide perfect readings, any robot design should try to shroud or shield the sensor from any external light to avoid variability when external light conditions change. If you can successfully shield the sensor from any external light sources, your sensor will work reliably and provide consistent readings for a surface no matter how the overall scenery where the robot is operating is lit.

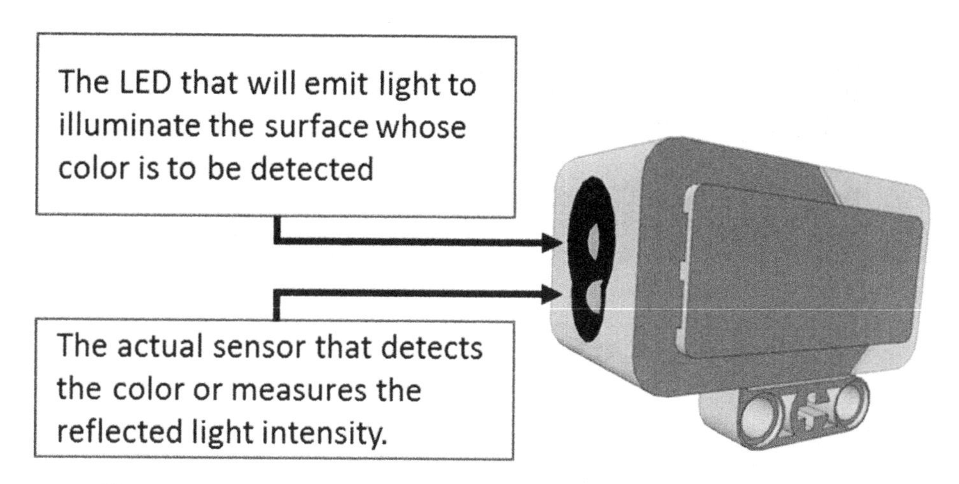

Figure 168: EV3 Color Sensor. The sensor generates all the light it needs to detect color or measure reflected light intensity and should be shielded against external light.

You should make every attempt to completely shield the color sensor from any external light sources. The color sensor produces any light that it needs for detecting colors or reflected light intensity and external light only introduces variability and confuses the robot.

Although one might think that the color sensor could face horizontally to detect colors, this is not the recommended way since it exposes the color sensor to external light and there is no reliable way to shield the sensor when it is placed in such a manner. Instead, the color sensor should be mounted facing down and usually would perform the best when it is no higher than 1-2 FLU (Fundamental Lego Units - space between the center of two consecutive holes in a Lego beam) from the surface on which the robot is traveling.

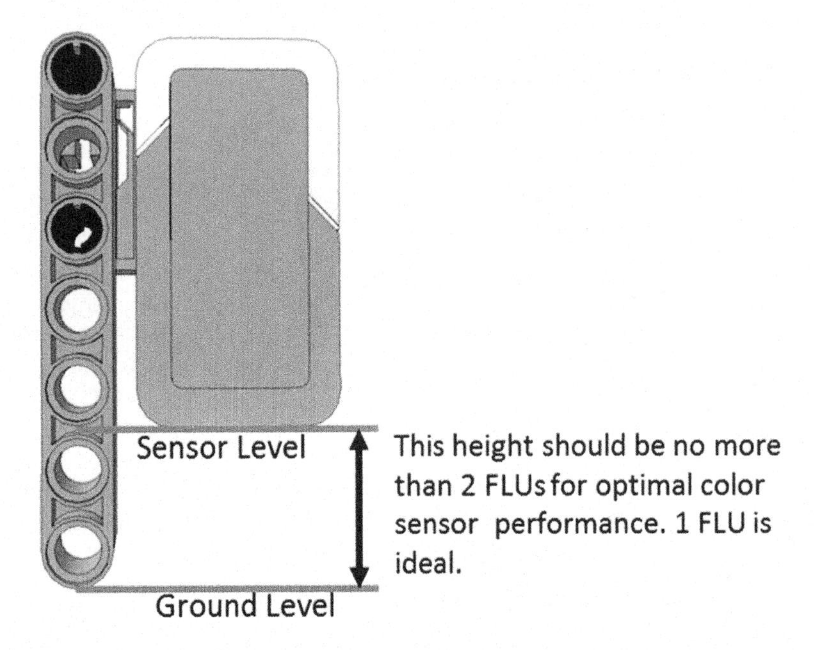

Figure 169: Optimal mounting of the Color Sensor. The sensor should be facing down and ideally no more than 1 FLU from the ground. Sensor does perform reasonably well till 2 FLU but after that the accuracy suffers significantly.

The color sensor should be mounted on your robot facing down and in a way so that the surface of the sensor is within 1-2 FLUs of the surface the robot is travelling on. A good place to mount the color sensor is under the body of the robot.

Now that we know how to mount the sensor in an optimal way, we can discuss the color sensor further.

Wait for color

The *Wait for color* program is similar to the *Wait until* programs for both touch as well as the Ultrasonic sensors. It works great on surfaces that are pure white and have nicely colored lines. Unfortunately, in the recent years, FLL mats are filled with full color pictures with many hues which has decreased its utility quite a bit. This program simply makes the robot move until the sensor sees a colored line. In the 2014 FLL World Class season, programs like the one shown in Figure 170 were used to make the robot travel to colored lines. The FLL Mat in that season (Figure 171) had a white surface with colored lines making it a perfect candidate for a *Wait until Color* paradigm.

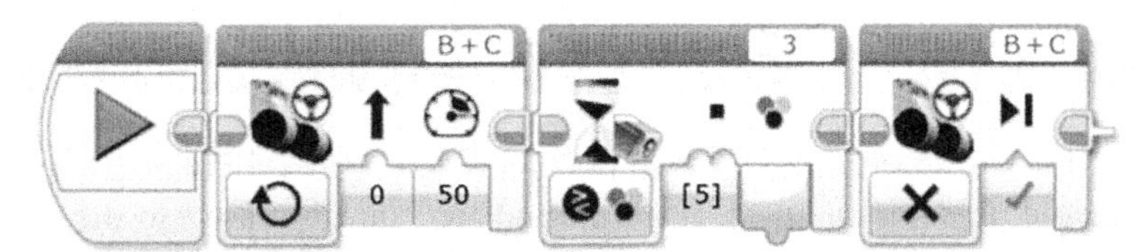

Figure 170: A simple program to keep the robot Moving straight until the color sensor senses red color. Such programs can be quite useful if the entire path of the robot is clear of the red color and is preferably white to avoid causing any confusion to the sensor.

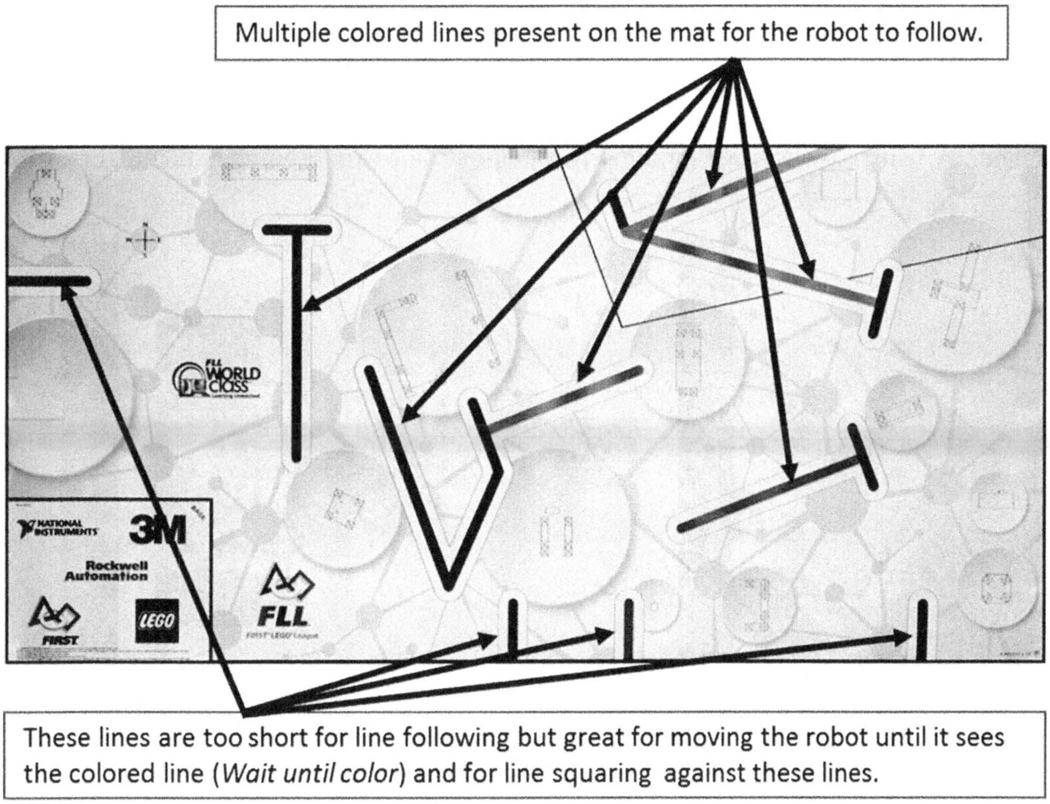

Multiple colored lines present on the mat for the robot to follow.

These lines are too short for line following but great for moving the robot until it sees the colored line (*Wait until color*) and for line squaring against these lines.

Figure 171: The FLL 2014 World Class Mat had a white surface with lots of clearly marked color lines that we could use with the "Wait Until color" block. Sadly, in recent years, the mat is full of graphics and pictures making this approach impractical for most part.

Line squaring to color lines

Line squaring to a color lines is a technique used to make a robot perpendicular to a line. This is a technique similar to *aligning to the wall with the bumper* technique that we had discussed in the chapter on Basic Robot Navigation. This is an extremely useful technique in FLL given that pretty much every FLL table in the past few years has had a lot of lines that you can use for this technique. Line squaring, in short uses two color sensors to detect a line on the table and then orienting the robot in a perpendicular direction to it. This is demonstrated in Figure 172 where the robot uses two color sensors and approaches the black line at an angle. The goal is to make the robot perpendicular to the black line. The way this is done by stopping the driving motor close to sensor 1 as soon as it detects the black line and only moving the motor close to sensor 2 until it too detects the black line. The intermediate positions of the color sensor 2 are shown for clarity. Once the alignment is done, the robot may decide to move back a few centimeters and redo the alignment a few times for better alignment. With every iteration, the alignment will get better. In our experience, 3 alignment attempts usually position the robot fairly perpendicular to the black line. If you are having troubles with the line alignment, try using slower speed for the driving motors since this is an operation that requires a lot of precision.

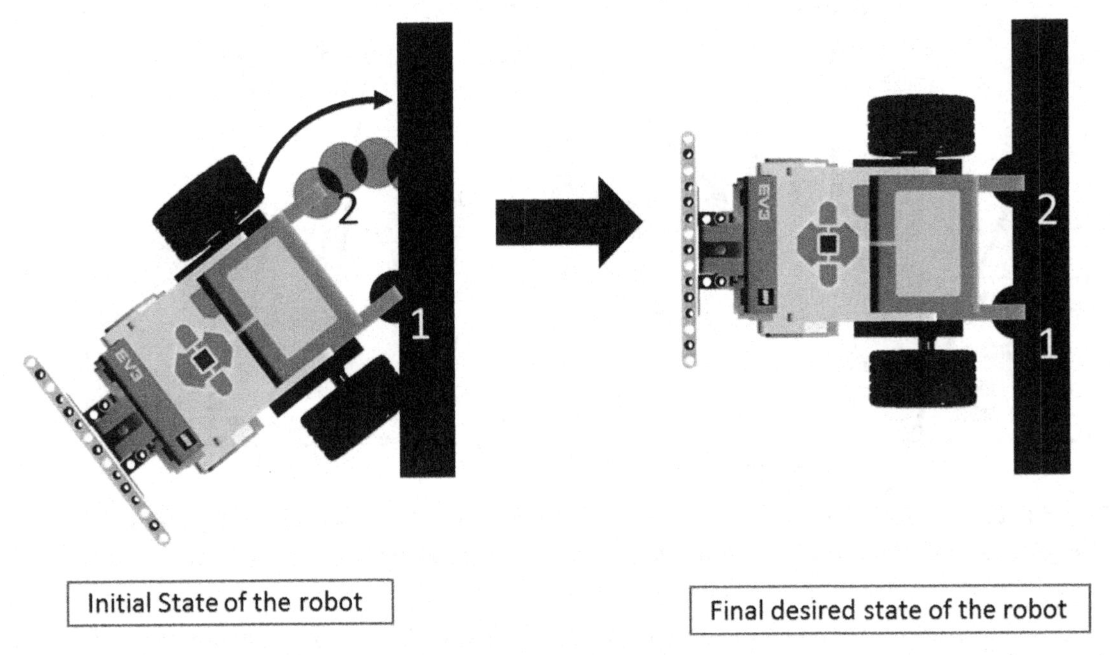

Figure 172: Aligning to colored line. Align to colored line is similar to bumper based alignment except that instead of the bumper we use two color sensors and instead of the wall, we use the color or the light intensity.

Here are two programs (Figure 173) that perform line squaring against a black line when the robot approaches the black line with the color sensor 1 hitting the black line first. Note that in this program, we have assumed that the left color sensor on the robot is connected to port 1 and the one on the right is connected to port 4. The two programs show the line squaring approach by using the color mode of the color sensor as well the reflected light intensity mode. If you are line squaring against black lines, the reflected light intensity mode works a lot better since it stops the robot when the two sensors are seeing about the same amount of reflected light intensity. Both the programs contain a small time-based wait in the middle of the program. The delay is there to ensure that the second motor starts moving only when the robot has stopped completely. Adding a small delay between successive motor operations quite often increases the accuracy of successive move blocks since it means that the operations don't impact each other. Additionally, the programs will need to repeat in a loop a few times (2-3 times) for a good alignment and between the loops you will need to back up a few centimeters.

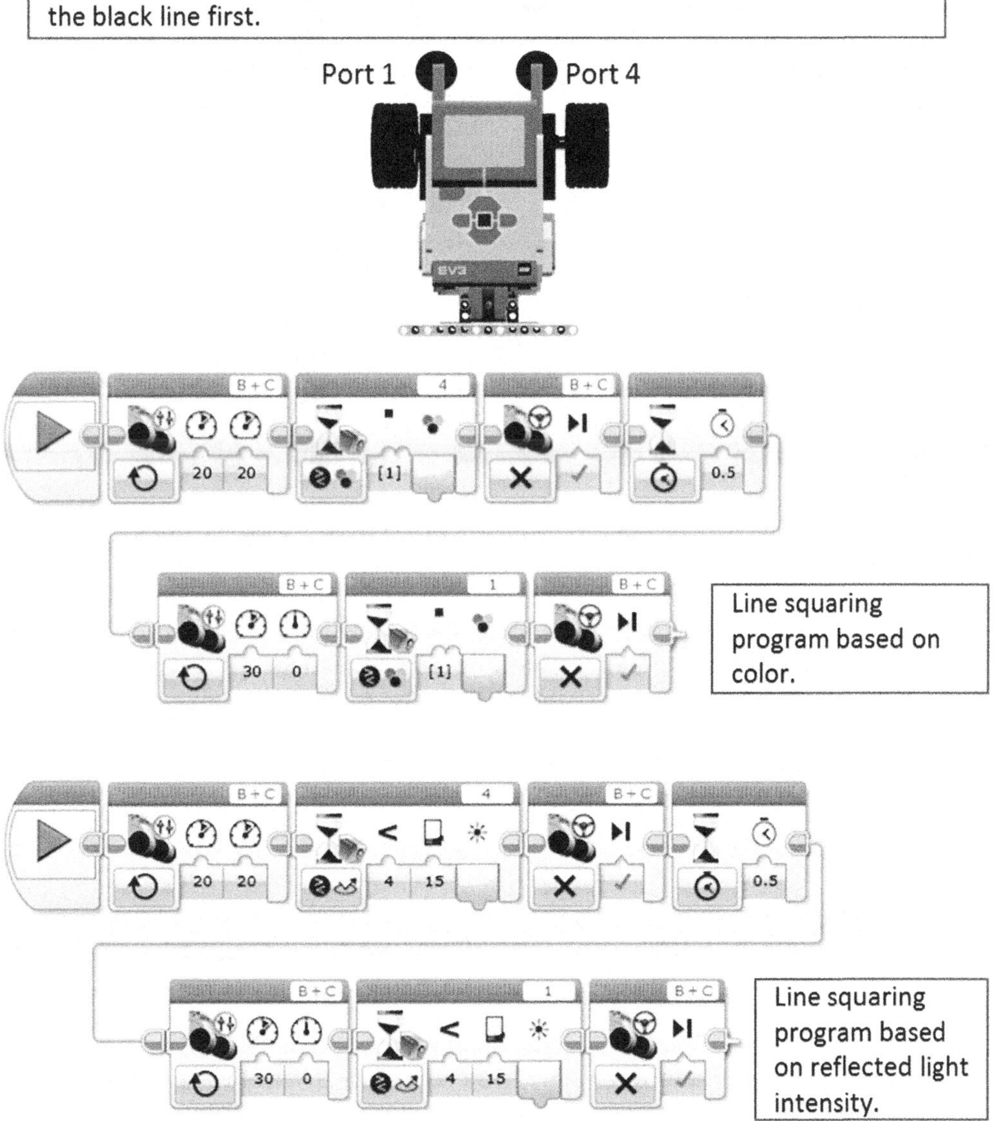

Figure 173: Line squaring shown using the color as well as the reflected light intensity modes. For optimum results, you will need to repeat the operations a few times in a loop with the robot backing up a few centimeters between attempts. Note that in general

reflected light intensity values of less than 10-15 mean really dark surfaces indicating black color.

Many teams used the line alignment technique to great effect in the FLL 2016, Animal Allies season to ascertain the robot's position at multiple points. An example from the Animal Allies season is shown in Figure 174.

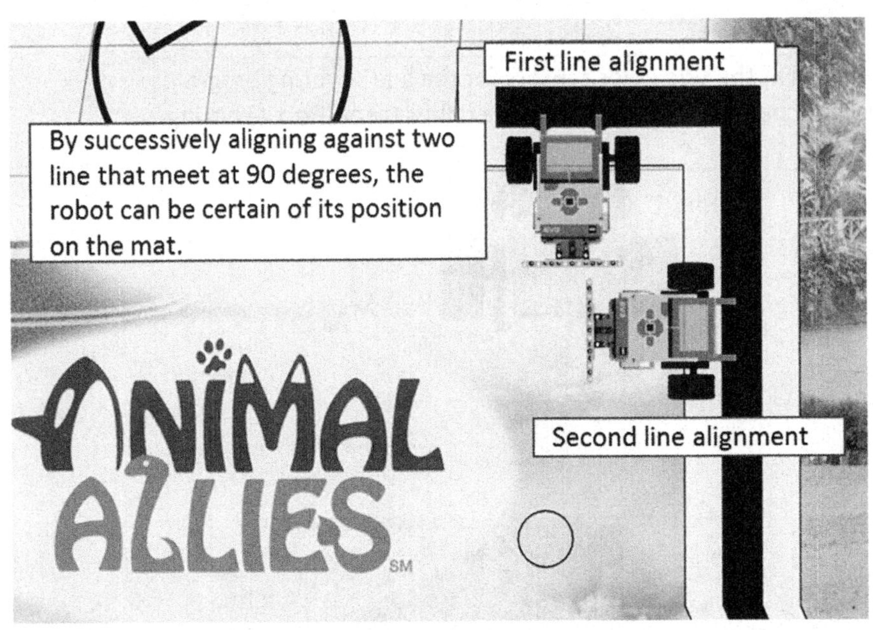

Figure 174: Great use of line alignment to ascertain the robot's position on the mat in the FLL 2016 Animal Allies season.

Even though we have shown line alignment when the robot is approaching with the color sensor on Port 4 hitting the line first, you can easily modify our program to line align when the robot approaches with the color sensor on Port 1 first. Although many teams may be happy with two separate line alignment programs, it is indeed possible to create a single program which aligns to the black line no matter which color sensor sees the black line first. Finally, you should convert the line alignment program into a MyBlock for ready access and use in any FLL programs.

Simple Line following with a color sensor

If you look at the FLL mats from the past few years, you will notice that they have many colored lines zig-zagging across them and quite often they are leading up to a model that the robot needs to interact with to fulfill a mission. Thus, being able to follow a line offers a significant advantage to a team since it allows them a great deal of certainty that they would reach the destination. Let's now discuss how to follow a line using the color sensor.

Even though the Line Following term is quite pervasive, it is a misnomer since it is not possible to follow a colored line itself and instead, what we follow is the edge of a line. One way to understand this is use the example of a real-life person on a road in an hypothetical scenario. Let's say if the person is standing on a colored road so wide that they could not see its edges, then for all intents and purposes, it becomes a huge colored ground that you will not be able to navigate without any additional markers. If, however you could see the edge of the road, you would know which way the road leads. The same principle follows with the robot. If the robot can see the edge of the line, it can successfully traverse the edge. Now that you understand that line following is really edge following, let's look at ways to perform this operation. Just for sake of consistency, we will be calling the edge following with its common name i.e. line following. There are two ways of following a line, we simply call them the *Simple Line following (also called bang-bang or two state line following)* and the PID line following. We will explain the *Simple Line Following* in this chapter and delegate the PID line following for the later chapter in the book named *PID algorithm for wall following using Ultrasonic and line* following using Color Sensors. Let's now look at the Simple Line Following program.

We may interchangeably use the words Program and Algorithm in the rest of the text. The difference between the two words is that an algorithm is the descriptive way of explaining how to perform a certain operation and Program is a formal specification. In simpler terms, an algorithm is simply the concept of how to perform operations whereas a program is an actual implementation that can be run on the robot.

In its simplest incarnation of the line following algorithm, we place the robot's color sensor on the edge of the line and turn the wheels of the robots one at a time. When the robot sees the color of the line, the robot turns the wheel closer to the line so that the robot is moving away from the line. When the robot has turned too far, and the color sensor cannot see the line, we turn on the wheel further from the line and the robot moves forward while tilting and placing the sensor back on the line. By moving in this slightly crooked, jiggling motion, the robot follows the edge. This is illustrated in Figure 175.

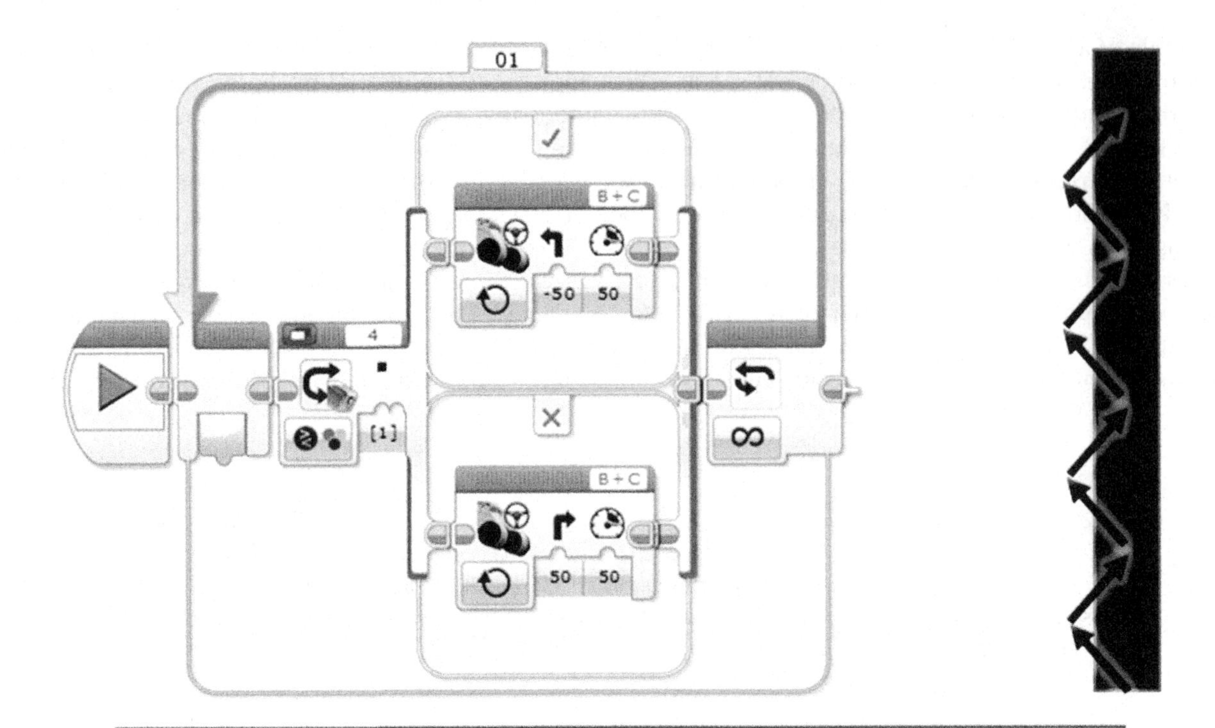

The robot will move in the zig-zag motion when following the left edge of the black line using the program shown.

Figure 175: The simple line following algorithm. Notice the use of the move steering block that is simply set to "On" and the use of -50 and +50 as the turn amounts

Using a color sensor for following a line is a great idea in FLL and in general works very well. You MUST, however, supplement the line following algorithm with a way to determine when to stop following a line. This can be done using a second color sensor in case the line that your robot is following ends in a T-shaped line as illustrated in Figure 176. Having a T-shaped termination line is quite common in FLL and thus having a second color sensor to stop line following on reaching the end of the line is extremely helpful.

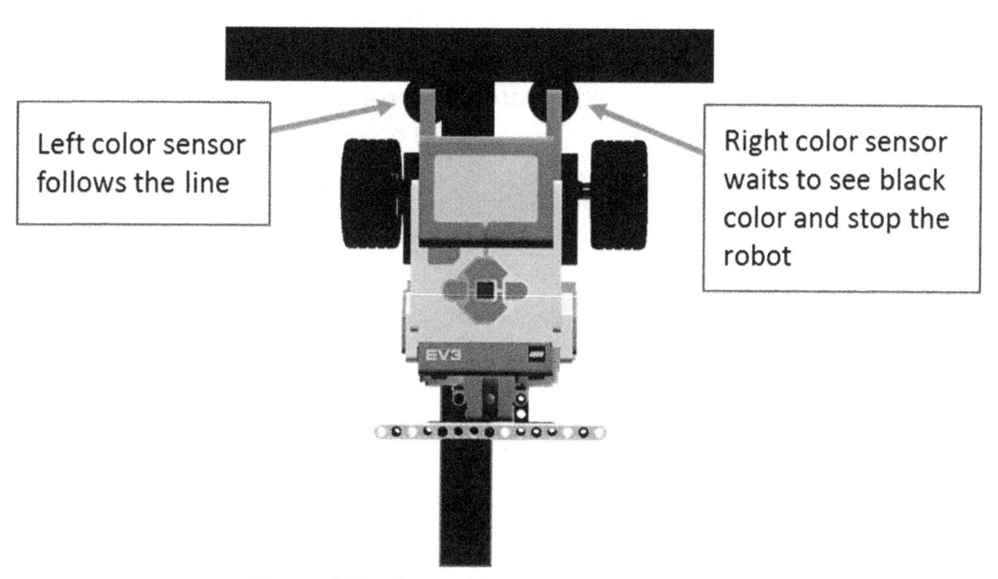

Figure 176: Showcasing a termination condition in a line following program.

A program illustrating line following with a termination condition based on a second color sensor is shown in Figure 177.

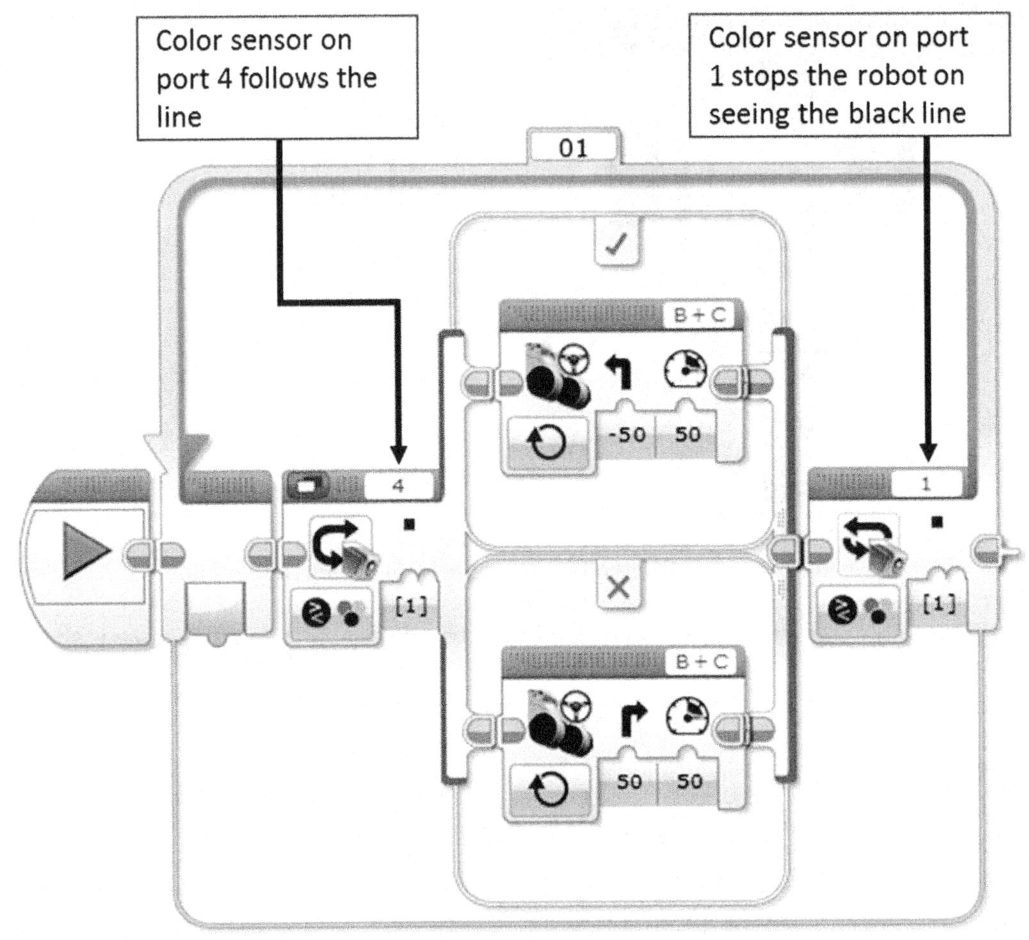

Figure 177: A second color sensor used while line following can provide the termination condition

Calibrating the color sensor

It is important to realize that in spite of your best efforts to shield the color sensor from external light sources, sometimes, it is impossible to do so. Changing light conditions will throw your line following, line squaring or simply the Wait for Color programs off. Fortunately, to tackle this, you can use the technique of *calibrating* your color sensor.

Calibrating your color sensor simply means that you reset the values that the sensor assumes to be black and white. The intuitive way of understanding calibration is to understand that under brighter and brighter light, all colors start reflecting more light. Prime example of this is when on a bright sunny day, white reflective surfaces become extra reflective and even black surfaces become a bit shinier. In such cases, we have to tell the sensor how much light is being reflected when the sensor is being pointed over white as well as black. Given these two values, the sensor can infer all the intermediate values for all the other shades of black.

The program to calibrate the color sensor is shown in Figure 179. The program uses the color sensor block available in the yellow tab in EV3 programming studio. This block has the options required to calibrate the color sensor as shown in Figure 178.

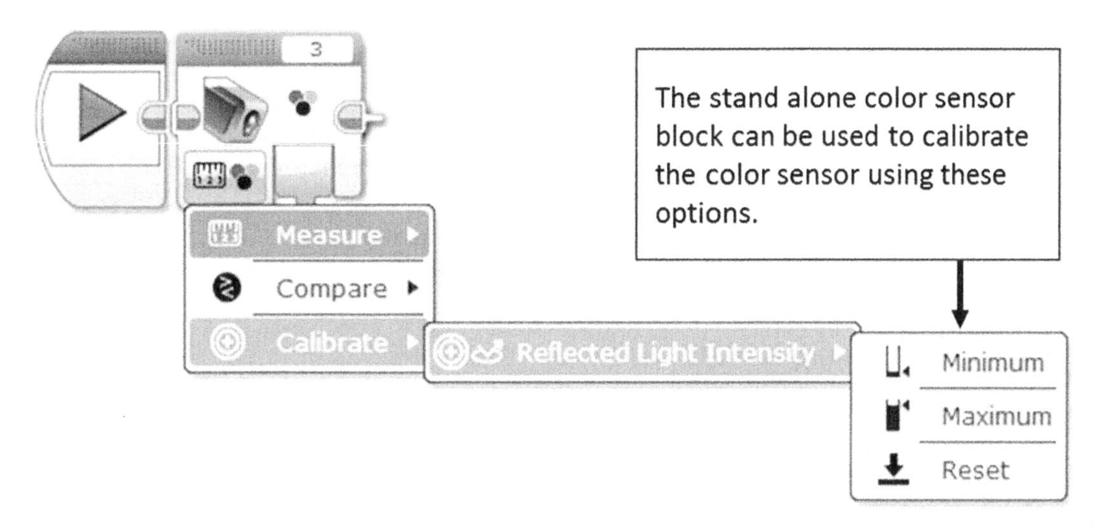

Figure 178: The stand-alone color sensor block with the options shown that allow you to calibrate the color sensor.

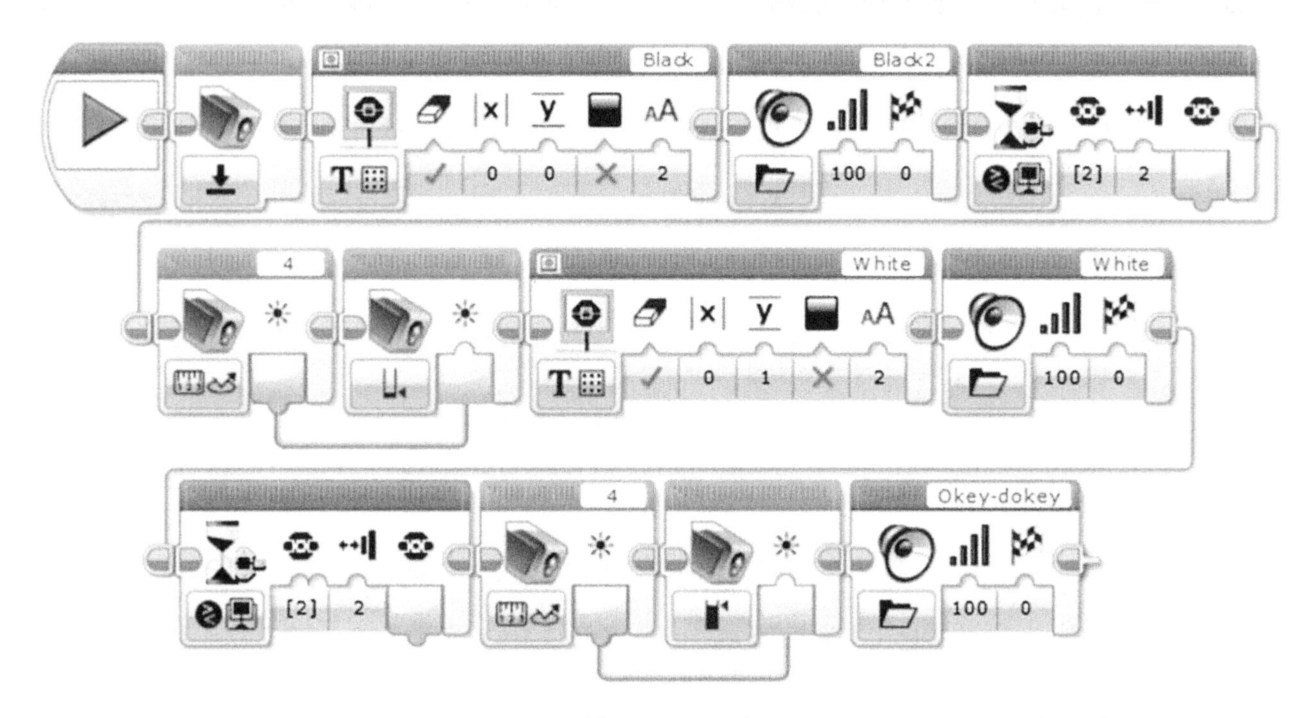

Figure 179: Color Sensor Calibration. In this program, we assume the color sensor is attached to port 4. The program uses the stand-alone color sensor block from the yellow/sensor tab in EV3 which provides the ability to reset the color sensor and to set the minimum and maximum intensities under any given light conditions. When run, the

program displays Black on the EV3 brick screen and speaks the word black. At this time, the color sensor should be placed completely on a black line and the center brick button pressed. The sensor will prompt you to do the same for white color. Once you do that and press the center brick button, the sensor will be fully calibrated and the EV3 brick will say "Okey-Dokey" to signify that everything is done.

You should convert the above program into a myblock so you can include it into your final list of FLL programs. This is one program that should always be present on your FLL robot.

Summary

The Color sensor is one of the most useful sensors in the EV3 kit. The color sensor can be used to detect color on the mat and it can additionally detect the reflected light intensity. The FLL mat design in the past few years leads itself very strongly to the use of color sensor for line squaring and line following. To score highly in the Robot game, the use of the color sensor is a near must.

Gyroscope Sensor

The LEGO Gyroscope - also simply called as the gyro - is a sensor that allows a robot to measure its rotation or rate or rotation. This can be used to turn a robot in a fairly accurate manner. Although it seems like a fairly useful sensor, the gyroscope is one of the most unreliable sensors in the EV3 kit and many teams have hit its pitfalls at the worst times. Given that the key reason for using the gyroscope is the robot turn, which we can perform using calibration and the turn MyBlock we created, we strongly discommend this sensor for its use in FLL. However for learning purposes and getting familiar with the sensor, in this chapter we will explain how to use this sensor. Although, reiterating, we strongly recommend against any teams for using the sensor in FLL.

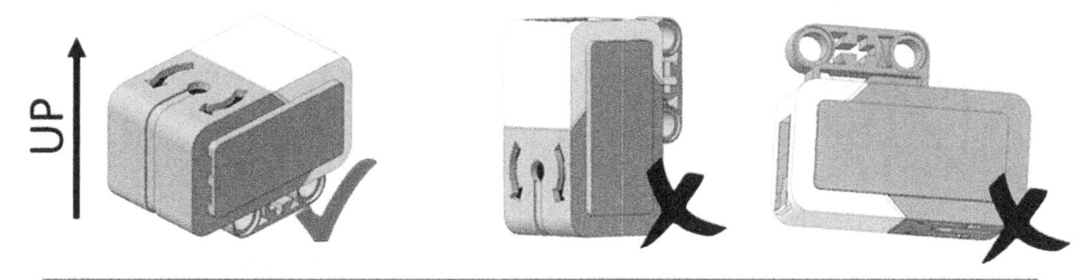

The arrows at the top of the sensor indicate the directions it will measure. This is the correct orientation for mounting. It will not measure correctly when the arrows are vertical and if mounted upside down, it will output inverted values.

Figure 180: Gyroscope Sensor

Basic Functionality

The Gyroscope measures the degrees the gyroscope has turned and we can use it to measure the degrees the robot has turned. Before we delve into how the gyroscope sensor works, we want to point out a fact that may be obvious to you i.e. when the robot turns, any point on the robot turns the same number of degrees. If this is not clear to you, take a look at Figure 181 where we show that two different points on the robot turn the same amount as the robot turns. In practice, what this implies is that you can mount the gyro sensor anywhere on the robot body and still measure the degrees turned accurately.

The Gyroscope is a finicky sensor, and we have found it unreliable since it needs a perfect startup as well as careful monitoring to ensure it is working perfectly. We have seen many robots fail at FLL competitions at crucial times due to these idiosyncrasies and thus we do not recommend using or relying on the gyroscope sensor in FLL.

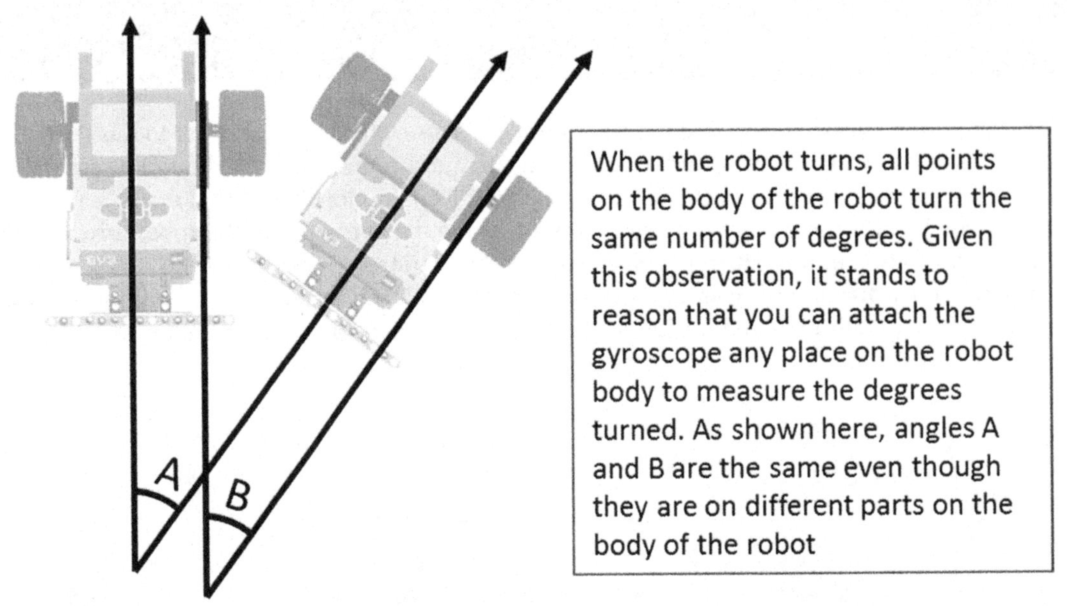

Figure 181: You can mount the gyroscope on the body of the robot at any place

The Gyroscope sensor works in two modes: the angle mode and the rate mode. In the angle mode, it measures the number of degrees from its previous known position and in the rate mode, it can measure the rate of turn i.e. how many degrees per second the sensor is turning at. Of these two, in FLL the measurement of the number of degrees is the more useful mode since most of the robot navigation operations simply require the robot to turn a certain amount and do not need the information about turn rate.

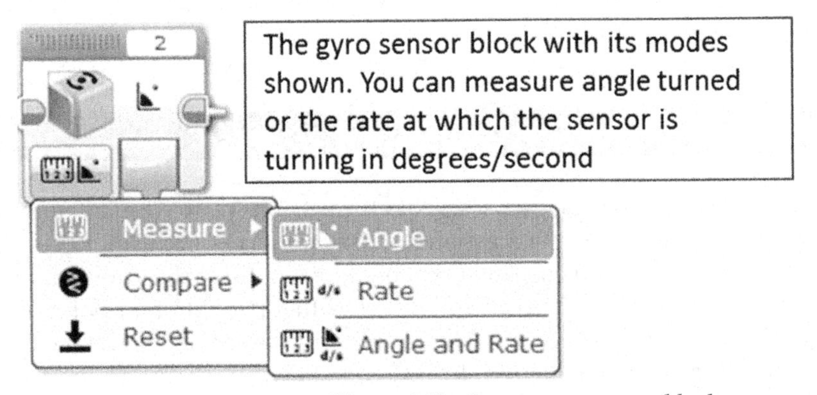

Figure 182: Gyroscope sensor block

Recall from the Chapter *Fundamentals of Robot Movement* where we had to calibrate our robot to calculate how many degrees of the motor to turn so that we could turn the robot by a specific amount. Using the information about how gyroscope can measure the degrees the sensor and hence the robot it is mounted on has turned, we can turn the robot by a specific number of degrees without needing any calibration at all. This can be done using the program shown in Figure 183 that shows how to turn the robot 90 degrees to the right using just the gyroscope.

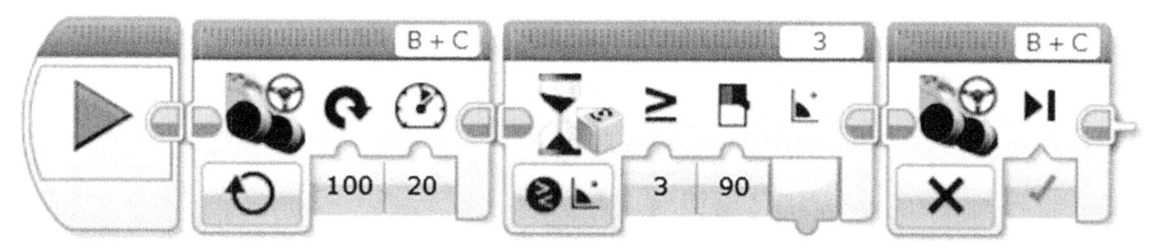

Figure 183: Turning the robot 90 degrees to the right using the gyroscope. Notice we use the greater than or equal to 90 degrees in the wait condition. Depending on the amount of slippage while applying the brakes once the gyroscope reaches 90 degrees, the robot will overshoot 90 degrees. You will have to correct this in your programming.

One of the few places where we saw the gyroscope truly be useful was when a wall following robot had to leave the wall. Using the gyroscope, the robot kept turning until it had fully accumulated 90 degrees on the gyro sensor even though parts of it were hitting the wall for a while as the robot struggled to get to 90 degrees. A calibrated turn block would likely not have done the job in that scenario.

If you look for it, you will likely find many resources that claim that they use the gyroscope for making the robot travel perfectly straight. In our testing, this hasn't borne fruit, especially to counter robot drift which is what we aim to solve. Even over the course of 100cm where the robot drifted almost 1-3cm to either sides, the gyro sensor did not register any changes in our experiments. You can however use the gyro to create a fun *unturnable* robot that resists attempts to turn it by hand. This *unturnable* robot, although cool to play around, is in fact not very useful in FLL since it cannot keep the robot precise to the degree where it can meet required to go straight. We provide an example of the unturnable robot program in Figure 184.

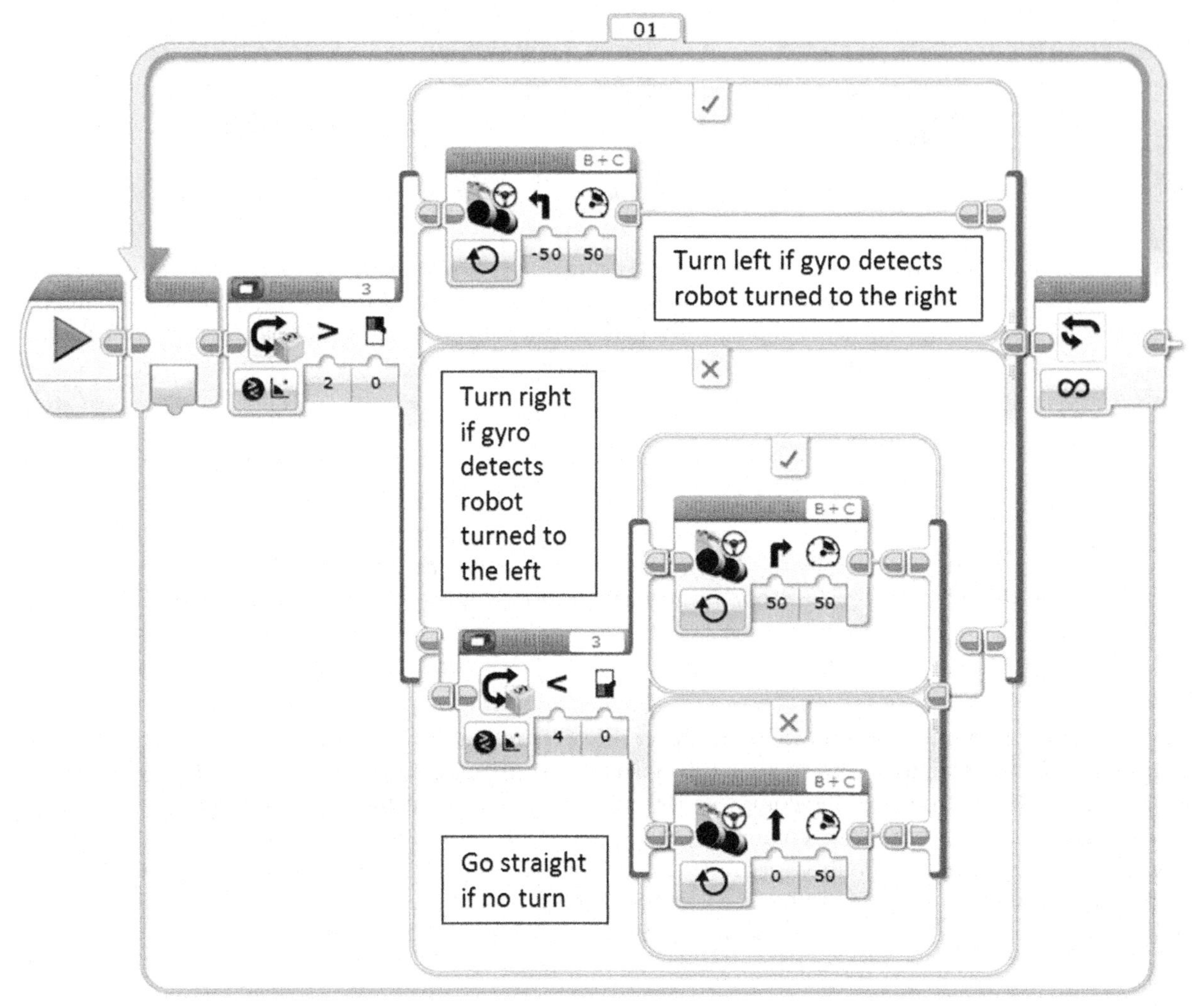

Figure 184: A program that resists the robot turning. The robot tries to go forward in a zig zag fashion. As you turn it away from its path, it will try to turn back on its original path.

If for some reason you must use the Gyro sensor, you can maximize your chances of success by shutting down the robot, placing it on a rigid stationary surface, such as a floor, pressing the EV3 start button and not touching the robot till the EV3 brick has finished starting up completely. This will initialize the Gyro sensor correctly since the Gyro sensor requires the robot to be perfectly still during startup. If the robot moved during startup, there is a good chance that the gyroscope will start drifting. Drifting is the process where a stationary gyroscope will behave as if it is turning at a steady rate. There are multiple other methods where it is claimed that resetting the gyroscope or waiting a few seconds after resetting will help alleviate the gyro drift. In practice, we have found most of these unreliable and thus our recommendation of leaving this sensor in the box.

Summary

The Gyro sensor allows you to measure the angle that the robot has moved from its last known position. Although the gyro sensor can presumably make your life easier by providing accurate turn values, in practice the gyro sensor does not work very well and can cause errors at the least expected time. Even though we explained the gyroscope in this chapter, we recommend FLL teams to leave this sensor in the box.

Motor Rotation Sensor and its use in conjunction with other Sensors

The Motor Rotation Sensor, or the tachometer or the encoder is a sensor that is built into the EV3 motors and is the sensor responsible for counting the degrees as the motor spins. The tachometer is the sensor that is internally used by move steering, move tank and single motor move blocks to determine when to stop the motors once the rotation has started. So far you have only used the motor rotation sensor implicitly when you used the move steering or move tank blocks. However, the motor rotation sensor is also available in the EV3 programming environment as a standalone block that can retrieve the number of degrees or rotations that the motor has turned. The standalone Motor Rotation Sensor block with its options expanded is shown in Figure 185.

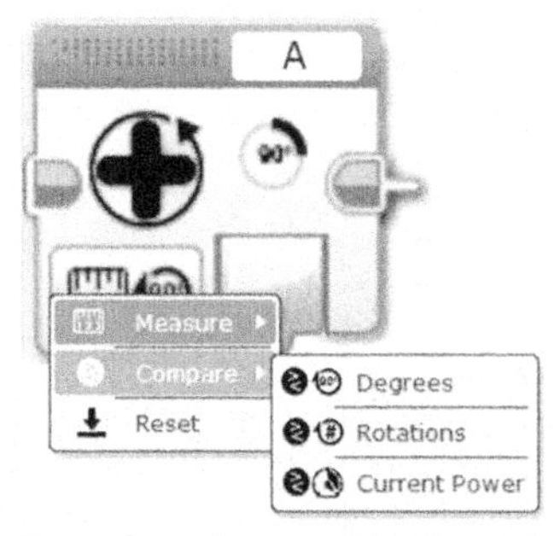

Figure 185: Motor Rotation Sensor block with expanded options. Note that the motor rotation sensor tells you the amount turned for the motor selected at the top right corner of the block. In this figure, the selected motor is the one connected on port A.

The motor rotation sensor is one of the least used sensors in FLL if you don't count the implicit use in all the motor blocks. This is not a fault of the motor rotation sensor, but rather how little it is understood. Getting to use the motor rotation sensor turns the usual way of thinking about the robot completely around.

Usually, we know how far we need the robot to travel, so we calculate the number of degrees to supply to the motors, so it can travel the required distance. In this scenario, you would think that there is really no reason to be able to know how many degrees the motor has turned because you already know it and in fact you supplied it. That is where you need to think out of the box and look at scenarios where the distance the robot needs to travel are not known and the robot needs to *measure* the distance it needs to travel.

As discussed in previous chapters let us say we have to travel *D* centimeters and the *Moving straight Ratio* is *GSR*, then to travel *D* centimeters, the number of degrees required by the motor is *D* x *GSR*. Now let's consider the scenario of wall following using the ultrasonic sensor or line following using the color sensor. In either of these situations, you may need to follow the wall or the line for only a fixed number of centimeters instead of all the way to the end of the line. In such cases, you can use the motor rotation sensor to measure the degrees that the robot has travelled and compare it with the motor degrees that you calculated (*D* x *GSR*) to travel the required distance. Then use the calculated value as a termination condition for your line or wall following. A program and a myblock derived from it, that follows a line for 50cm is illustrated in Figure 186. Keep in mind that we are only using the motor degrees from the B motor. The program assumes that while following a line, both the driving motors i.e. the B and the C motors will turn the same number of degrees and hence relies on the degrees turned by only one of the motors, the B motor.

The Motor Rotation sensor is an extremely useful sensor when used in conjunction with a color sensor or ultrasonic sensor to help in stopping the robot after travelling a known distance.

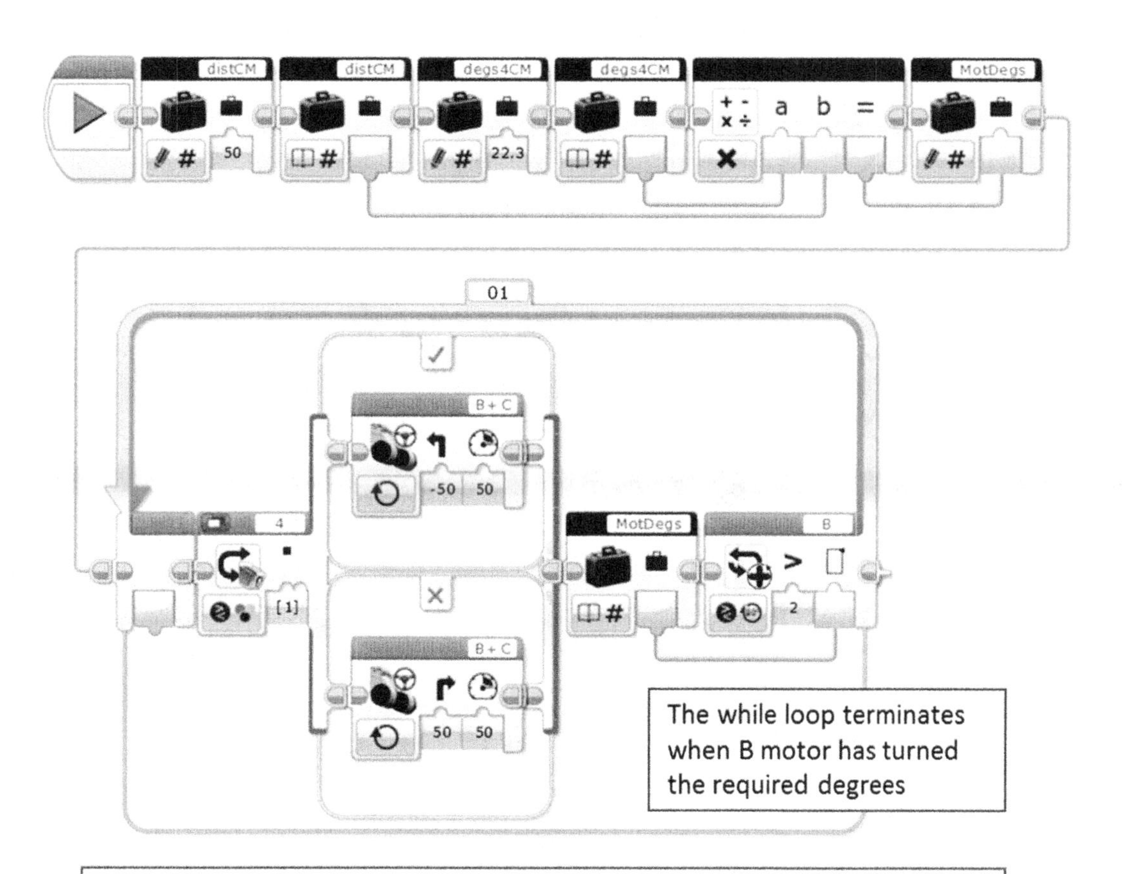

Variables used in this program:
distCM: distance in Centimeters that you want the robot to move
degs4CM: motor degrees per CM of movement also known as the going straight ratio as described in the *Robot Navigation* chapter.
MotDegs: Calculated motor degrees for going straight

Figure 186: A simple line following program that follows a black line for 50 centimeters. We use the distCM and degs4CM variables as input to the math block to calculate how many motor degrees that distCM distance translates into and set that as the initial value of the MotDegs variable. The MotDegs variable is later used as the input to the motor rotation sensor to ensure that the line following algorithm inside the loop block finishes when the motor rotation sensor detects the degrees turned as greater than MotDegs. This program can easily be converted into a myblock for use in FLL programs.

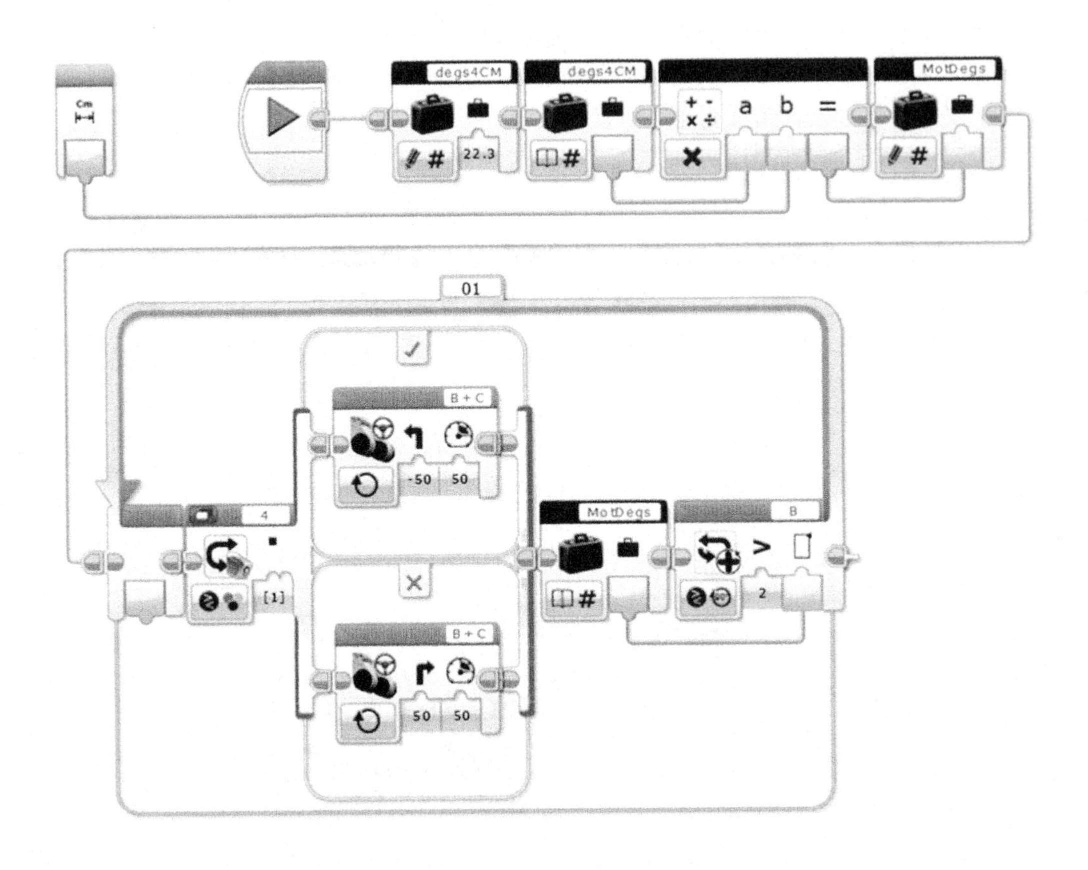

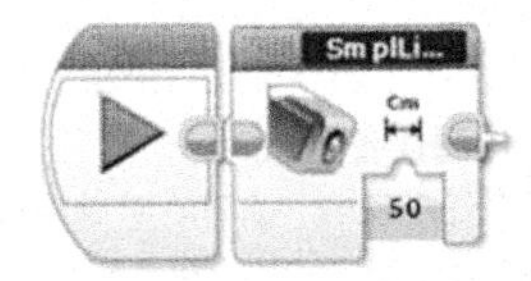

Figure 187: The program in Figure 186 shown as a myblock and how its use will look like as a myblock. You can convert the program in Figure 51 by selecting all the blocks excluding the distCM and start blocks at the beginning and then using the menu options Tools->MyBlock Builder

Another example of the use of the motor rotation sensor along with color sensor can be seen in Figure 188, which is the *Milking the Cows* mission from the 2016, FLL Animal Allies season. The robot had to traverse a circular path for only part of the path. There were no other discerning reference points to stop the robot and many teams relied on the motor sensor to stop at the correct location to finish the mission.

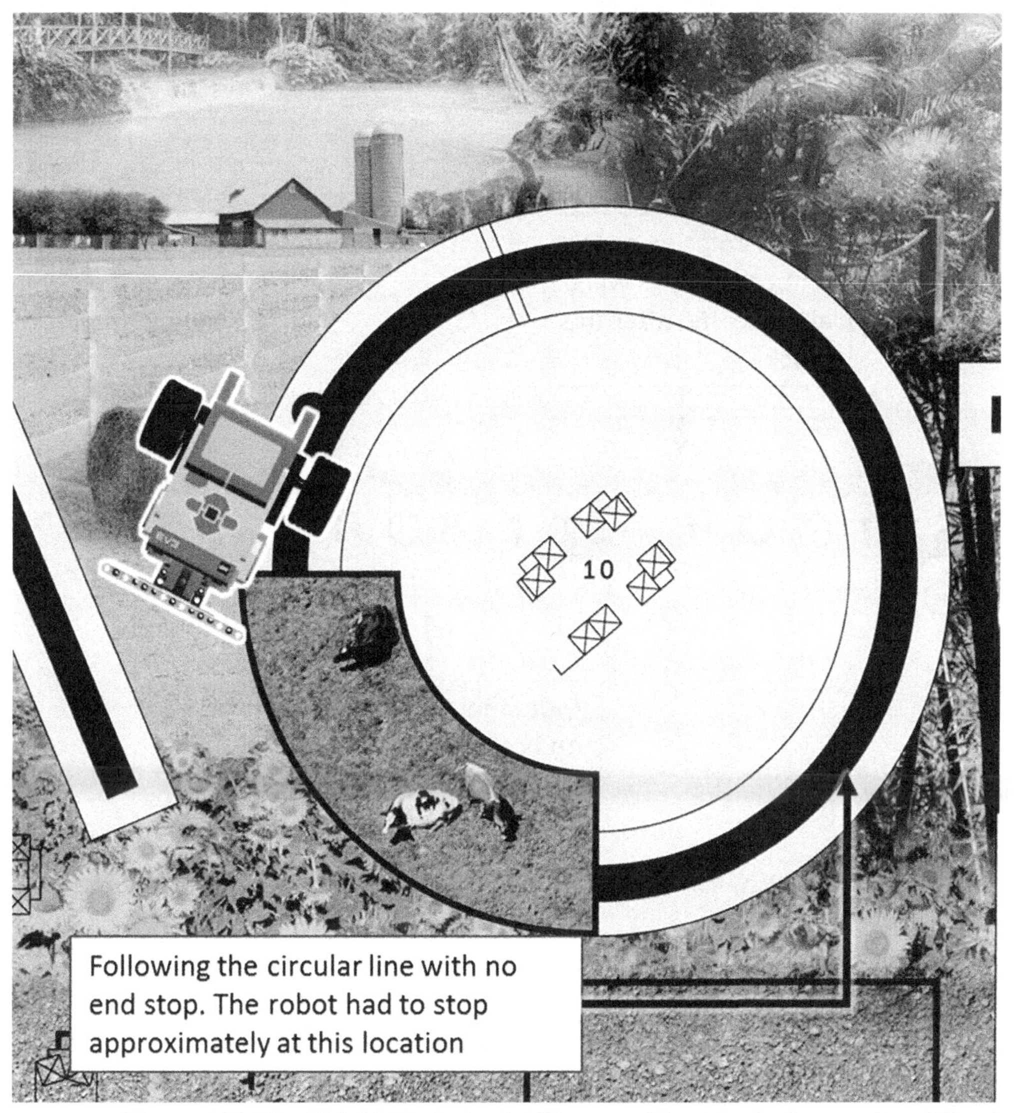

Figure 188: The FLL 2016, Animal Allies mat. The robot had to use line following to traverse part of a circular partway to finish the Milking the Cows mission. This would have been exceedingly hard without measuring the distance travelled with the motor rotation sensor.

FLL programs are rarely ever comprised of a single step. In general, a robot when it goes out on a mission, might intend to perform multiple tasks. This can save time for multiple trips. If possible plan for a mission to include multiple sub-tasks within an area. The navigation for such a mission may be somewhat complex since robot has to perform multiple navigation steps of movement and turns. You may need to make use of the motor rotation sensor for the robot to identify its position between tasks. Thus, it is key to remember that the motor rotation sensor accumulates the degrees as the robot travels and uses motor blocks. That is if you have multiple move blocks one after another, then starting at the first block with zero, robot continues to accumulate the total turns at every block. An example of this is provided in Figure 189. As described in the figure, the First block drives the motors connected to Ports B and C to spin 360 degrees, the second block drives the motor 530 degrees. The rotational value stored by rotation sensor will be accumulated 360 degrees and 530 degrees hence 890 degrees. After the execution of third block the stored value will be 990 degrees

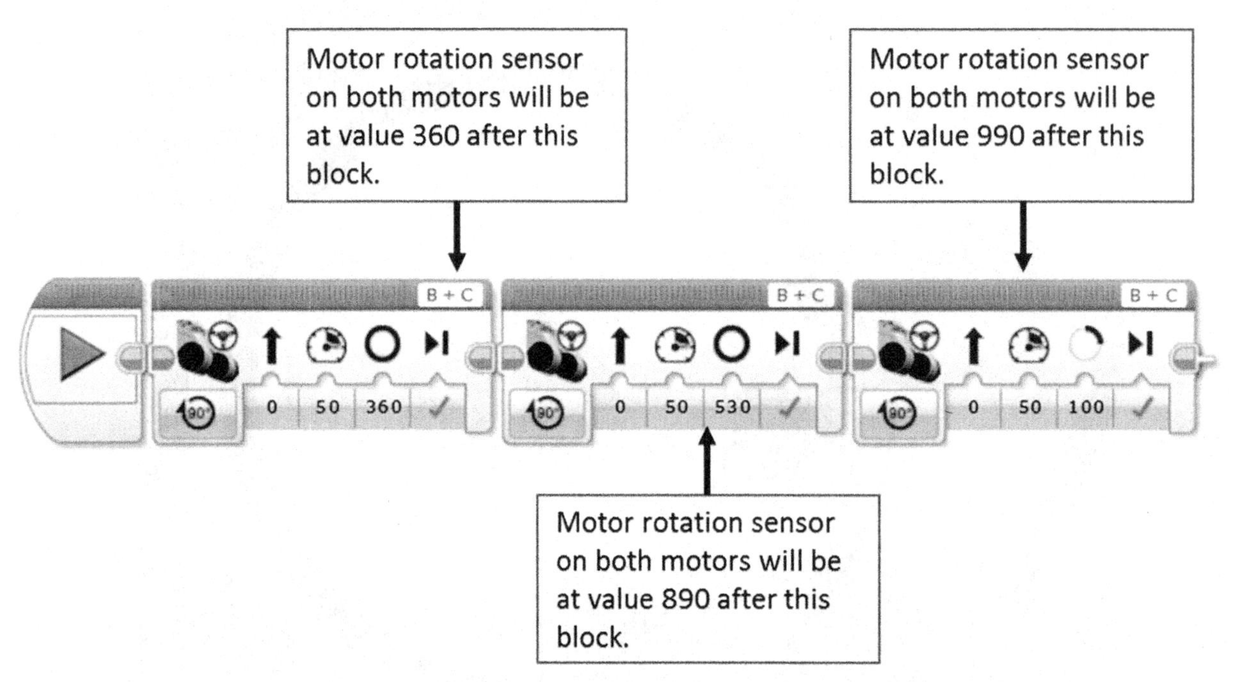

Figure 189: Motor rotation degrees keep on accumulating in a program as the robot moves around.

Given that the motor degrees keep on accumulating, you must reset the motor degrees to zero before you start measuring the degrees for a specific purpose. You can reset the motor rotation degrees using the motor sensor block in the yellow/sensor tab.

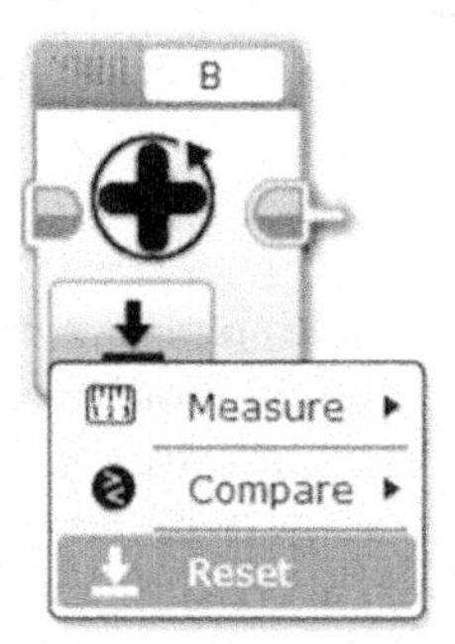

Figure 190: The Motor Rotation Sensor with the Reset option illustrated

Here is an example where we can use the motor rotation sensor to reset the motor degrees.

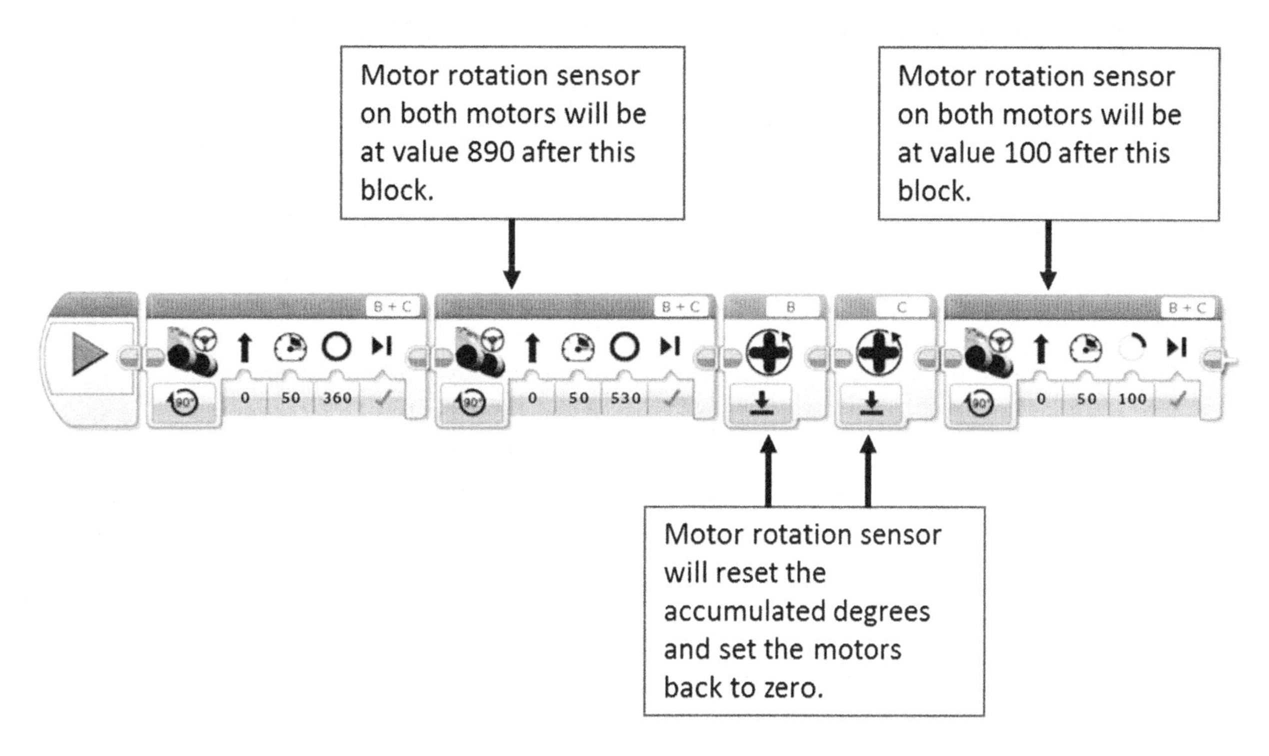

Figure 191: Resetting the motor rotation degrees using the motor rotation sensor block before the last move steering block.

Exercises

Here are a few other exercises with the motor rotation sensor.

Exercise 1: With a robot where the touch sensor is mounted as shown in Figure 148, have the robot start at one corner of the table and move until the end of the table. There it moves back a bit, turns left 90 degrees and hits the other wall. At the end of the process, the robot stops and waits for a brick button press. While waiting, it displays the distance that it measured using the motor rotation sensor for both dimensions.

Exercise 2: This is a maze solving exercise and challenges the kids in multiple ways. The program needs to use variables, switch conditions, loop condition and alignment at key locations to be able to navigate the maze correctly.

1.

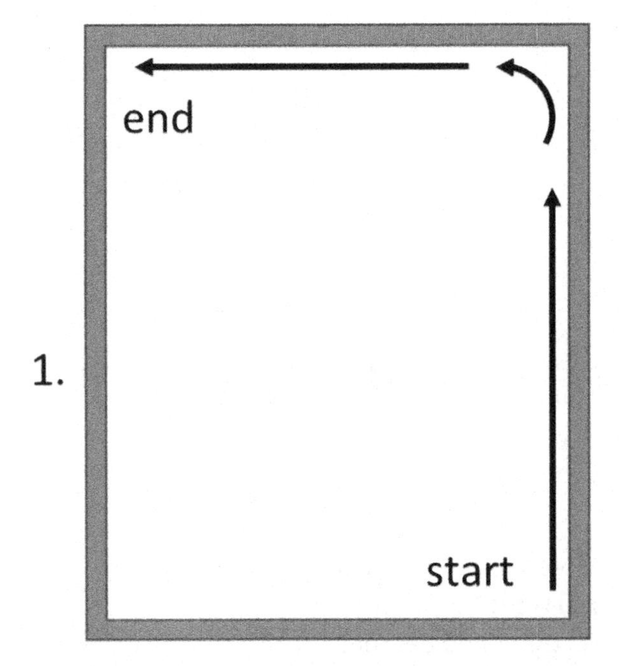

The robot should have a touch sensor mounted in the front. The robot has to start at the start location, move forward until it hits the wall, move back a bit and turn left. The robot then needs to move forward again until it hits the wall next to the end point. Robot should stop at the end point and display the dimensions of the table on the EV3 screen as measured using the motor rotation sensor and variables to remember the intermediate values.

2.

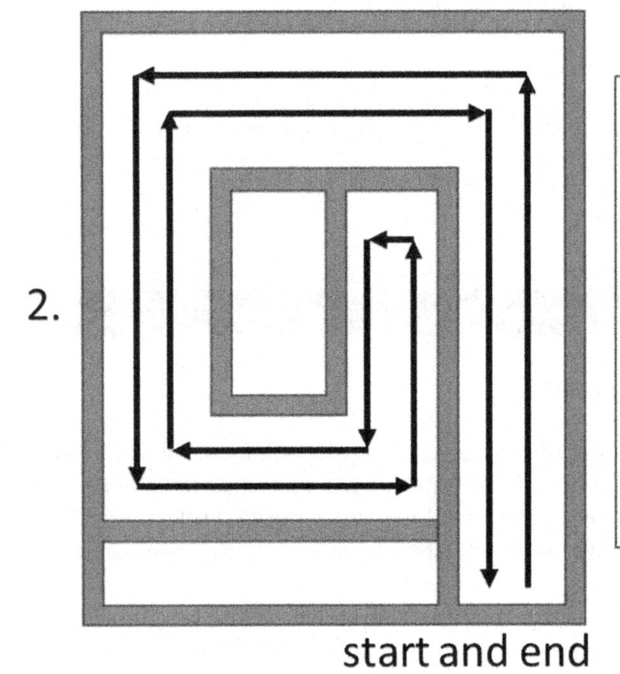

The robot should start at the beginning of the maze and use a touch sensor in the front to find the walls and the motor rotation sensor to measure the distances and store them in variables. Once the robot reaches the point of U-turn, it should use the distances it measured earlier to traverse the path backwards without using the touch sensor.

Summary

The motor rotation sensor provides us the ability to measure the distance that the robot has travelled based on the number of degrees that the robot's driving motors have turned. This provides you the ability to perform distance measurements without needing any additional sensors and is quite useful when following lines using the color sensor and following walls using touch and ultrasonic sensors.

PID algorithm for wall following using Ultrasonic and line following using Color Sensors

We covered line following using the color sensor and wall following using the Ultrasonic sensor in earlier chapters. Our programs for the line following and the wall following using ultrasonic sensor suffer from the issue that at any point of time while following the line or the wall, we are not certain exactly which way the robot is pointing. Unfortunately, this is caused by the very design of the program. The program works by turning the robot left or right based on the condition that it is monitoring. So, for example, for the color sensor, if the sensor is seeing the colored line, it turns away from the colored line and otherwise, it turns towards the color. Although this lets the robot follow the line, the robot motion is not smooth and instead looks jerky with the robot orientation switching between left pointing and right pointing. In FLL, where the robot orientation needs to be precise to be able to perform many tasks. This puts us in a predicament because at the end of the line following, Robot needs to reorient with the help of other reference points which might or might not be available. Thus, either we need a line following algorithm that works without the zig-zag motion of the robot or we need other reference points at the end of line following. Luckily, there is another method that we can use to follow lines and walls that overcomes these shortcomings by severely reducing the zig-zag motion.

To solve this problem, we will use an algorithm called PID control that you hear about often in FLL circles. PID control primarily refers to line following or wall following using a mathematical equation commonly used in control theory. The full form of PID is proportional-integral-derivative and the algorithm is complex to understand. However, you don't need to understand all the details and the math of the PID algorithm to be able to successfully use it. In the rest of the chapter, we are going to be explaining the portions of the algorithm that you need to understand to be able to use it successfully in FLL. To differentiate the line following algorithm that you have seen so far and the PID line following, we will call the former as the Simple Line Following and the new one as PID or proportional-integral-derivative line following.

PID line following

To start with, instead of thinking of PID as a completely new concept, it is easier to think of it as an extreme case of the simple line following. To better understand, let's discuss the Simple Line Following program again. The Simple Line Following Algorithm is a two-state algorithm i.e. either the color sensor is seeing the color or the color sensor sees background color. The Robot keeps following the colored line by moving from the *seeing the color state* to the *not seeing the color state* while still travelling in a forward direction. Moving between the two states by turning the robot on one wheel is what causes the jerky behavior and hence the uncertain orientation of the robot at any point of time. What if instead of the robot jumping between two states, the robot could increase the number of states i.e. instead of simply taking action on *seeing the color* and *not seeing the color*, the robot could have other states like *seeing the color 0%, 20%, 30%, 40%,* all the way *100%*. If we could have more than two states and turn the robot less or more depending on how much color the color sensor is seeing, then that would certainly smooth out the robot motion. This is exactly the idea behind the PID line following. PID line following works on the basis that instead of relying on two states, the robot could look at a very large number of intermediate color sensor states and adjust the amount of turn with respect to that. That is exactly what the complex equations of PID line following convey, although in a more rigorous fashion.

PID line following works on the basis that instead of relying on two states, the robot could look at a very large number of intermediate color sensor states and adjust the amount of turn with respect to that.

To better explain, the differences, the two-state line following program is illustrated in Figure 192.

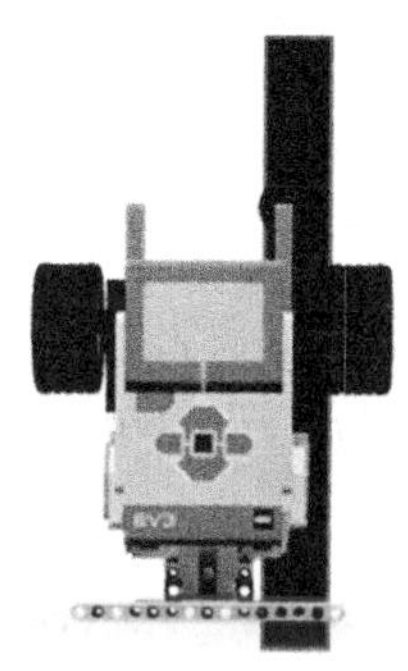

The color sensor denoted by the red dot, fully on the black line.

The color sensor denoted by the red dot, fully off the black line.

Two State, Simple Line Following

Figure 192: The two state Simple Line Following Algorithm

By contrast, a multi-state line following program is illustrated in Figure 193.

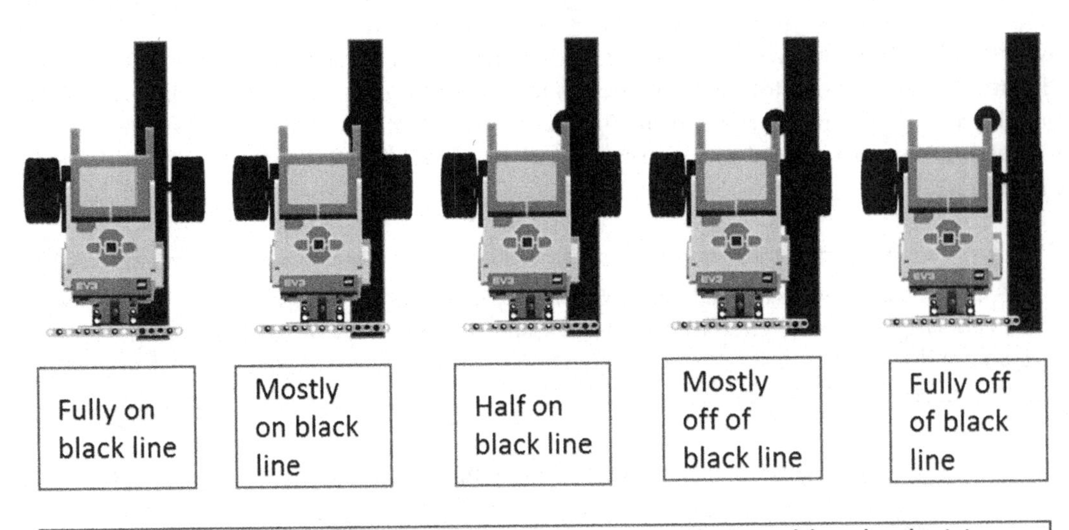

One could potentially write a multi-state program that could make decisions based on how much color from the black line the color sensor is seeing at any moment.

Figure 193: Multiple States of the Color sensor, from fully seeing the black line to not seeing it at all. This can be used to create a more sophisticated line following algorithm.

It is important to note that when the color sensor is in the mode where it detects colors instead of reflected light intensity, it does not provide information on how much color the color sensor is seeing. The color sensor, when sensing a color provides only a *yes/no* answer i.e. it can only tell whether it is seeing a specific color or it is not seeing that color. This throws a wrench in our multi-state color following program that we would ideally like to create. To solve this problem, we need to use the *reflected intensity mode* of the color sensor. As discussed in the chapter on the color sensor, in the reflected intensity mode of the color sensor, the color sensor measures the amount of light entering it and presents it as a number between 0-100. In this range, 0 means perfect black where none of the light emitted on the surface by the color sensor is reflected back and 100 means perfect white i.e. the color sensor sees all the light it shines reflected back. By contrast, the other colors such as red, blue and green reflect different amount of light back and thus reduce the range available to the algorithm. Thus, using black lines on white or vice versa, you get the maximum range in the reflected light intensity i.e. from 0-100. Other colors will provide a number between 0 and 100 and thus will not have as many states and will lose effectiveness to varying degrees.

Thus, unlike *simple line following* which works on most colors, multi-state line following works best for black lines on white mats or vice versa. The robot still places the sensor on the line and moves left or right depending on whether the color sensor can see the color. However, unlike the previous case where the robot tries to completely turn the sensor away from the line and then place the sensor completely back on the line, in the case of multi-state line following, we adjust the robot turns in smaller amounts depending on how much reflected light intensity is being observed by the robot.

Coming back to the PID line following, it is important to understand that the PID line following is simply a multi-state line following with practically infinite states. Instead of having a fixed number of states and then jumping from one state to another in the multi-state line following, the PID algorithm adjusts the amount of turn based on the amount of reflected light seen by the robot as frequently as the EV3 brick can read the sensor values and run the line following loop. This causes a fine-grained turning of the robot and hence a much smoother line following. With a good implementation of the PID line following, the robot seems to magically follow a line with nearly no jiggly movements, even across curves and thus when the robot stops, it stops in a consistent position.

Now finally coming to the actual line following algorithms, we would like to point out that the PID line following, even though considered the holy grail is sometimes unnecessary complex for FLL and hard for kids to understand. A simplified version of the PID algorithm, which is simply called the P or the Proportional algorithm and ignores the Integral and Derivate portions or the PID still works extremely well and is easier for students to understand and implement. For sake of comprehensiveness, we are listing out the three algorithms that you should consider in your FLL strategies:

- The two-state, simple line following algorithm, explained in the chapter on Color Sensor. The great thing about the two-state line following algorithm is that even with a bit improper positioning, the robot can find the line edge and follow the line. On the flip side, this line following algorithm has the maximum uncertainty in the robot orientation when the robot stops.
- The proportional line following algorithms which is a simplified version of the PID algorithm. This line following algorithm works very well with smooth line traversal, is reasonably easy to understand but if the robot positioning is a bit suboptimal, may not work as well.
- The PID algorithm. This line following algorithm works the best with extremely smooth motion but is quite hard to understand and requires complex tuning. If the robot is in a suboptimal position when beginning to follow a line, this program can completely throw the robot off and the robot may start going in circles trying to find the line. Depending on how you tuned the algorithm, once the robot starts going in circles, the robot run might not be recoverable and user may have to manually stop the program.

Getting back to the algorithms, it is key to remember that when following the line based on how much reflected light is seen by the color sensor, the color sensor is trying to be exactly between the middle of the black and white lines. The reason for straddling the line right in the middle is because in that case, the sensor is on the perfect edge and the robot can simply move straight without needing to turn left or right to correct the robot's path. Unfortunately, due to mechanical limitations, the robot will inevitably move a slight bit left or right and thus we keep striving to keep moving the robot with slight turning motions so that the color sensor gets in the ideal position of half black and half white.

The Proportion and the PID algorithms strive to place the color sensor exactly on half black and half white and adjust the robot turn in small amounts to keep the color sensor in this ideal position.

We are going to start with a straightforward algorithm (Figure 194) that follows the line based on how much light the light sensor is seeing at any point of time and evolve it into the Proportional and later the PID algorithms. This algorithm first assumes that the ideal state the color sensor should be is on exactly half black and half white i.e. the reflected light intensity the color sensor should see should be exactly *midpoint* between 0-100 i.e. a value of 50. Thus, we call numerically denote this ideal state as the MidPoint variable and assign it a value of 50. Next, this algorithm calculates the difference between the reflected light intensity that the color sensor is seeing at any point and the MidPoint, which in other words is the difference between where we want the color sensor to be (MidPoint) and where the color sensor currently is. In a way, you could say that the robot is in an erroneous state when the color sensor is not exactly seeing 50% reflected light intensity. Additionally, if we had to quantify the error from the ideal state of having the color sensor on exactly the MidPoint position, it would be the difference between the color sensor current reading and the MidPoint. Larger the difference between the two, higher the error or displacement of Robot is. Correspondingly more the robot needs to turn towards the ideal state.

Quantifying the above, when line following,

Error = Color sensor's current reflected light intensity reading - Midpoint

Next, note that the error ranges from -50 to +50 since the Reflected Light Intensity can only have values between 0 (seeing pure black) and 100 (seeing pure white.) Next, it is time to pay attention to the move steering block's turn amount again by referring to Figure 89. Note that when the turn amount is set to -50, the robot turns left with only the right wheel moving, at turn amount set to 0, both wheels move at the exact same power and at turn amount set to value 50, the robot turns right with only the left wheel moving.

Drawing conclusions from the above, we can say that when the error is -50 i.e the robot is completely on black, then to correct this error, the robot should stop its left wheel and turn only the right wheel i.e. use the turn amount of -50 until the color sensor starts seeing white. When there is no error i.e. error value is 0, the robot does not need to turn at all and so it should use a turn value of 0. When the error is 50 i.e. the robot is completely on white, the turn amount should be set to 50. Since the robot turn amount varies linearly from -50 to +50 for full left turn to full right turn, the intermediate values represent partial left turns to Moving straight to finally partial right turns. This coincides very nicely with the error values range from -50 to +50 and thus, the error values can directly be used as the amount of robot turn in the move tank block. This is exactly what we have shown in Figure 194.

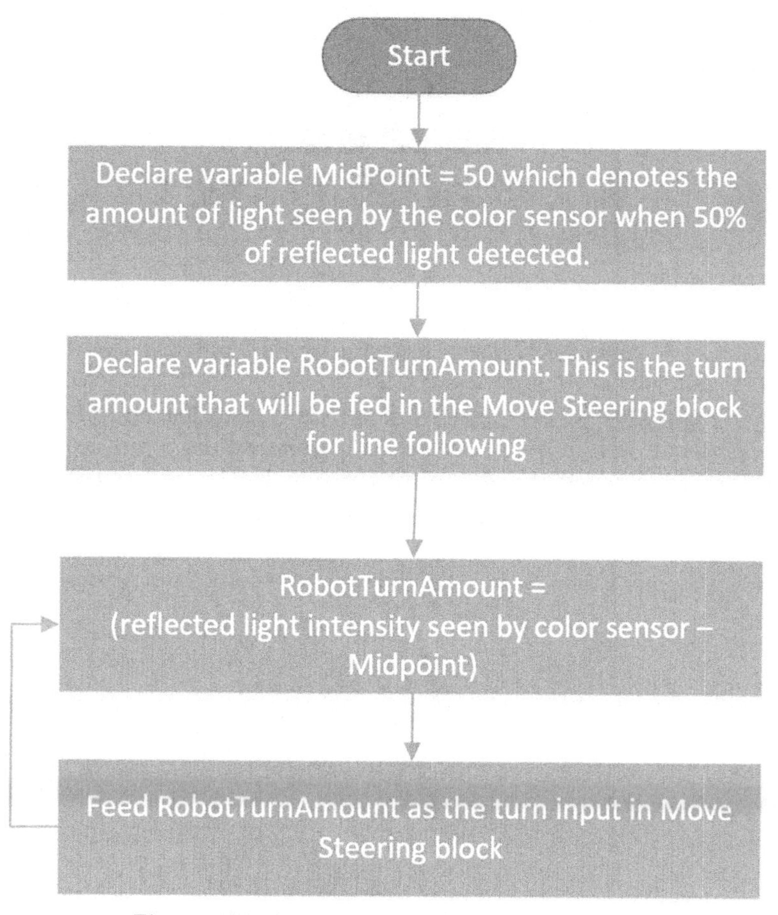

Figure 194: An intuitive infinite state algorithm for line following.

The implementation of simple proportional is described in in Figure 195. Please not that this algorithm is strongly dependent upon the diameter of Robot wheels, speed, as well as and how far apart the wheels in your robot are. For certain size of the wheels and wheel-spacing, the robot might move smoothly whereas for another combination of wheel diameter and wheel-spacing the robot might move in a zig-zag fashion like the simple line following algorithm. On may adjust the three parameters that is wheel size, wheel-spacing and speed to arrive at a sweet spot. However, an alternate approach used to counter this problem is to introduce proportional factor. Use of proportional factor is described in next section.

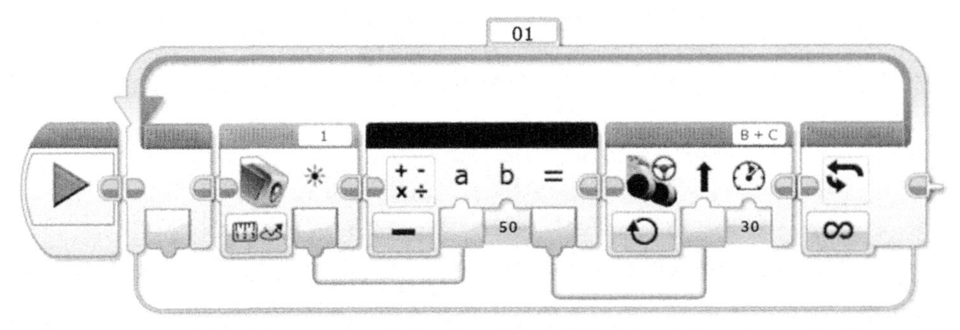

Figure 195: Implementation of the algorithm shown in Figure 194

The algorithm shown in Figure 194 is very close to a *proportional algorithm* and the key difference between it and the *proportional algorithm* is that in the *proportional algorithm* we use a multiplication factor, also known as the *tuning constant (Kp - K* for constant and *p* for proportional.) The updated algorithm with *Kp* is shown in Figure 196.

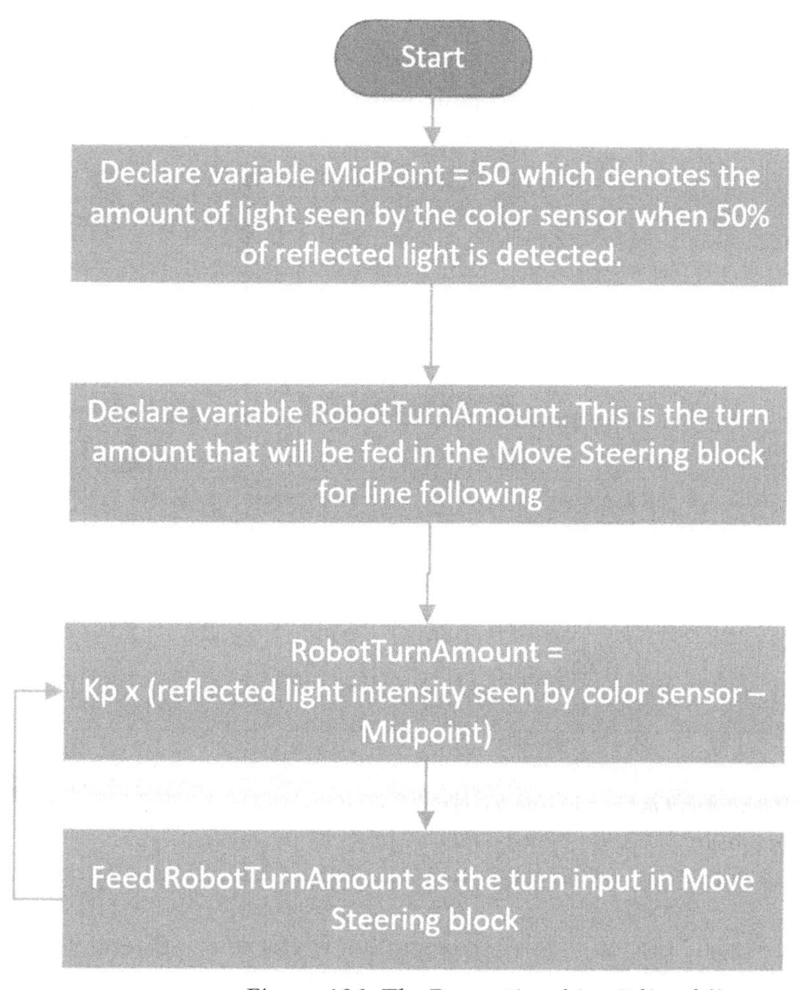

Figure 196: The Proportional i.e. P line follower.

The value of *Kp* is experimentally determined which is a fancy way of saying that you need to try values between 0 and 1 and see at what point your robot starts moving the smoothest. Determining the value of *Kp* is known as tuning the proportional algorithm using the *Kp* parameter. We are showing the updated algorithm in Figure 196 and the program in Figure 197. For our sample robot, a *Kp* value of 0.5 produces smooth line following.

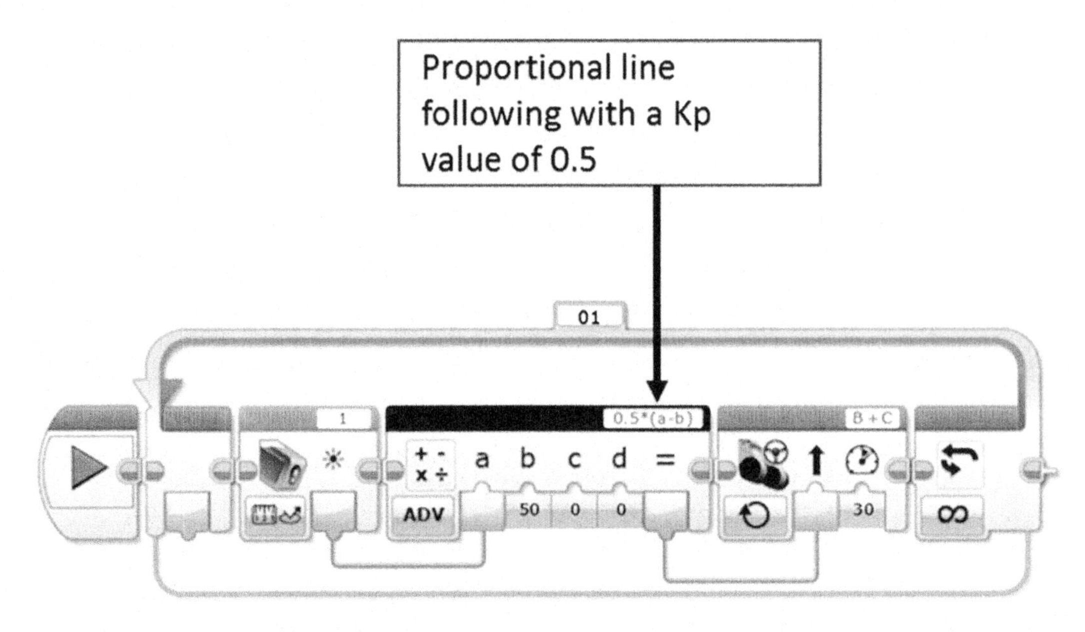

Figure 197: The Proportional Line Following Program

Large distance between wheels and higher power to motors are both enemies of the accuracy of line following. If your robot is zig-zagging too much, first decrease the distance between the wheels and the power and finally star tuning the P and PID program parameters.

For all practical purposes in FLL, the *proportional* algorithm works very well, and it is relatively simpler to understand.

The program in Figure 197 will follow the left edge of a black line. If you want to follow the right edge of a black line, you would have to switch the order of the a and b inputs so that instead of computing Kp(a-b), you are computing Kp*(b-a). The switching of a and b, reverses the motor power and causes the robot to follow the right edge of the line. This program is easily convertible to a MyBlock. However, you will need a termination condition for the loop based on another sensor such as the color sensor or the motor rotation sensor. We leave this as an exercise to the reader.*

For the readers who may be interested, we are going to explain the PID. Given that we just went over the proportional algorithm and understand that we need a tuning parameter *Kp*, the PID algorithm can be explained in a relatively straightforward manner using two more tuning parameters called *Ki*, the integral parameter and *Kd* the derivative parameter.

As we mentioned earlier, the difference between the measured reflected light intensity and the midpoint is called as the *error* and the *Kp* term allows you to tune the error per your robot design. The other two terms, the *integral* and the *derivative* are both related to error as well and their related parameters *Ki* and *Kd* are used to tune those terms.

The integral term controls the error by accounting for the cumulative error and the time that the error has been occurring. On the other hand, the derivative term looks at the rate of the change of error and aims to bring it close to zero via anticipation. The PID flowchart is provided in Figure 198 and you can create a program based on this that will be functional. The tricky thing with the PID algorithm is that there are three parameters, Kp, Ki and Kd to tune and they impact the robot differently, especially when the robot misses the line or if the line starts curving. Please bear in mind that non-optimal values will likely make the robot spin in large circles or even keep it circling in one spot. Choice of the Kp, Ki and Kd is strongly dependent on different scenarios hence needs adequate tuning effort. We will leave the PID controller as an exercise to you.

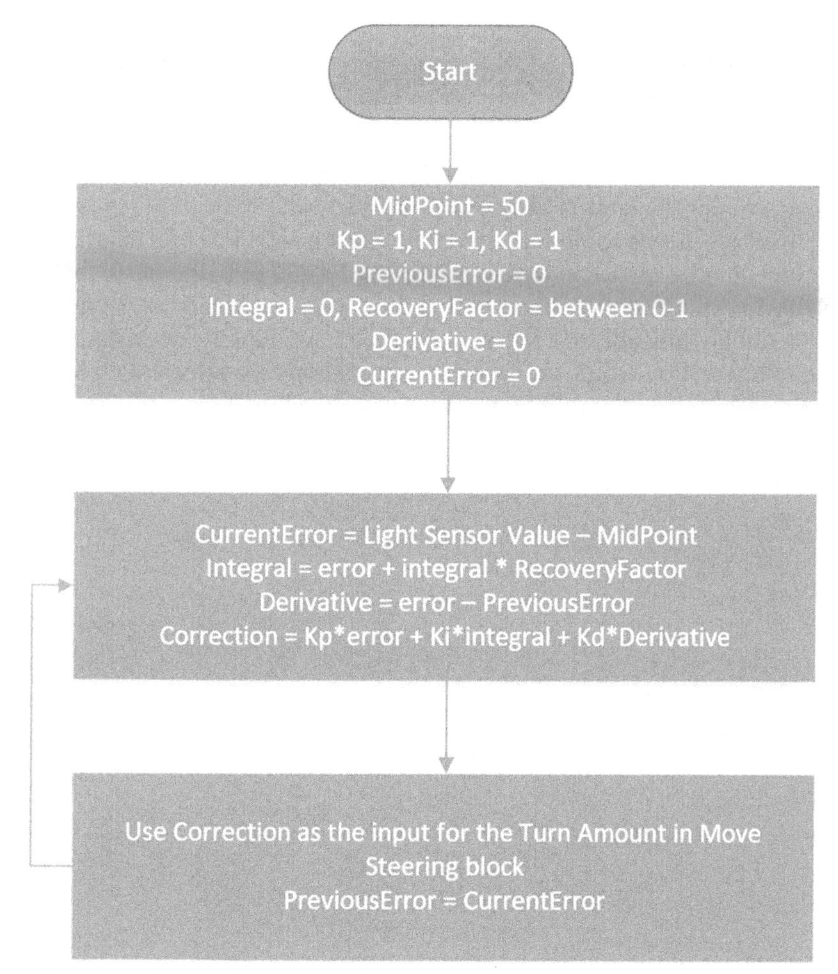

Figure 198: The PID algorithm as applies to color line following. Note that the program starts with an assumption that Kp, Ki and Kd are set to 1. These parameters must be tuned for your robot for it to have a smooth motion. The Recovery factor must be set to an

appropriate value between 0-1 to ensure the integral does not grow unbounded and make the robot spin on the spot.

Even though we have explained the P or the PID algorithm with respect to only the color sensor, it can easily be modified to work with the Ultrasonic Wall Following. We are leaving this as an exercise to the reader.

Summary

The Simple two state line following program works for the simplest of the cases but is hard to use due to it leaving the robot in uncertain orientations. However, for nearly all practical purposes, the proportional program works very well and needs some tuning. Algorithm for the PID controller is described, although we don't really recommend it because it is difficult to tune. Since in FLL, the participants should well understand and be able to explain all the concepts that they have been using, it is better for them to use and algorithm that they would be fully able to explain and implement. Finally, do keep in mind that even with the proportional line follower, you would still need to use a secondary sensor such as a second color sensor or the motor rotation sensor to stop the robot once it has traversed the line to the desired place.

The Master Program

The Master program which, is also referred to as the Sequencer sometimes, is a program that can simplify the final execution of your FLL robot game. The master program is a program created to solve the problem of confusion during the competition. The EV3 startup screen contains a list of all the recently run programs. During the competition, the participants need to be able to select and run the correct programs as fast as possible and without having any confusion. The EV3 startup screen makes this extremely hard because it moves the most recently used program at the top of the program list. Even if you start with a perfectly ordered list, after you have run a few programs, the list becomes jumbled and it takes precious seconds to find and run the correct program. This issue is identified in Figure 199 and the master program solves this problem by creating a fixed order and having a single program in the recent program list. Another disadvantage of using the EV3 startup screen is that, the font size is somewhat difficult to read.

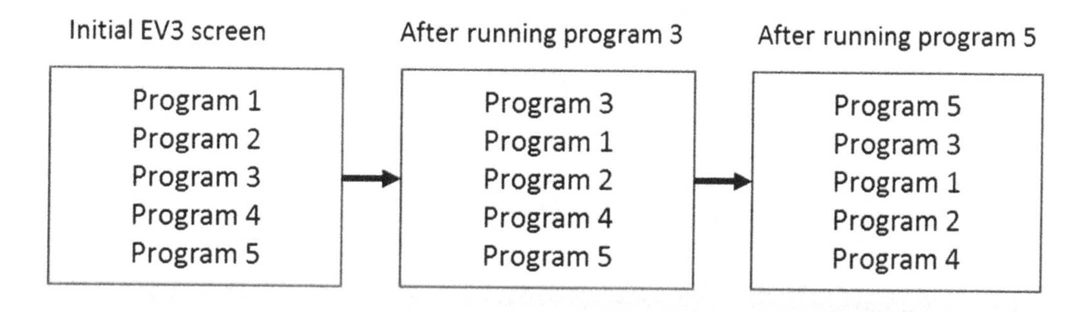

The EV3 screen shows a list of recently run programs. As you run a program, that program goes to the top of the list and is removed from under. After a few programs have been run, the ordering of the programs is messed up and causes a lot of confusion as kids try to find the right program during the competition. The master program solves this by creating a single program containing all the other programs and preserving the order.

Figure 199: The EV3 recent program list keeps getting reordered causing confusion and consuming precious competition time.

The master program is essentially a *single* program that encompasses all your programs and presents the user a menu on the EV3 screen. The menu shows the current program that you wish to run and activates the program using one of the brick buttons. As soon as the program is finished executing, the master program automatically increments the menu to show the next program in sequence. The user has the option to press back and front EV3 buttons to navigate the programs available through the master program. Additionally, one may enlarge the font size for better visibility. To better understand this, refer to Figure 200.

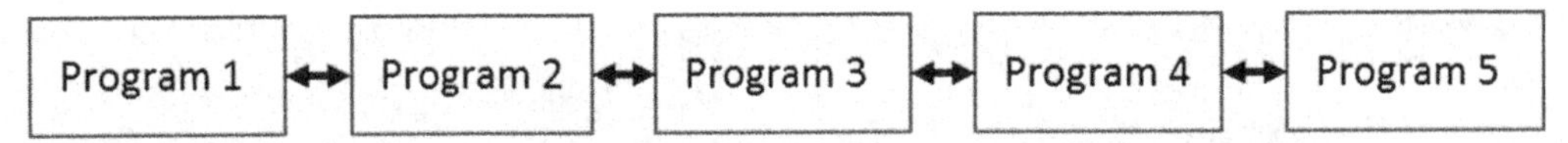

The user needs to select and run programs 1-5 on the brick during the competition. The Master program encapsulates each of programs 1-5 as myblocks within it and exposes them via a menu as shown below. User clicks the EV3 left, right buttons to move from one program to another and the master program displays the name of the program that will be run when you press the center brick button. Additionally, it is visually indicated that you may move left or right in the program order and select the correct program to run. The user has exactly one program that they need to run and a consistent order on the screen, thus removing all confusions.

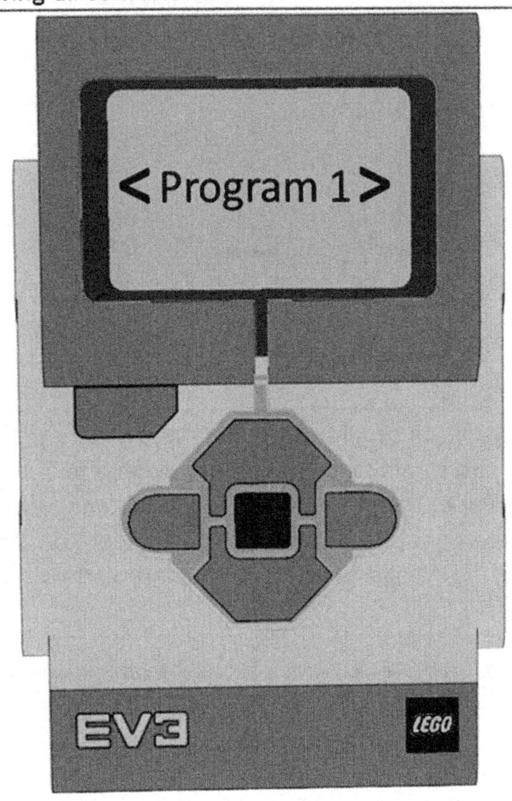

Figure 200: The master program shown running on the EV3 brick.

A master program is a somewhat complex program, so it is a good idea to look at a flowchart that explains how it works. A flowchart for the master program is provided in Figure 201.

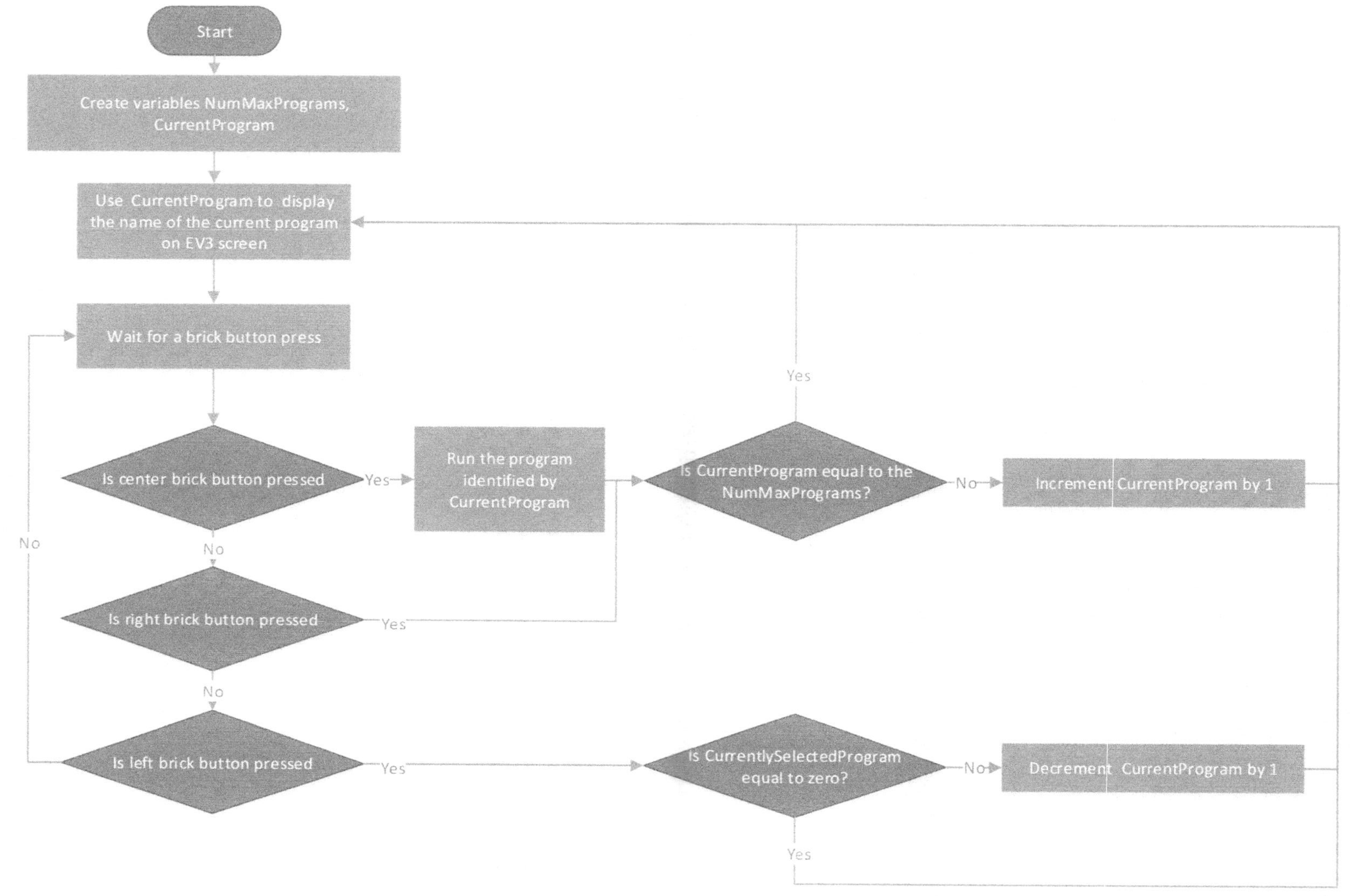

Figure 201: Master Program flowchart

To convert this flowchart into a program, we need to use:

- Variables
- Myblock for displaying the name of the current program
- Multi-State switch statement for the Brick Button
- MyBlocks for each of the program options

Suffice it to say that showcasing a full Master Program in this book would be an extremely cumbersome process as the program would take multiple pages and even then we would not be able to show all the conditions in a meaningful manner. A full downloadable version of the Master program based on the above flowchart is present at www.FLLguide.com. Even though we cannot showcase the Master Program in its entirety here due to space constraints, we will showcase some key features of the Master program that will help you understand and implement it.

Display Program Names MyBlock

Let's say we have a master program that has three subprograms called *MazeProgram, Wall Following* and *Color Line Aligning.* We want to display the name of these programs on the EV3 brick in the manner shown in Figure 202.

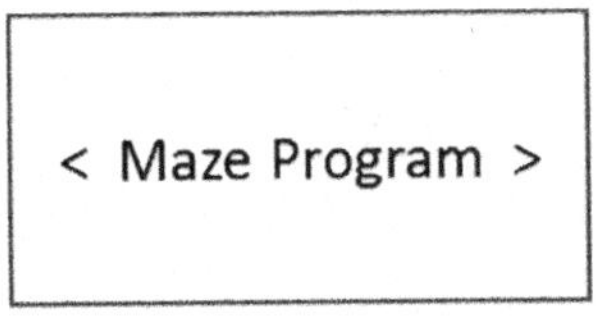

< Maze Program >

Figure 202: The desired display for the master program

To display the names of these subprograms, we can use the program shown in Figure 203. You can easily convert this program into a myblock that only takes the CurProg numeric variable as input. The program with the myblock is shown in Figure 204.

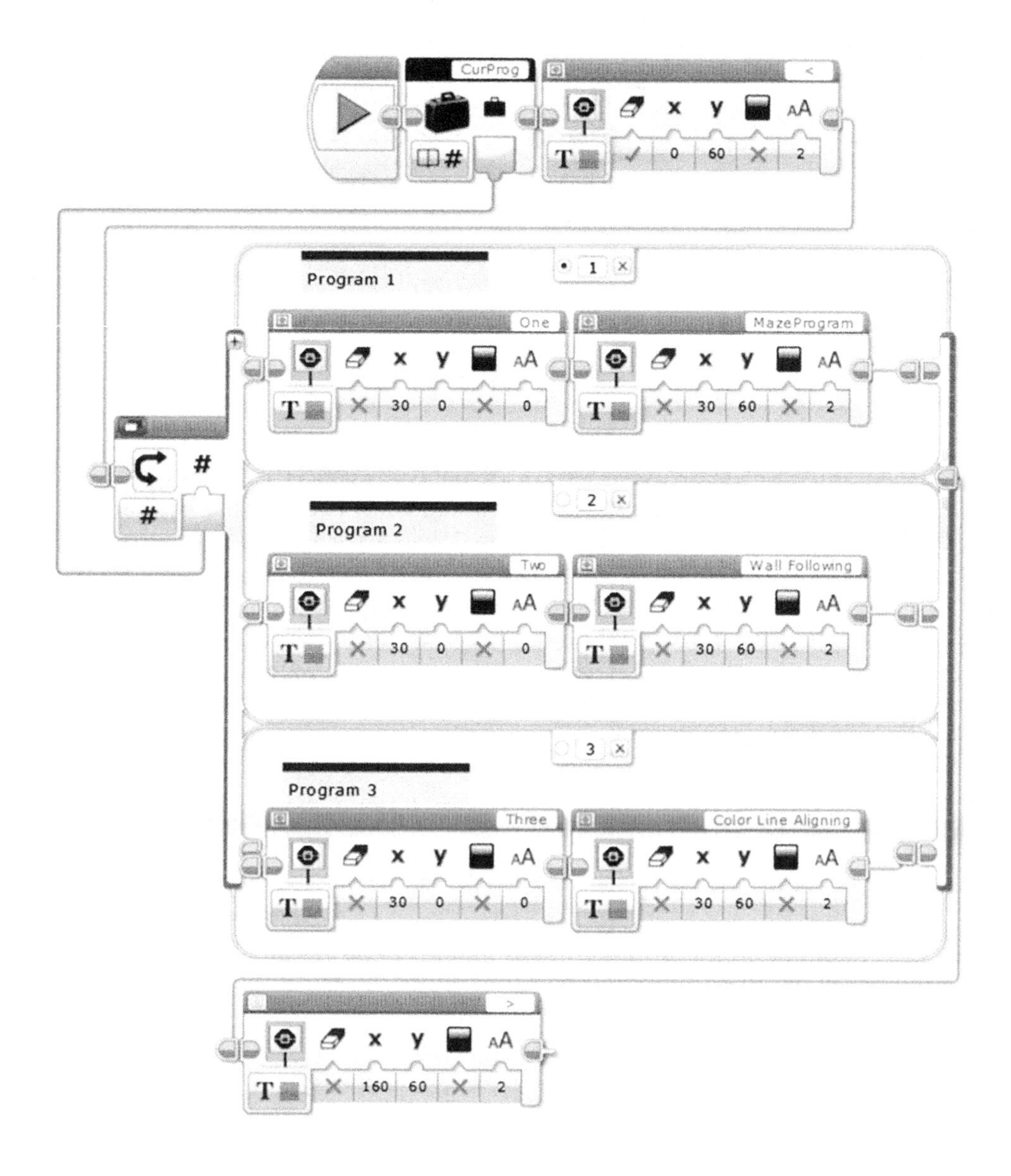

A program that takes the Current Program Number numeric variable (*CurProgram*) as input and displays the correct program name on the EV3 screen. It additionally displays the Numeric Option above the program name.

Figure 203: The Display Program for showing program names

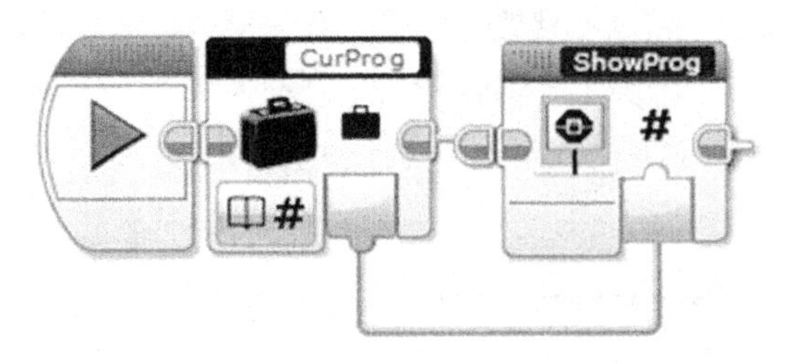

Figure 204: The program in Figure 203 shown as a Myblock

Next you need to create three programs i.e. the MazeProgram, Wall Following program and Color Line Aligning program and convert them to myblocks. We are not really concerned with what goes in these programs as we are simply using them for sake of illustration. Thus, you may place just a sound block with a unique sound in each of those programs for this exercise. The idea is to have three separate programs that do three different things so we can convert them into unique myblocks. Next, you need to create a program that uses the CurProg variable as the input and uses a switch block to run MazeProgram, Wall Following Program and the Color Line . The usage of the values 1,2, and 3 for the CurProg variable ties the three programs to respective numbers. Finally, you must convert the above program to a myblock that takes the value of CurProg as the input to it. This program is shown in Figure 205.

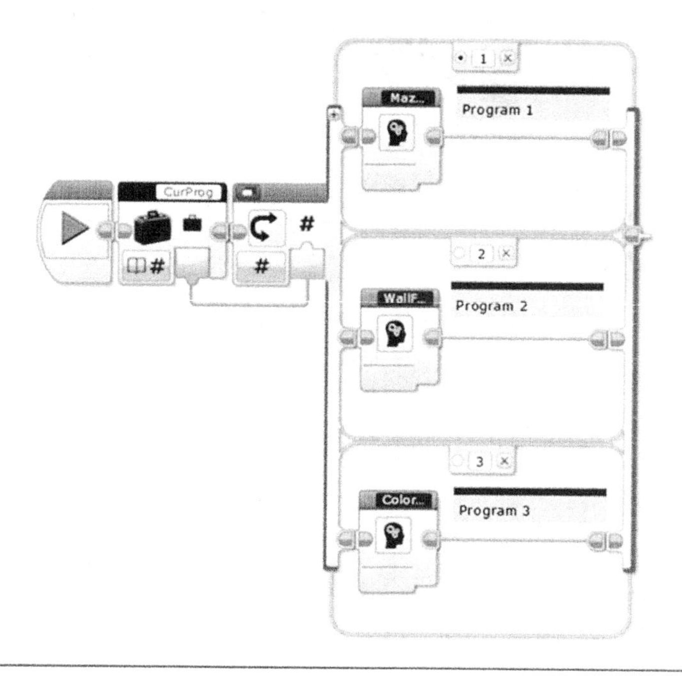

A program that takes the Current Program Number numeric variable (*CurProgram*) as input and executes the correct program. Each of the options denotes a program that has been converted to a myblock.

Figure 205: A program to select one among multiple programs (represented as myblock) and execute it.

Next we convert the above program to a Myblock. This MyBlock will take CurProg variable as an input and we will name this MyBlock the *ProgRunner* as shown in Figure 206.

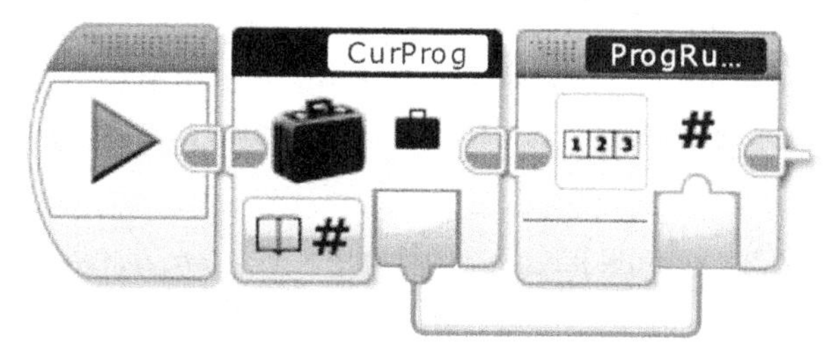

Figure 206: The program in Figure 205 shown as a MyBlock.

Finally, we get to the last piece of the puzzle that is how to use the brick button left to move to the previous program, the right button to move to the next program and center button for selecting the current program for running. This is shown in Figure 207. Using the three parts i.e. the Program display myblock in Figure 203, the myblock for selecting and running the correct program as shown in Figure 206 and finally the program for selecting the correct brick button and taking an action based on that in Figure 207, you can easily write a full-fledged master program using the flowchart in Figure 201.

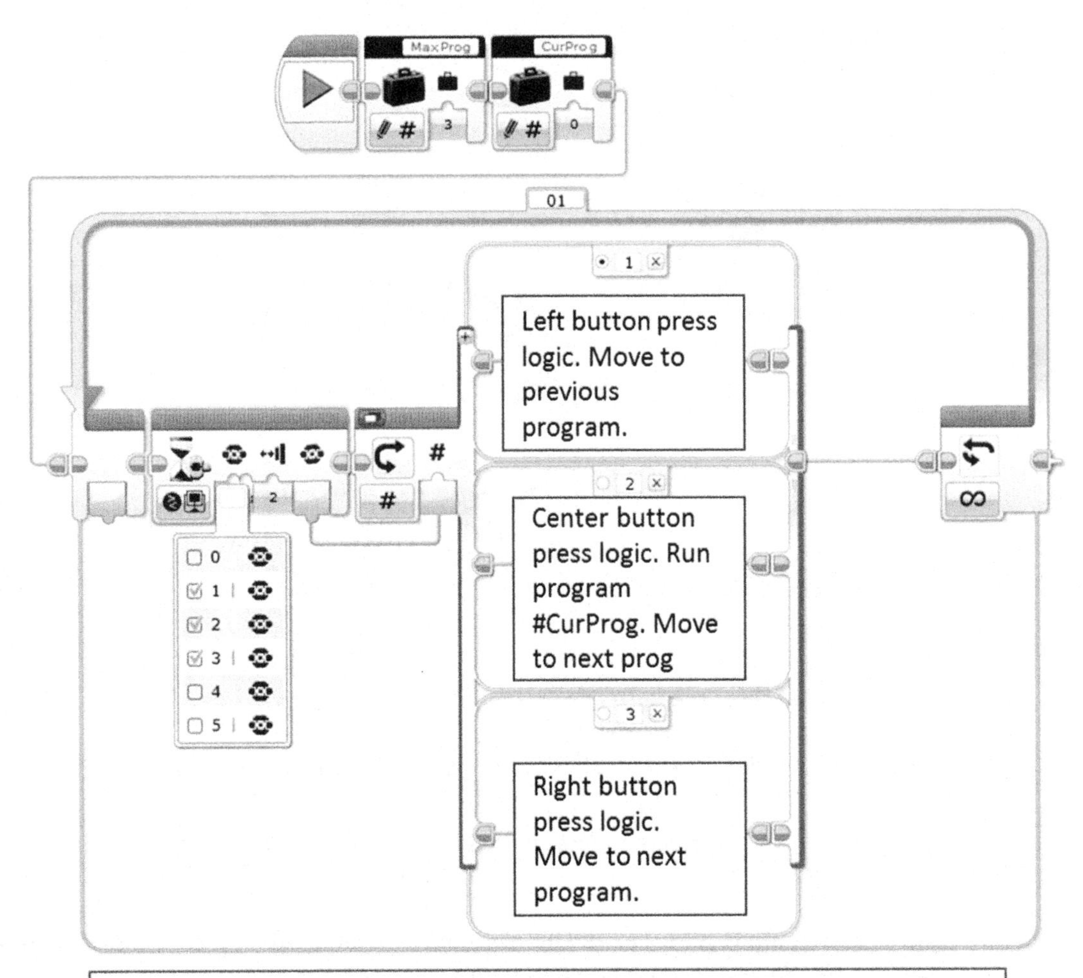

A program that takes the Current Program Number variable (*CurProgram*) and maximum number of programs (MaxProg variable) in the master program as input and depending on left, right or center brick button press performs the appropriate action. The brick button options are shown expanded.

Figure 207: The final piece of the Master program takes appropriate action based on the brick button press.

Summary

Master program is a program that allows you to take the uncertainty and complexity out of selecting and running the correct program in the heat of moment at the FLL competition. This is an advanced technique and can help your team tremendously by removing confusion and saving time during the competition.

Some Design suggestions to improve accuracy and timing

In this section, we are describing some designs based on out of box thinking to improve the navigation accuracy, robot alignment and designs to accomplish missions accurately and quickly. The designs serve as guidelines and the user may change the design as per need.

Self-aligning features

Despite best programing and design, the robot and its subassemblies are susceptible to errors. The mechanism performance may not be repeatable. In order to complete a mission, it is very important for the robot to not only arrive at a location but also maintain suitable orientation. Local structures and self-aligning features may be used for the robot to locate and align itself.

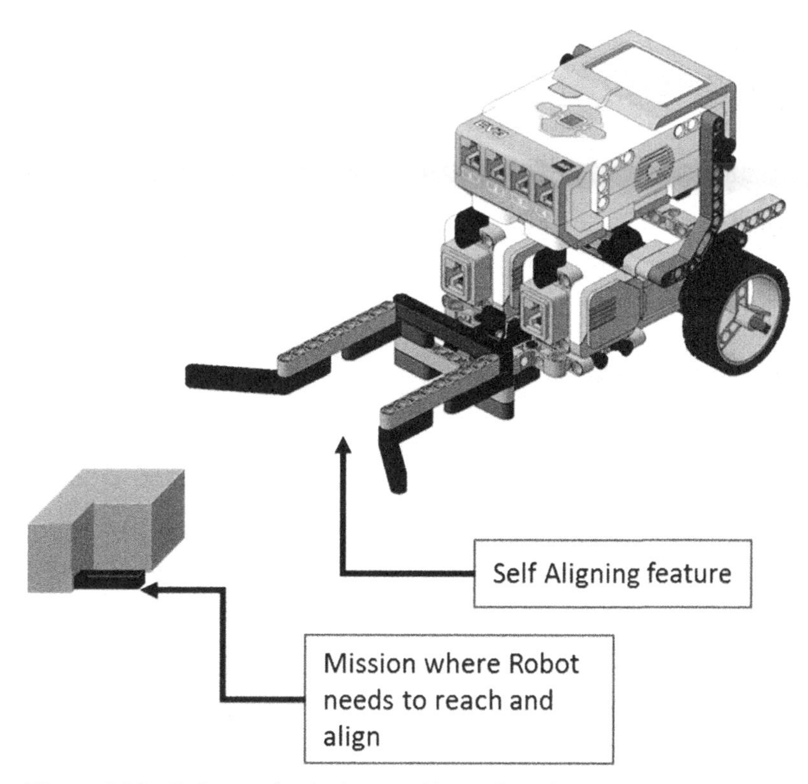

Figure 208 : Robot and mission needing robot alignment

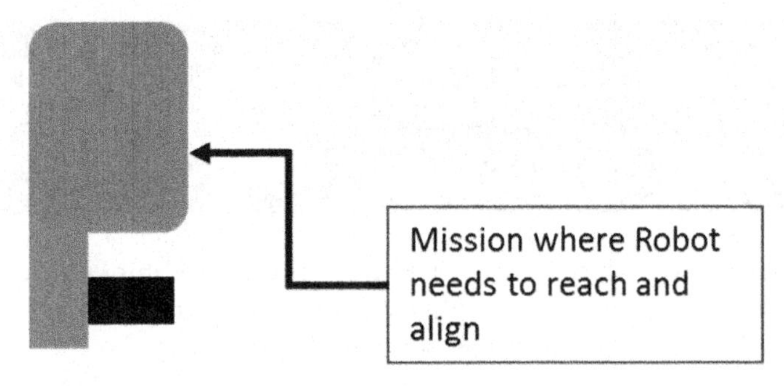

Figure 209 : Mission, top view

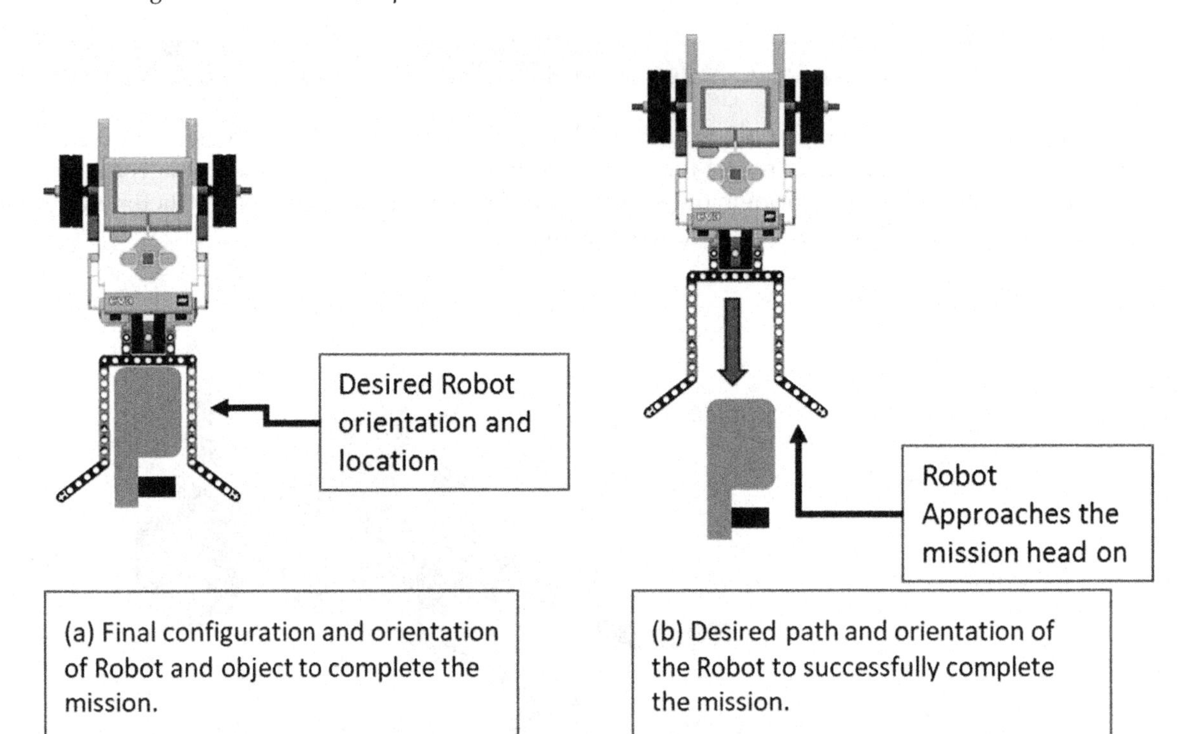

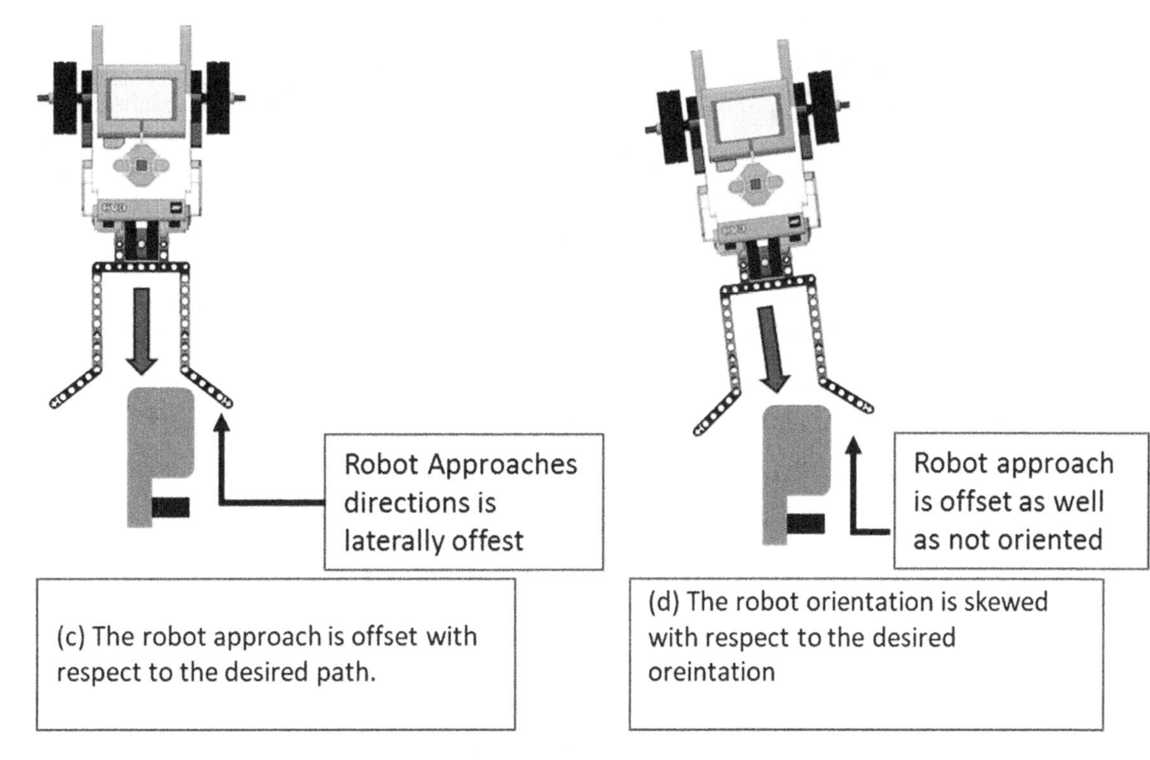

Figure 210 : Minor offset and disorientation of the Robot can be easily accounted by using a funnel like self-alignment feature.

Figure 208, Figure 209 and Figure 210 describe a mission that requires a robot to align with the mission pieces. Total robot alignment timing can be reduced by attaching a self-aligning mechanism. The self-aligning feature is a structure in the shape of funnel. The narrower part of the funnel is about the same width as the object where the robot needs to engage. The wider opening of the funnel allows the robot to access and establish contact with the object. When forced to move, the funnel will guide the robot to narrower area and slide into required position. The wider entry area of the funnel not only accommodates the offset but also fixes misalignment in case Robot doesn't approach the mission head on.

Optimize motor usage by eliminating motorized mechanisms and attachment by Robot Movement

EV3 brick has only 4 ports to connect the motors. Additionally, per FLL rules maximum 4 motors are allowed. However, with varied range of missions it certainly helps to reduce the dependency on the motorized arms and actuators. Since Robot can move back and forth and turn, design a suitable attachment to utilize the robot motion for simple the push-, pull, sweep- and collect type of problems. NOTE: All scenarios are not similar, therefore authors suggest that accuracy, force requirements and timing should be consideration in choosing between the two options.

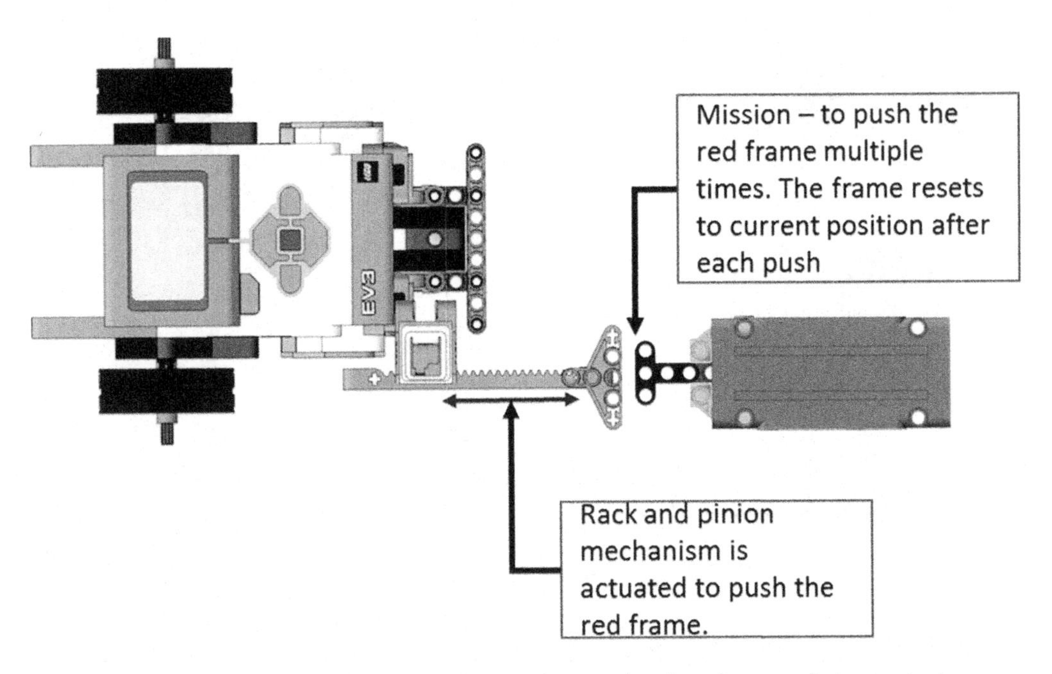

Figure 211 :Use of motorized actuation mechanism for a push type mission

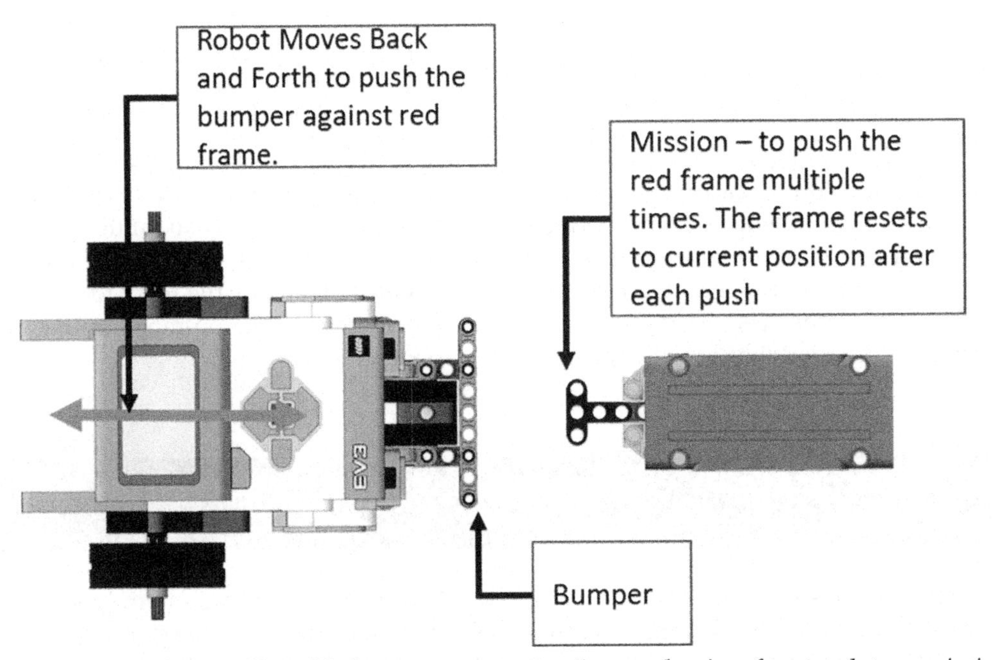

Figure 212: Use of Robot movement actuation mechanism for a push type mission

Figure 211 and Figure 212 describe such a scenario. As described in the figure, the mission requires a red frame to be pushed multiple times. The red frame resets to its original position when disengaged. A motorized rack and pinion attachment (Figure 211) can be used to actuate the red frame multiple times. However, this solution not only needs an additional motor but also a special mechanism. The same action can be accomplished by engaging a static bumper attached to the robot (Figure 212) and then moving the robot back and forth.

Use passive non-motorized end effector for collecting objects

For problems involving collection of objects that are placed on ground, static constraining features eliminate need for additional motors but such static attachments can also be easily switched. Figure 213, describes a motorized but hard to configure arm attachment to collect objects on the ground. The robot may start with the up position of the arm. When robot is within the vicinity of the objects, arm may be lowered to engage and collect the objects. A non-motorized passive end effector can be attached to accomplish the same task. However, the sequence of the robot movement (Figure 214) is important for successful object collection.

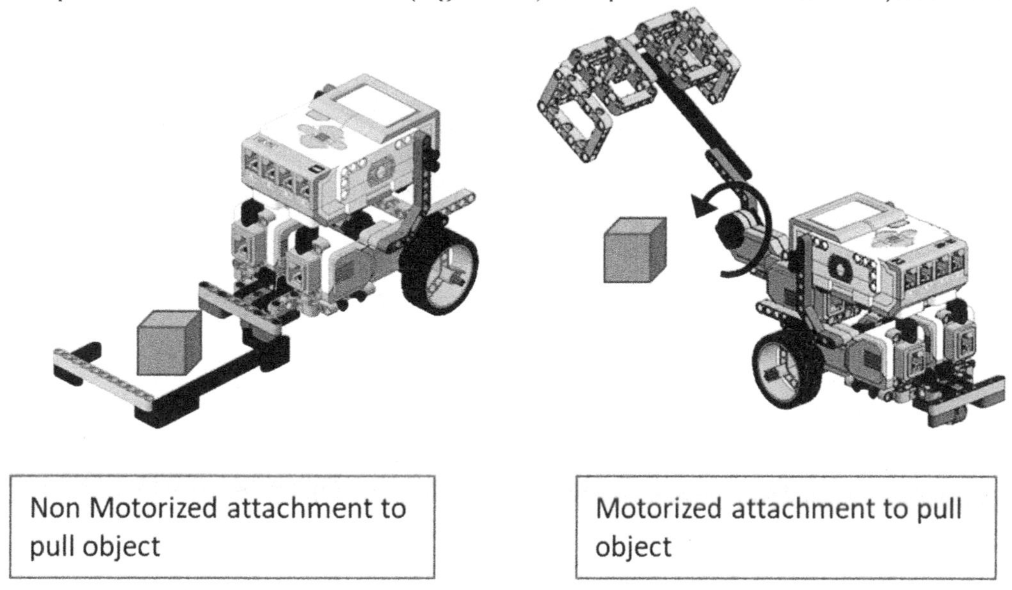

Figure 213:Using a motorized vs. non-motorized end-effector for pulling an object

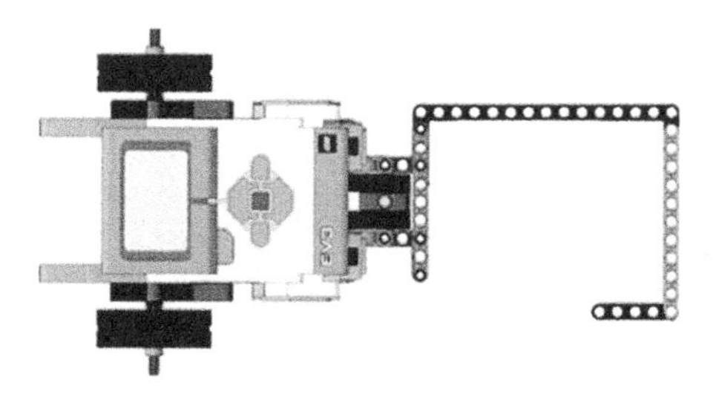

a. Initial position

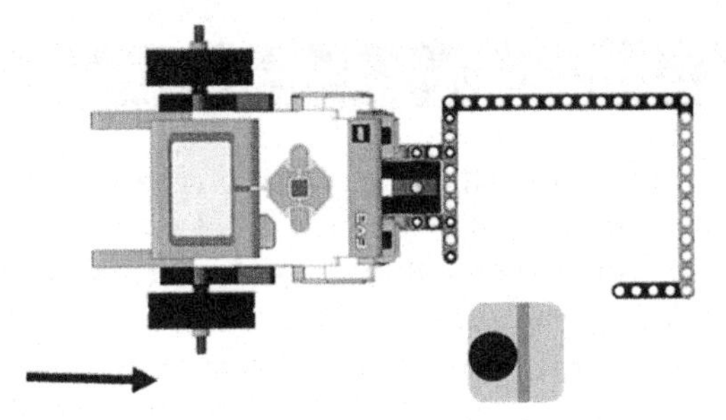

b. Robot aligns with object

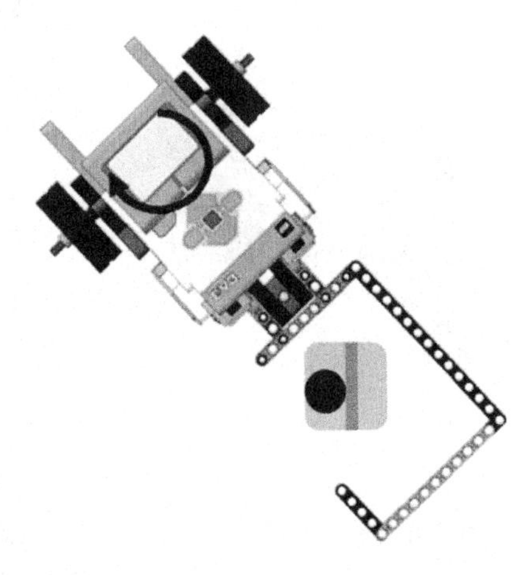

c. Robot turns to engage with the package

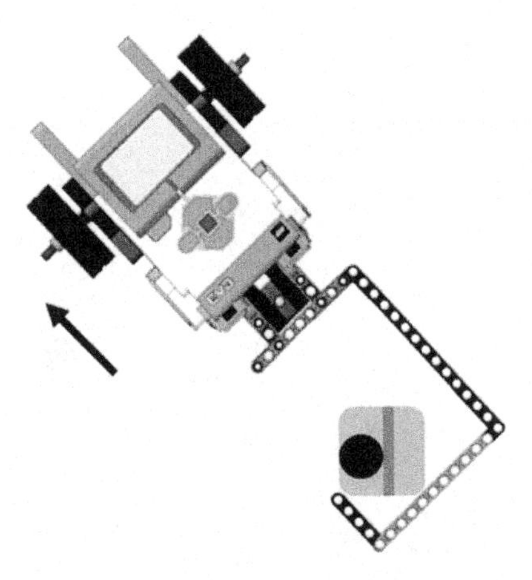

d. Robot moves back to fully secure package

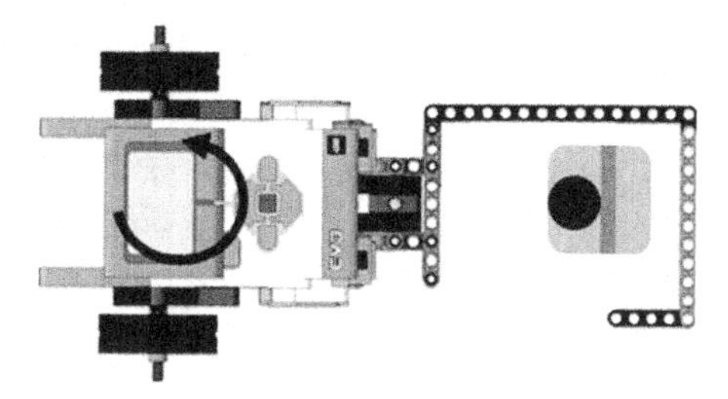

e. Robot reorients

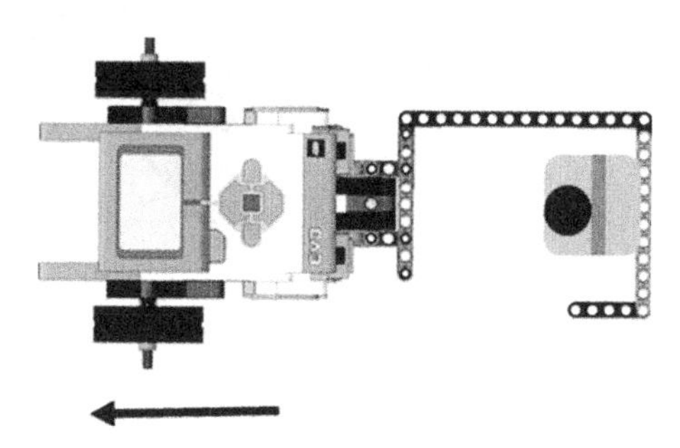

f. Robot moves back to the station

Figure 214: sequence of robot movement for collecting object with help of a passive attachment

As described in the Figure 214, robot may start the mission and move closer to the object. When in vicinity of the object, robot aligns itself and turns to engage the object. Following the engagement, robot turns back and moves backward to bring the object to base.

Designs based on Out of box thinking for objects placed on ground - effector for collecting objects

Figure 215 describes a design also referred to as one-way curtain. Such a design not only eliminates the need for motorized mechanisms but also eliminates the need for robot to traverse a complex sequence of motion to collect objects. The design comprises of freely swinging pieces that are constrained on one side. Figure 216 illustrates how the curtain is attached and used for collecting an object. The robot moves and overlaps with the object. Contrary to other passive/active end effectors, this design allows the robot to let the curtain slip over an object by robot movement. Once the curtain slips over the object and is within the confines of the curtain, robot moves in opposite direction. The constraining feature will contain the object and pull it along with the robot.

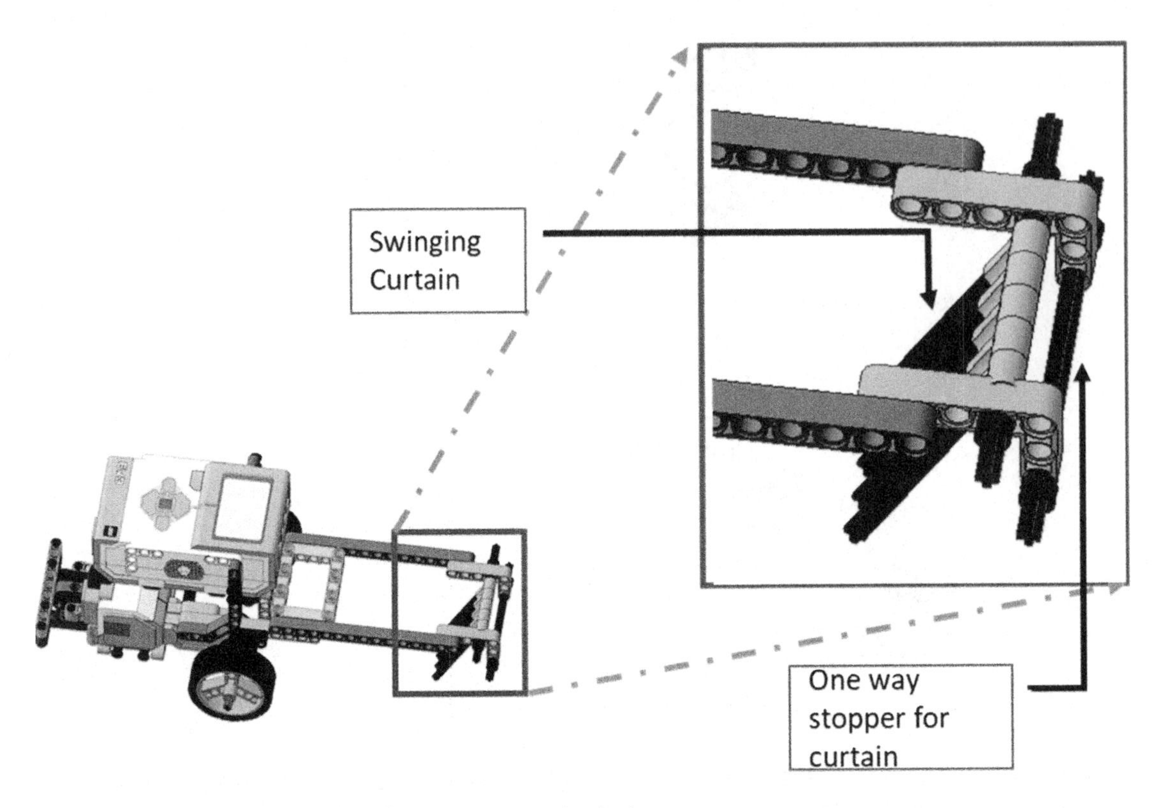

Figure 215: Details of One-way curtain design

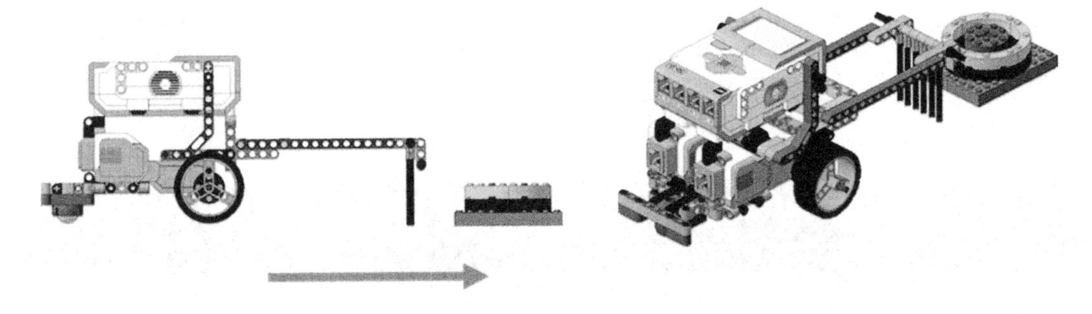

Step 1: Robot approaches the object to be pulled

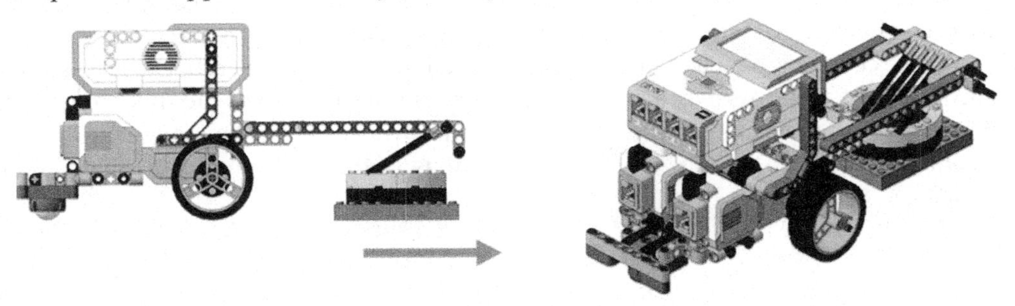

Step 2: Robot continues forward motion, and the curtain slips over the object

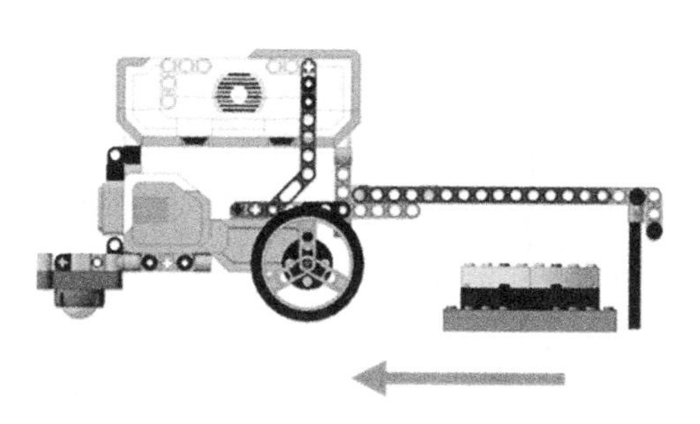

Step 3: As robot moves backward, the locking mechanism of curtain prevent the object from slipping out and therefore drags it along.

Figure 216: Steps for pulling the object with one-way curtain

Save time by using single fixture for simultaneous multiple deliveries

If multiple delivery missions are in close proximity, the total mission time can be saved by making multiple deliveries using a single fixture. As an example, let us refer to the part of a mission below described in Figure 217. There are three points in close proximity where certain items need to be dropped. Starting, orienting the robot and reaching to each location needs certain time. Additionally, Robot must sustain its bearings, otherwise deviation from its path at one location will affect rest of the mission.

An alternate approach based on a well-designed fixture that holds multiple items needs a much simpler path and saves time. A strong well-built fixture with multiple points of attachment can be built so that all the objects to be delivered are attached to the fixture. The points of attachment on the fixture are distributed per the location of the point of delivery. Instead of dropping pieces individually, the robot may reach within the vicinity of the delivery points, align itself and drop the fixture along with the delivery material since there is no penalty for that. Ensure that the fixture is not in the way for subsequent missions.

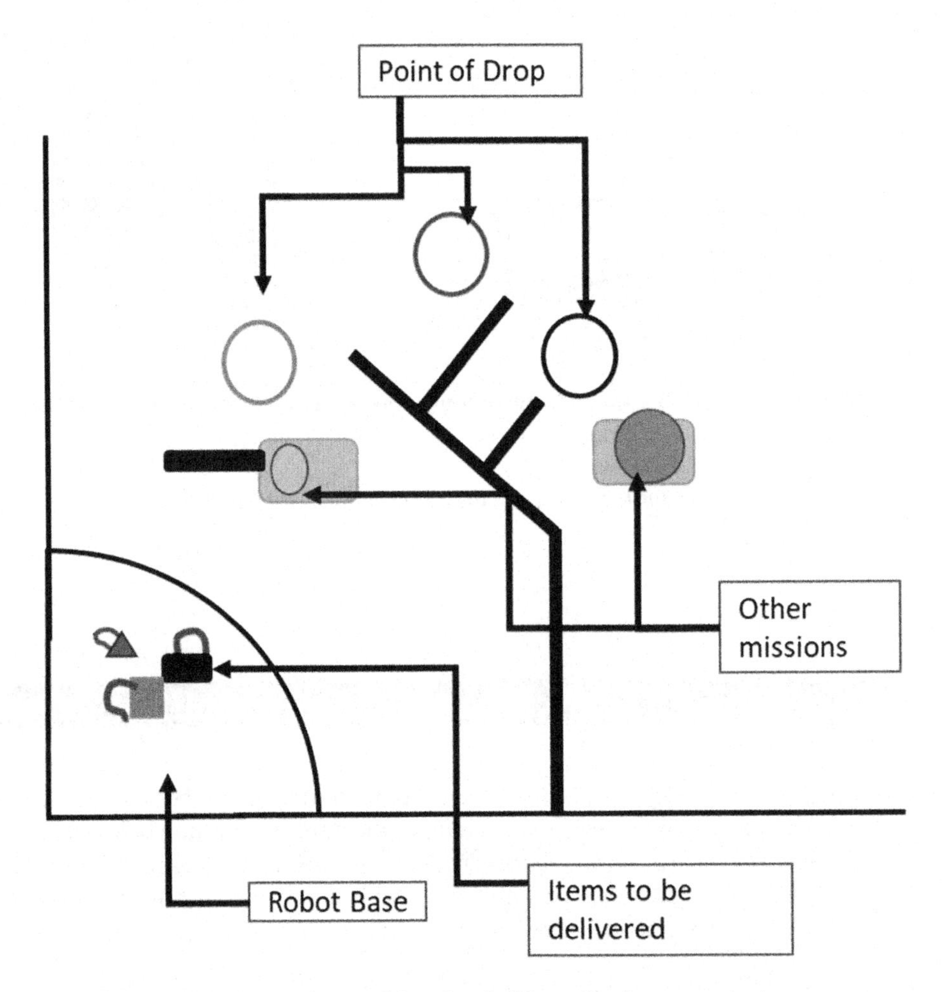

a. Challenge description for multi-point delivery in close proximity

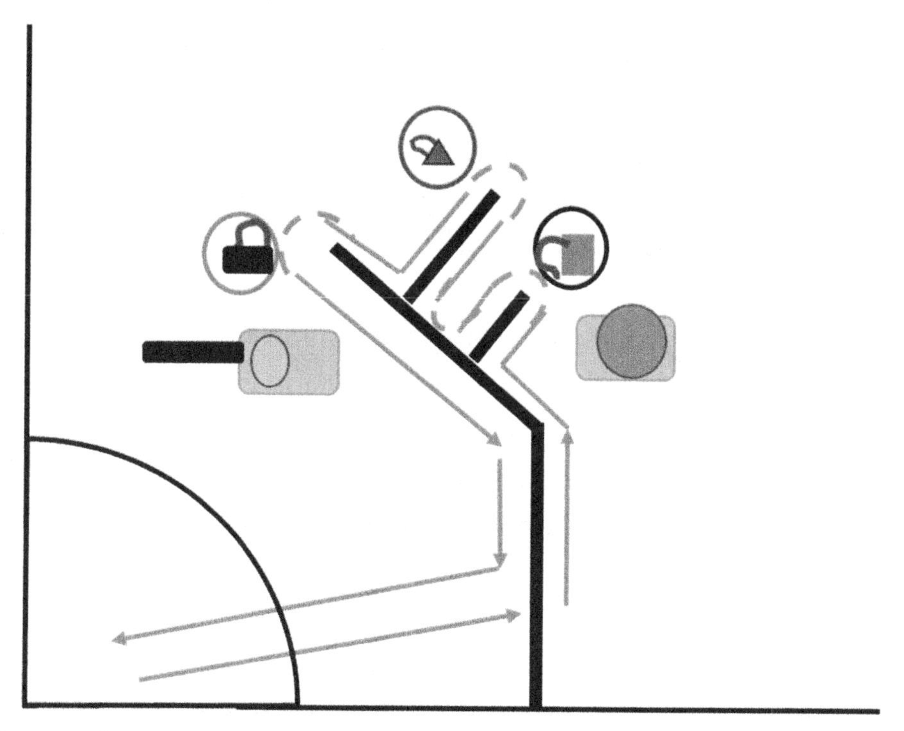

b. Longer slower approach to material delivery with any path segments and turns

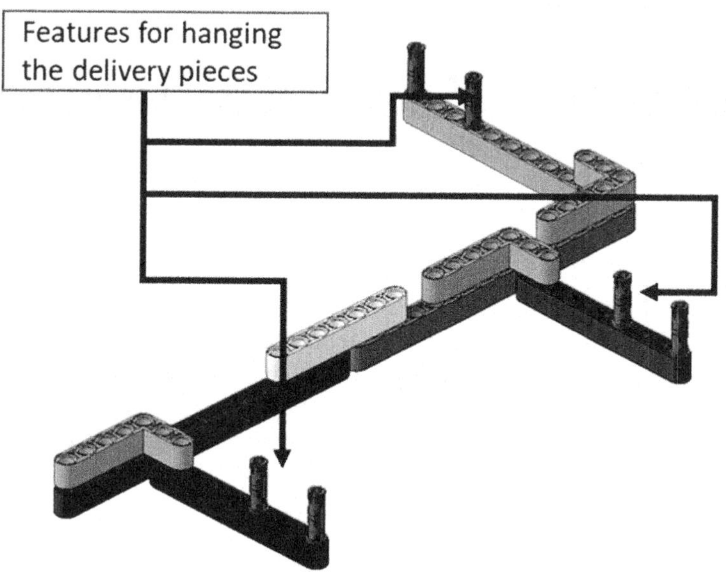

c. A suggested Fixture for multi-delivery

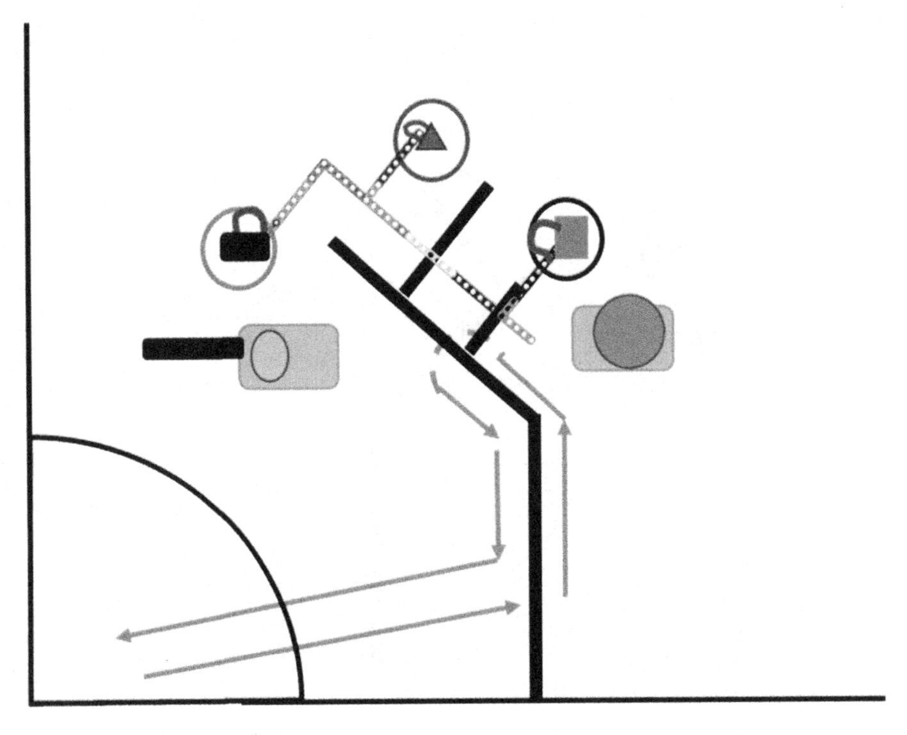

d. Much shorter and more reliable delivery with special fixture

Figure 217 : Use of a multi-point delivery fixture for quick and reliable delivery

Robot need not go behind the wall

In many a mission, the point of access for the Robot is in the far direction with respect to robot base, other obstacles or nearest point of access. One may device special attachments to go from above or from side with the robot still facing the mission from front.

For example, the challenge described in Figure 218. has an actuator on the far side. Other missions placed next to it, limit the access from one side. In order to access the actuator, robot needs to follow a complex, much longer path and then align itself for accessing the actuator (Figure 218 (b)). Instead, the robot may use a motorized actuator that comes from the top of the walls (Figure 218 (c), (d), (e)) so that the robot can access the mission from front. The path is much simpler, robot alignment is much more reliable and will consume much lesser time.

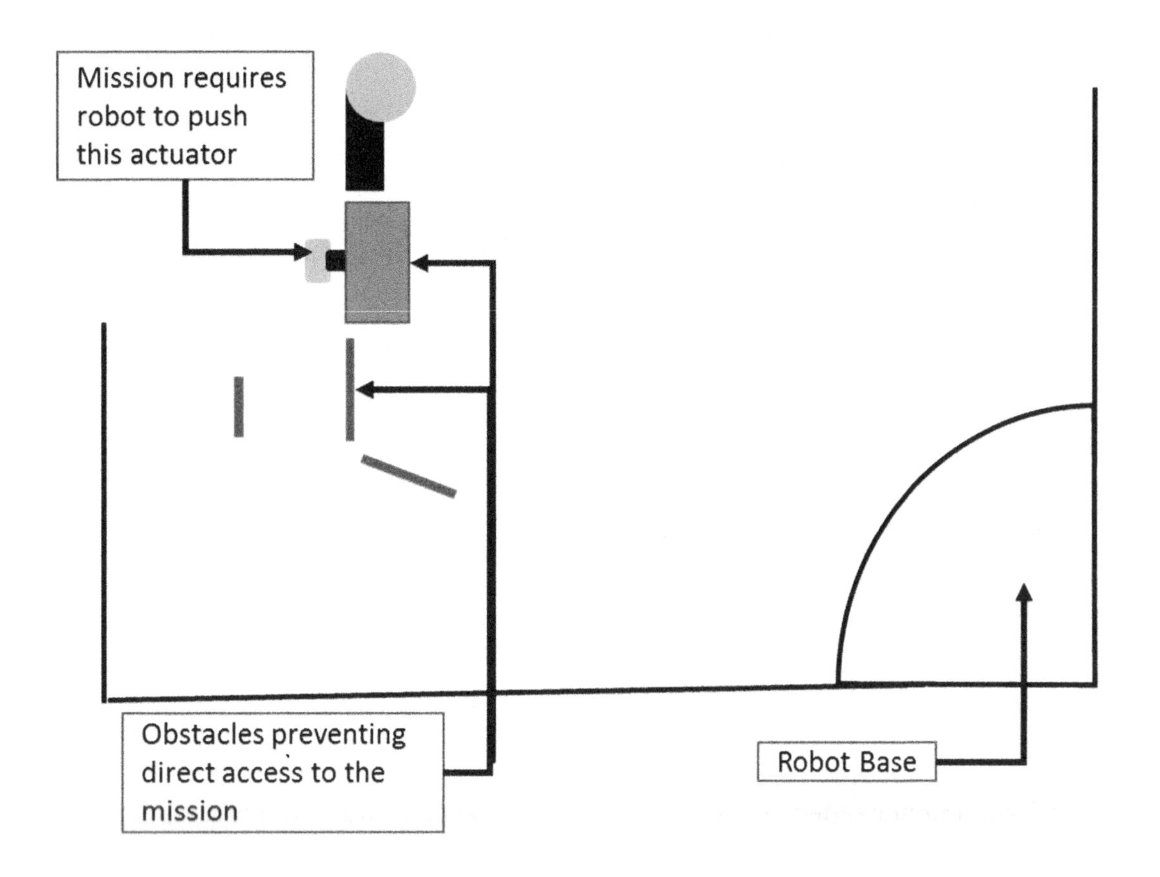

(a) Mission location that prevents Robot from directly accessing the actuator

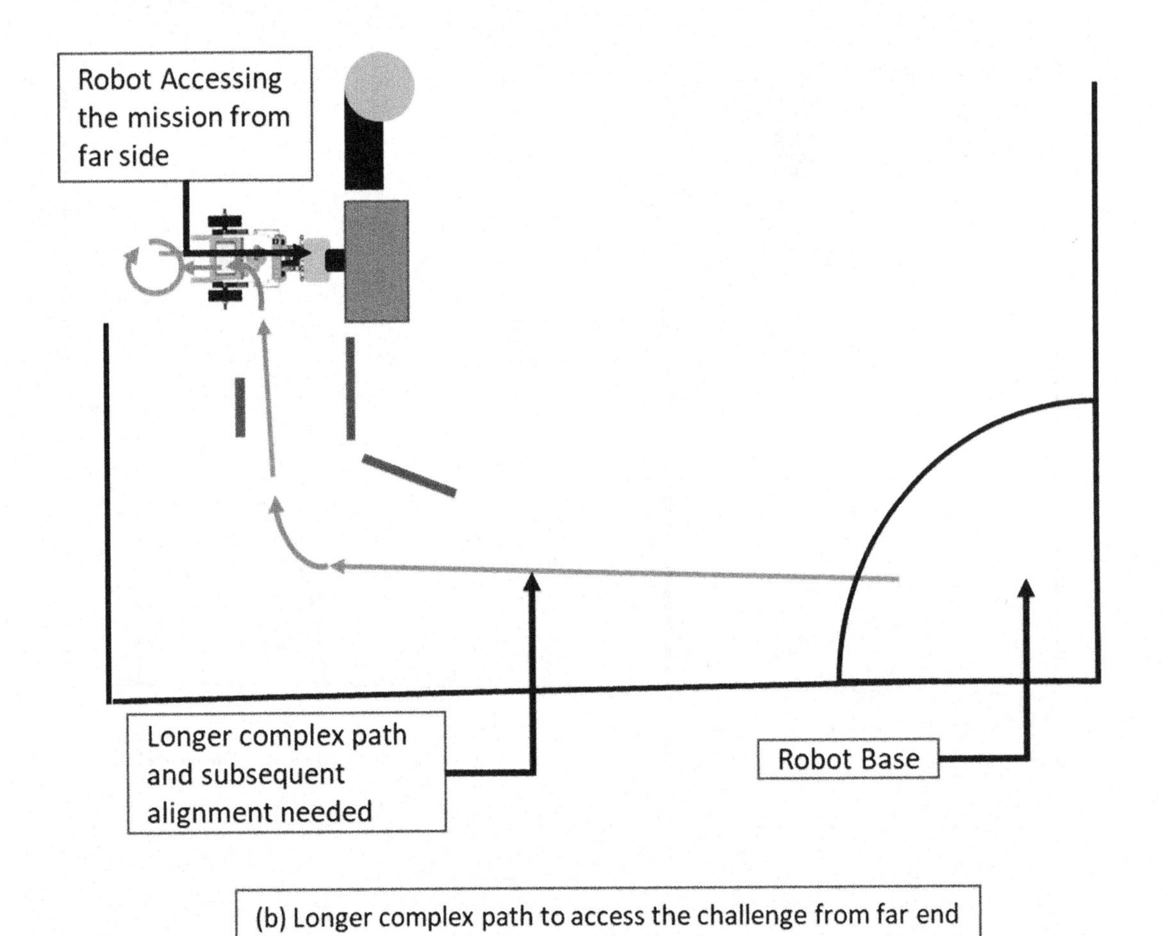

(b) Longer complex path to access the challenge from far end

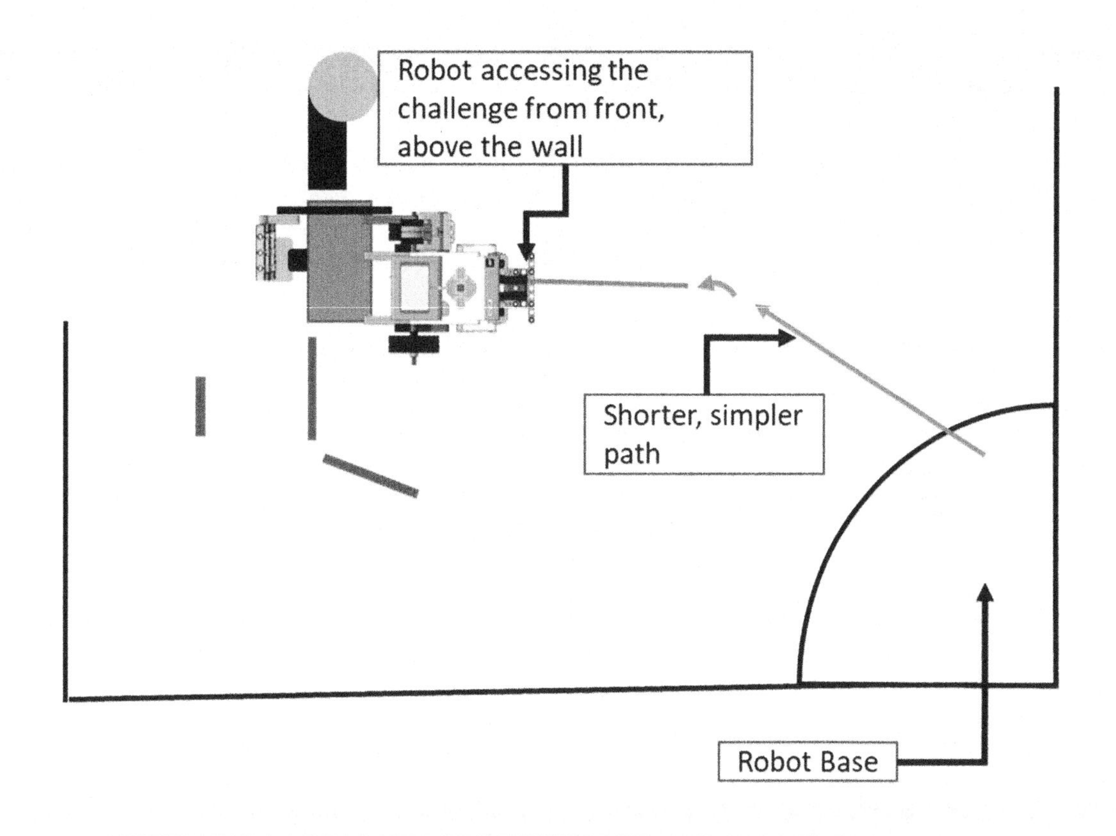

(c) Shorter simpler path to access the challenge from near end, above the obstacles

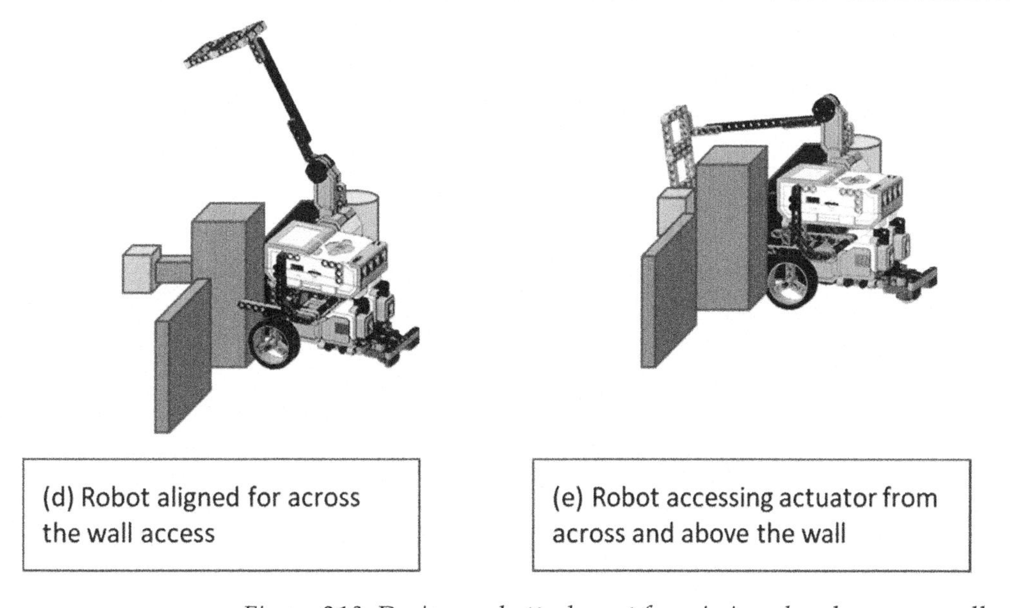

(d) Robot aligned for across the wall access

(e) Robot accessing actuator from across and above the wall

Figure 218: Design and attachment for mission placed across a wall

Axle based connectors for quickly switching static attachment

Pegs are used very frequently for semi-permanent attachment of two technic pieces together. When assembled as intended, parts are structurally robust. The pegs; however, when used to switch the attachments may be time consuming and may cause errors. Additionally, when participants are applying force for quick removal of an attachment, this may cause the robot assembly and subassemblies to detach in unpredictable manner.

Using long axles at connecting as well as receiving interface eliminates most of the problems.

Figure *219* describes an example of quick swap attachment that is based on an axle. The axle may be connected to the attachment itself with corresponding feature (hole) on the robot (*Figure 219*(a)). Alternatively, the axle may be connected to the robot itself with corresponding feature (hole) on the attachment (*Figure 219* (b)). It is very important that the length of axle as well as the span of engaging hole should be long enough to allow ample engagement of the two parts, else the parts may disengage and may fall off easily.

While it is not advisable to use the axle based attachment for moving parts; if the movement of arm is not very fast and if there is an opportunity, the axle based attachment may be used for moving parts as well. Ensure that the length of the axle allows ample engagement of the two parts (Figure 220.)

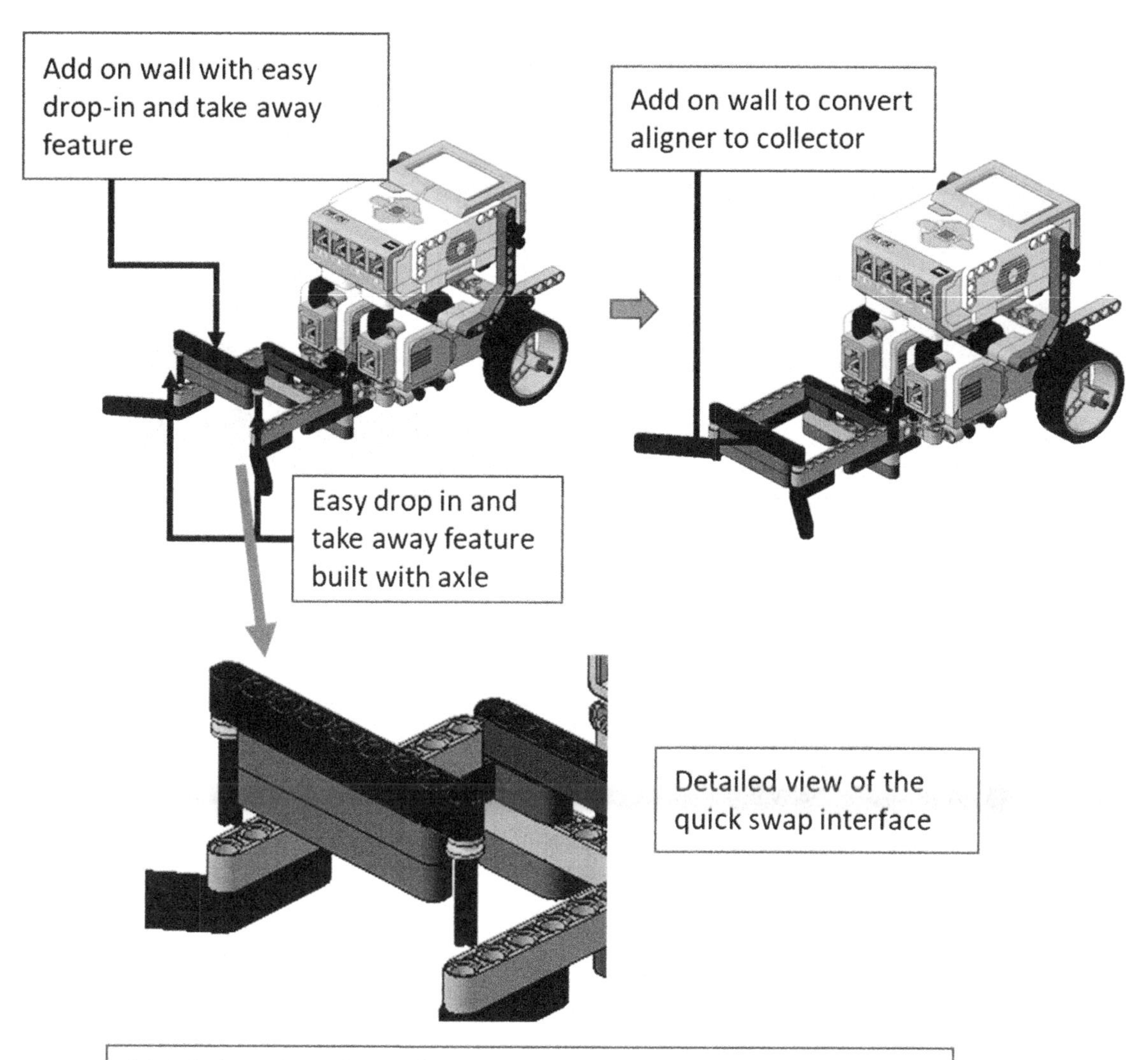

(a) Axle based quick switch mechanism with axle on the end effector

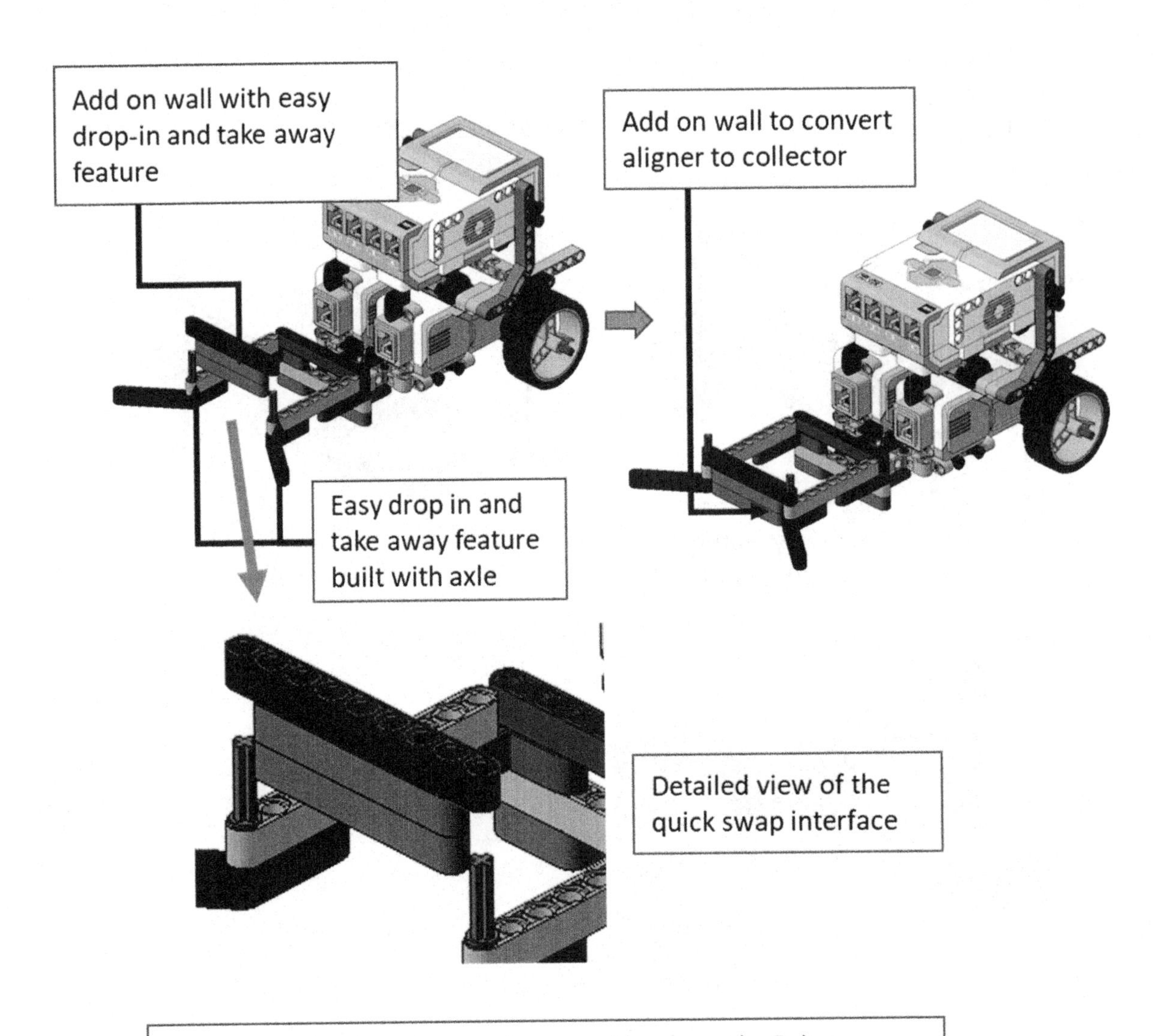

Figure 219 : Axle based quick swap interface for static end effectors (attachments)

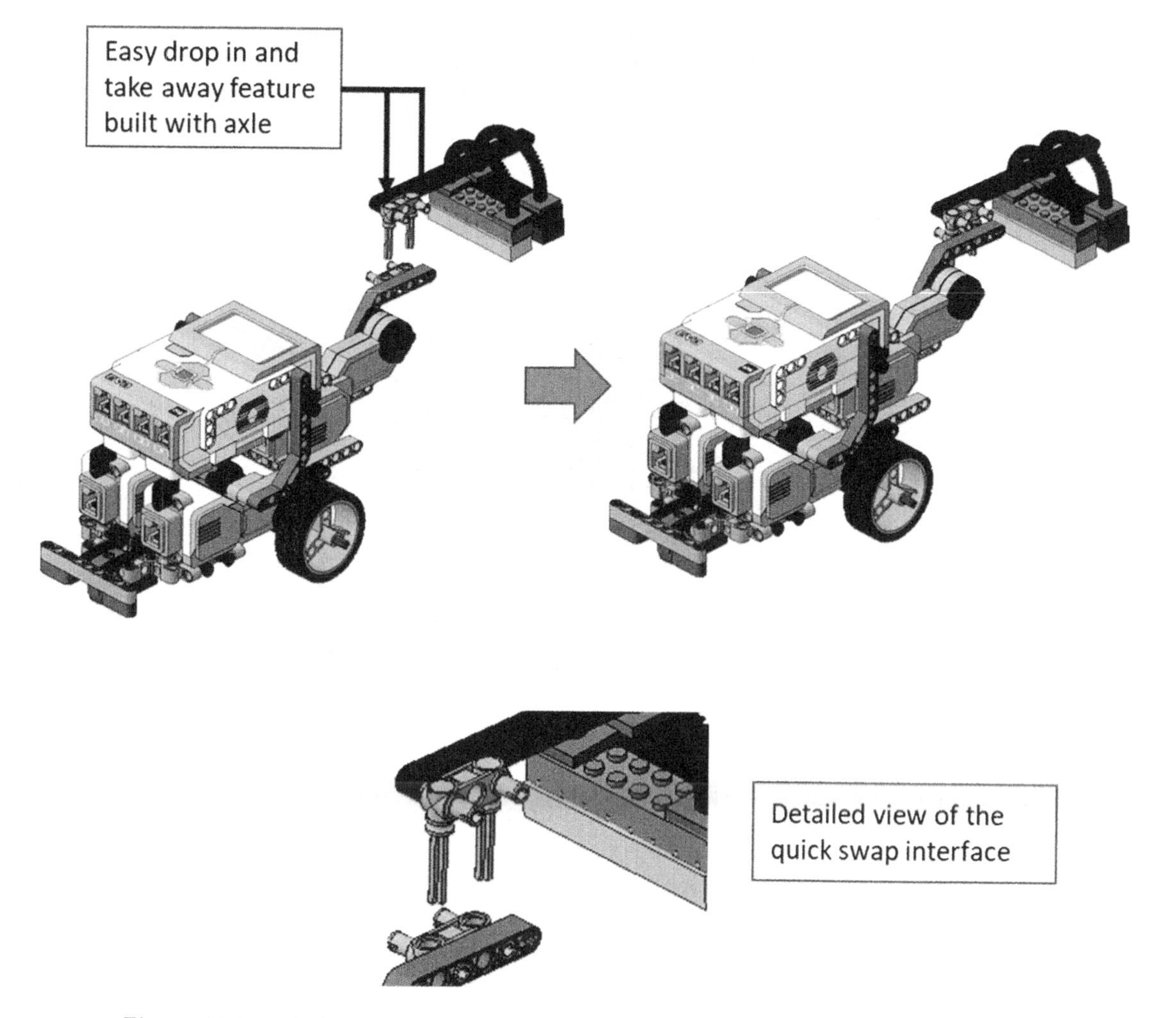

Figure 220: Axle based quick swap interface for moving arm

Touch sensor Engagement

While many missions may be accomplished by robot being within the vicinity, others may need very precise location of the robot. Touch sensor is very useful in precisely locating the robot as well as locating various missions. Contrary to the other sensors, detection with touch sensor is based on physical contact. Inherent to the physical contact is the possible effects of load and impact point. Ensure that the point of contact for touch sensor enables the application of force in-line with the red probe (Figure 221).

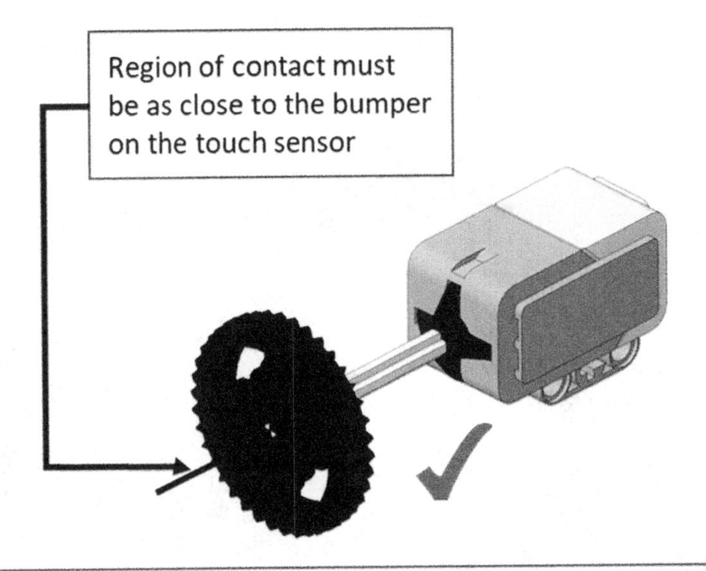

(a) Point of contact in line with the bumper for touch sensor enables more predictable performance.

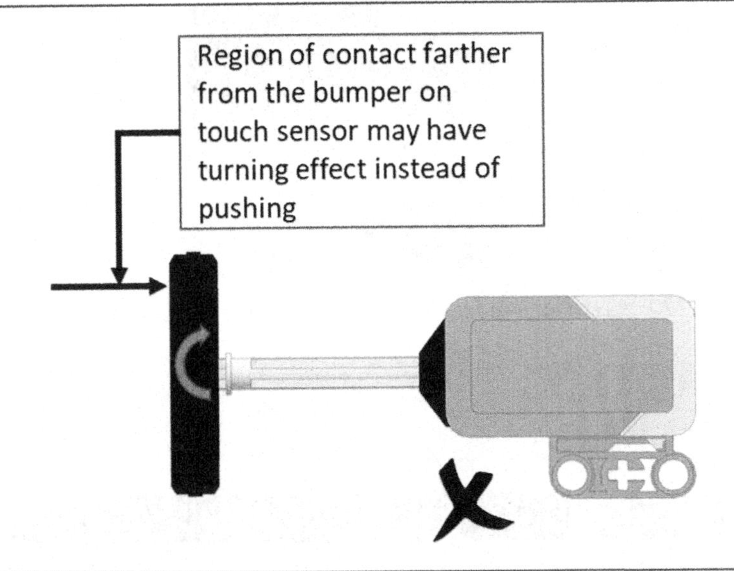

(b) Point of contact offset with respect to the bumper may have turning effect rather than pushing leading to unpredictable performance.

Figure 221 : The point of contact for touch sensor should be close to the bumper for more predictable performance

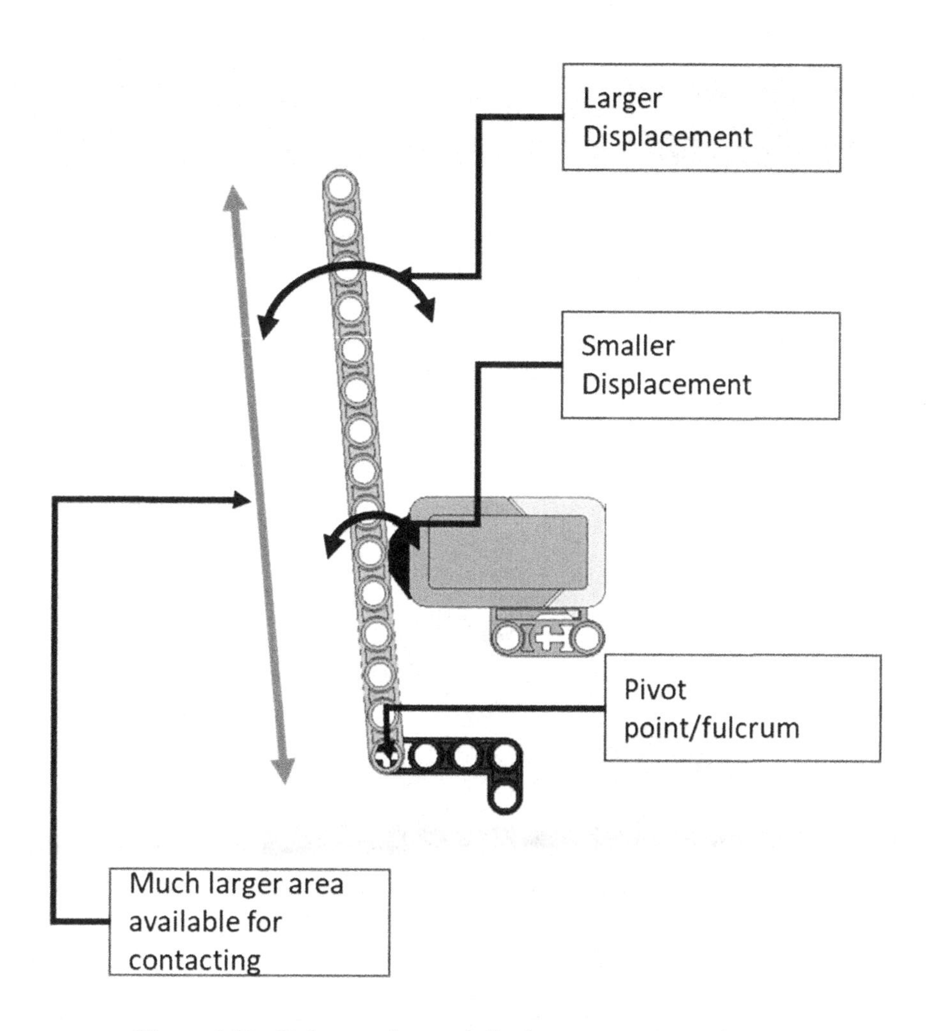

Figure 222 : Enhance the total displacement using a lever arrangement for touch sensor

As described in Figure 222, the overall displacement for touch sensor may be changed by engaging it with a lever arrangement. Additionally, the total area over which the touch may be registered is increased with the help of lever arrangement engaged with a touch sensor.

Use Level fix vehicle for aligning and make deliveries above the ground level

Certain missions require packages or objects to be delivered above the ground level as described in Figure 224(a). The package or the object to be delivered may have a loop or axle based feature to be used for engagement. While a motorized arm may be perceived as obvious solution, the motorized arm may deviate off its desired position and limit the total area available inside the loop (Figure 20 (b)). Maximizing the total available area within the loop is critical since it improves the chances of package transfer. It is suggested that we use a vehicle (Figure 20 (a,b)), with the object to be engaged placed in the vehicle to ensure the height of delivery material doesn't change. Additionally, the complete space within the loop is available for transfer.

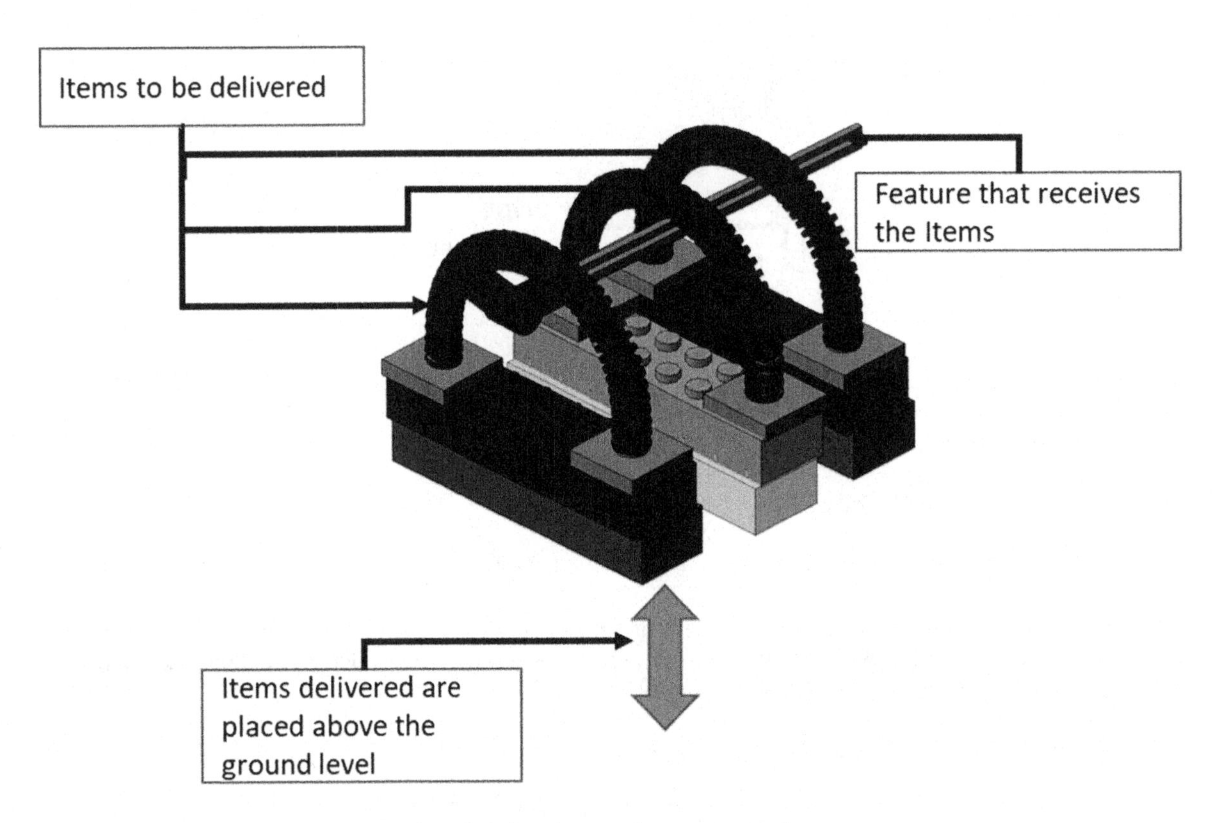

a. Packaged to be delivered in this format on the receiving feature

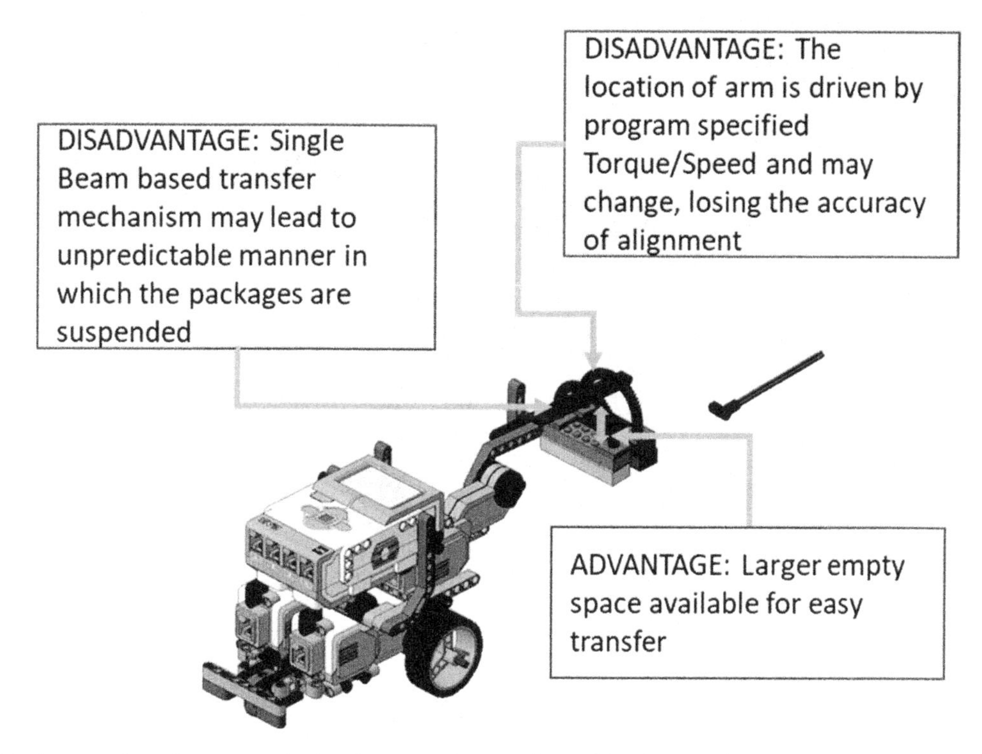

b. Objects are placed on a motorized arm and transferred to the receiving arm

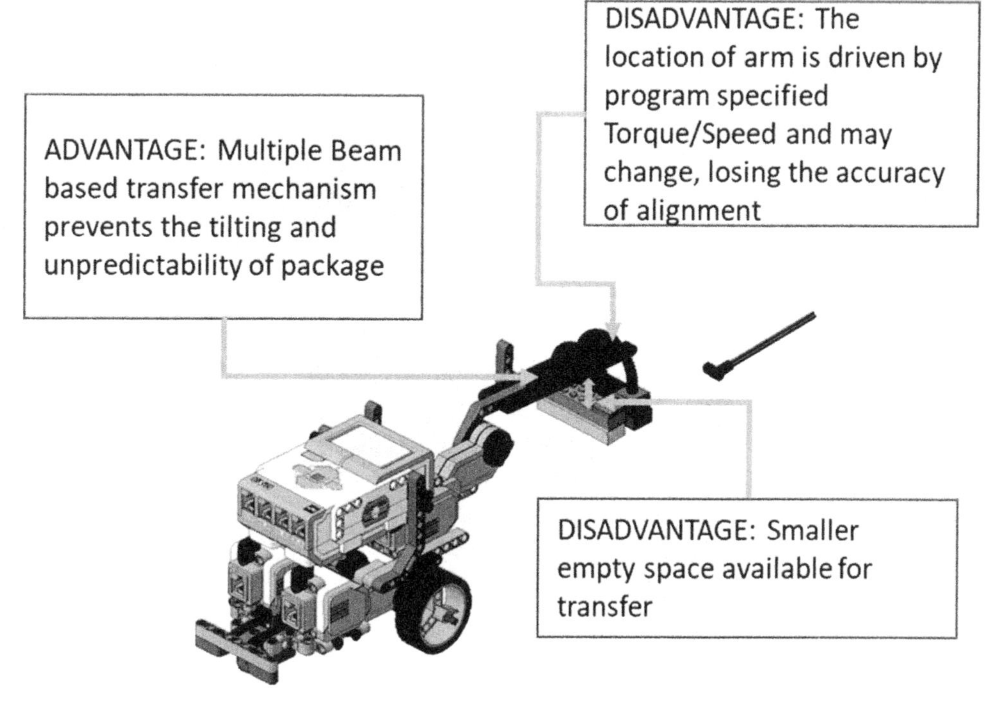

c. With two parallel arms the items are more stable but total empty space for transfer is limited

Figure 223: Using a motorized arm for transfer

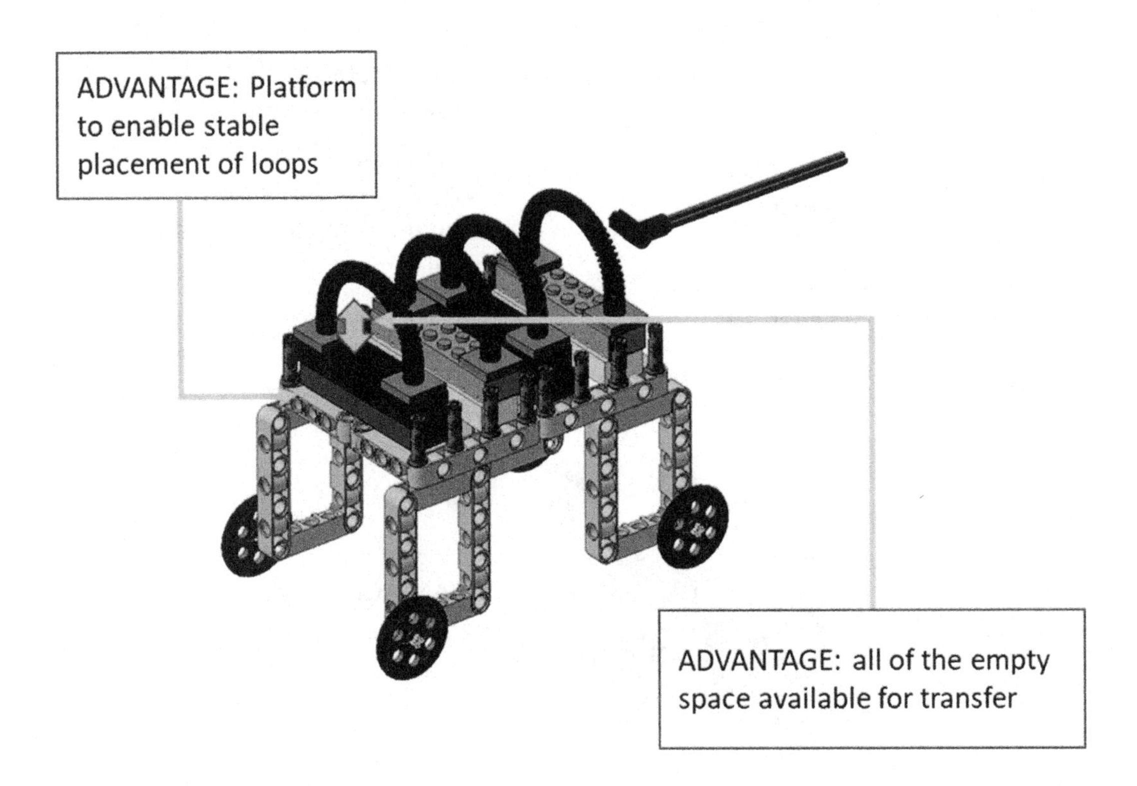

a. Vehicle for transfer of object onto receiving arm

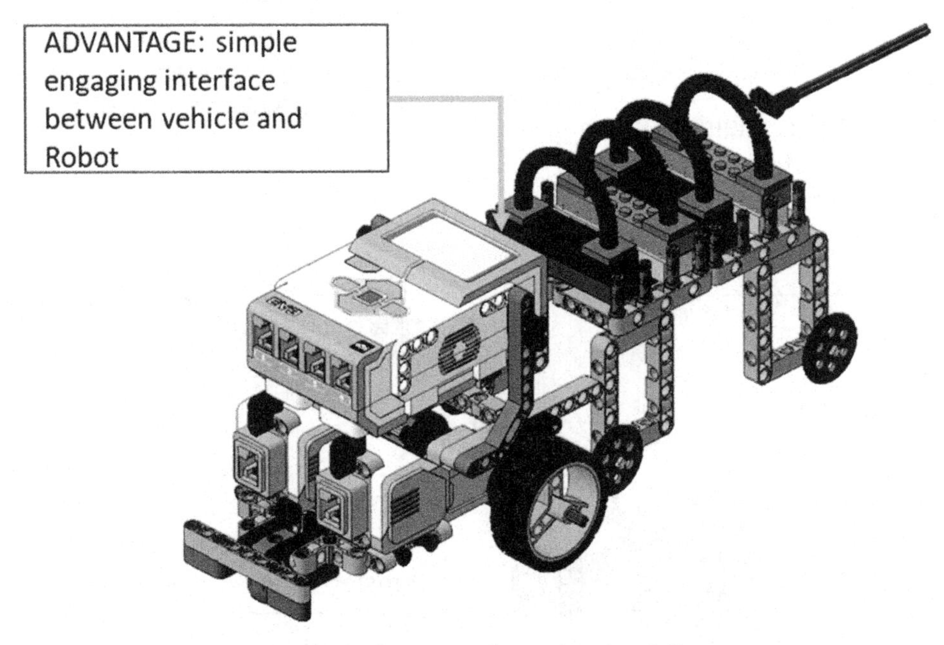

b. Vehicle engaged to robot for delivery

Figure 224: Using a vehicle for delivery of the loops onto receiving arm

Summary

In this chapter, we discussed multiple ways to optimize your robot to be able to perform missions with an efficient use of the resources available to you. We looked at designs such as passive mechanical designs that can grab and capture objects simply because of the way they work and mechanical designs that allow the robot to align against uneven objects using self-aligning features. We also looked at designs that used motors and sensors to perform certain tasks in a more intelligent and efficient manner. These techniques can easily allow your robot to do a lot more with relative ease and efficiency.

Robot Mission and Path Planning Strategies for FLL

In addition to various aspects of Robot Mechanics and Programming choice of mission strategies and order of executing missions are equally important to reduce the total time take and increase the overall score. In this section we will introduce you to some of the tips and techniques that are used by experienced team in planning the missions for the Robot Games.

Rookie teams that are competing in FLL for the first time typically tend to think about interacting with one model at a time and thus they have multiple missions to perform a single task at a time. It is an eye-opening moment for them to enter a FLL competition and see experienced teams tackling multiple missions in one single run.

Since time is of premium in FLL given that a team has to attempt as many missions as they can within a tight time limit of 2 minutes and 30 seconds, it is essential that each time the robot travels outside the base, it tries to accomplish as much as possible before returning back to the base. Given the constraint of time and doing as much as possible within a single outing, teams need to put thought into the path planning for their robot and which missions to attempt in single robot outing.

There are multiple schools of thought on how to perform mission planning and we will go over them one by one. We will provide a few examples from past FLL seasons to indicate what worked in a certain season and why a certain mission planning made more sense.

One at a time mission planning

This is the simplest of all mission planning. In this method, the teams plan the robot path to tackle one mission at a time. This is the most expensive path planning method as it requires the robot to return to the base after every mission. Additionally, this also means that the team is removing and adding attachments after every robot outing which costs precious time. This approach, although sometimes used by rookie teams, is not recommended as it wastes precious time for robot return and attachment change.

Zone based mission planning

One of the common ways teams consider mission planning for the Robot Games is using the Zone based mission planning as illustrated in Figure 225. The table is divided into multiple zones containing missions and plan the robot path so that the robot can complete one zone at a time. This method uses the concept of geographical proximity of missions to bundle them together and thus attempts to solve missions close to each other in one robot outing.

Although this approach has an element of simplicity since it is trying to optimize the number of times a robot travels the same path, there are times when duplication is simply unavoidable and you might score higher by traversing the same path multiple times. Additionally, one of the most crucial part of the Robot game that teams try to optimize on is the number of time an attachment must change. The reason for this is that the robot must return to the base for the team to change an attachment. This costs time when the robot travels back and additionally, changing the attachment is usually an expensive operation taking on the order of 10-20 seconds depending on the complexity of the attachment. Thus, teams try to reuse the same attachment for multiple missions if possible. This may force the team to either reorganize the zones repeatedly or have the robot travel across multiple zones forcing the team to reconsider their strategy.

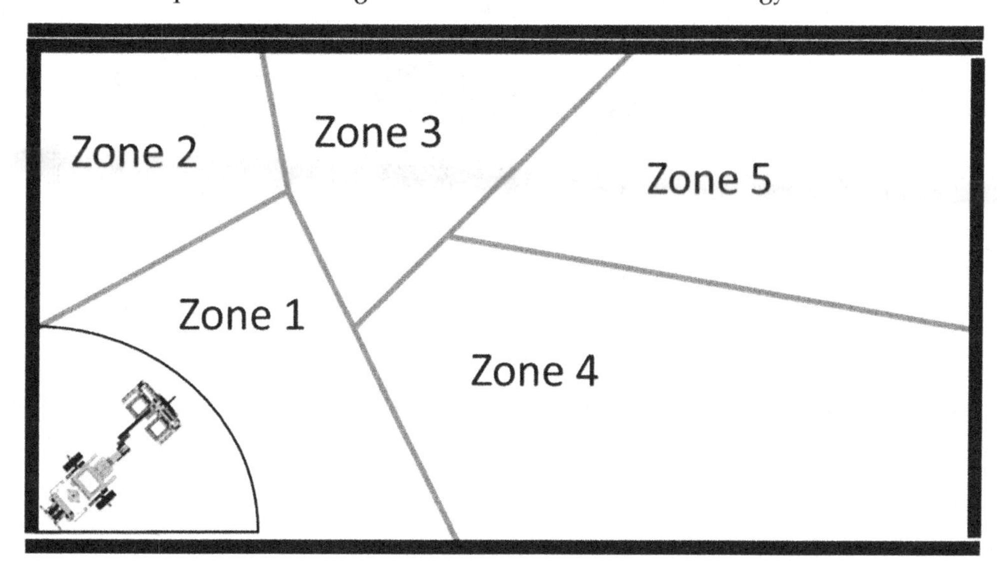

Figure 225: Zone based mission planning.

In Figure 226, we are showcasing the mat from the FLL 2016, Animal Allies season with two robot paths that many teams used to great success. Note that these paths have overlap, and it is not possible to cleanly separate out zones based on geographical proximity.

The filled in red shapes denote obstacles or missions. We are indicating only a few of them.

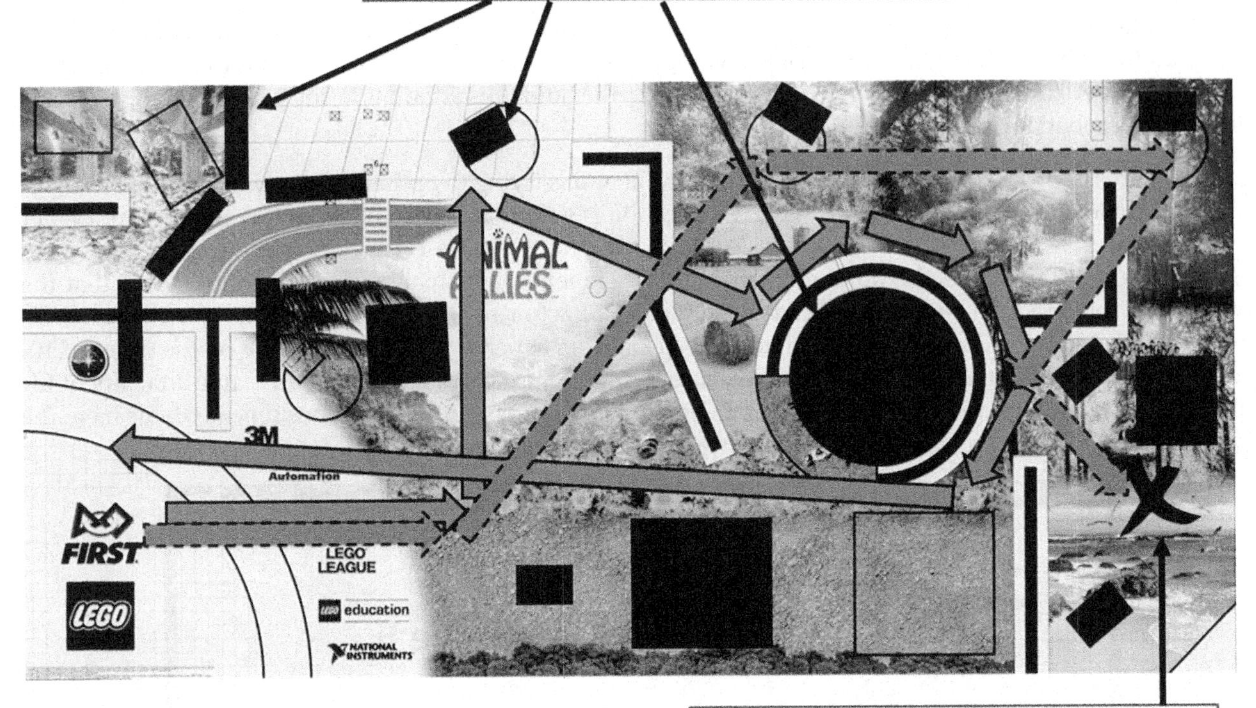

Green path (arrow with dashes) run abandoned here and penalty taken to bring robot back to base.

The blue paths (solid edge arrows) and the green paths (arrows with dashes) have overlap and it is not possible to cleanly separate the missions contained in them among zones.

Figure 226: FLL 2016, Animal Allies mission plans used by teams.

Path and Opportunity based Mission planning

There is another method of mission planning called Path and Opportunity based Mission planning. We identified many advantages of this approach and it has been used by our teams very successfully.

Let's start with the Path component of this kind of mission planning. The idea here is to figure out a robot path that the robot can travel consistently and with the least amount of uncertainty and around which the highest scoring models are located. In this method of planning, the team sits down and ranks all the missions based on how many points a mission is worth and what is the likelihood of being able to finish the mission successfully in a consistent manner.

Once the team has ranked the highest scoring missions in decreasing order of likelihood of success, the team maps paths that access the missions with the highest scores. A path, that can approach multiple high scoring missions with a higher cumulative score is picked over another path where even though individual missions may have higher score, the cumulative score over the path is lower.

Once a set of path has been agreed upon, the team creates attachments that can solve as many of the missions on the path as possible. In some cases, the robot design may become a limiting factor. Alternately, one may choose to attempt a mission that was previously discarded due to lower score but becomes available en route in alternate run without significant loss of time. In this case, we will *opportunistically* add this mission to the original path. We continue iterating over this process based on experimentation with the path and whether all the goals set at the onset are achievable.

To understand and make use of the *path and opportunity based mission planning,* we are going to introduce a few tips, tricks and concepts in the following sections.

Island of Certainty

Before we can plan a robot path on the FLL table, we need to understand a phrase - *Island of Certainty,* that we coined when explaining the mission planning concepts to our teams. *Island of Certainty* refers to the locations on the FLL mat, where the robot can, without any doubts, be certain of its position.

To understand the concept of *Island of Certainty,* consider a graph paper representing the *Cartesian Coordinate* system. In a *cartesian coordinate system,* every position on the graph paper is represented in terms of a unique *(x,y)* coordinate. Now think of the FLL mat as a graph paper that can be represented in terms of cartesian coordinates. When the robot is travelling on the FLL mat, it uses *odometry* or sensors to navigate the terrain. If it is using odometry, because of robot drift, the robot is not really certain of its exact location. Even if using sensors, the robot might or might not be certain of its exact (x,y) position on the mat. To gain certainty, the robot needs to find at least two reference points on the mat that it can use to ascertain its position. One of these reference points needs to provide the robot with its exact *x* coordinate and the second reference point needs to provide the robot with its exact *y* coordinate (Figure 227.) If you can locate a few points on the FLL mat where using any navigation, aligning or sensor based technique, the robot can be certain of its exact *(x,y)* position, then you can plan your paths around these fixed, easy to verify points. In our lingo, we call these fixed, verifiable reference points as the *Islands of Certainty.*

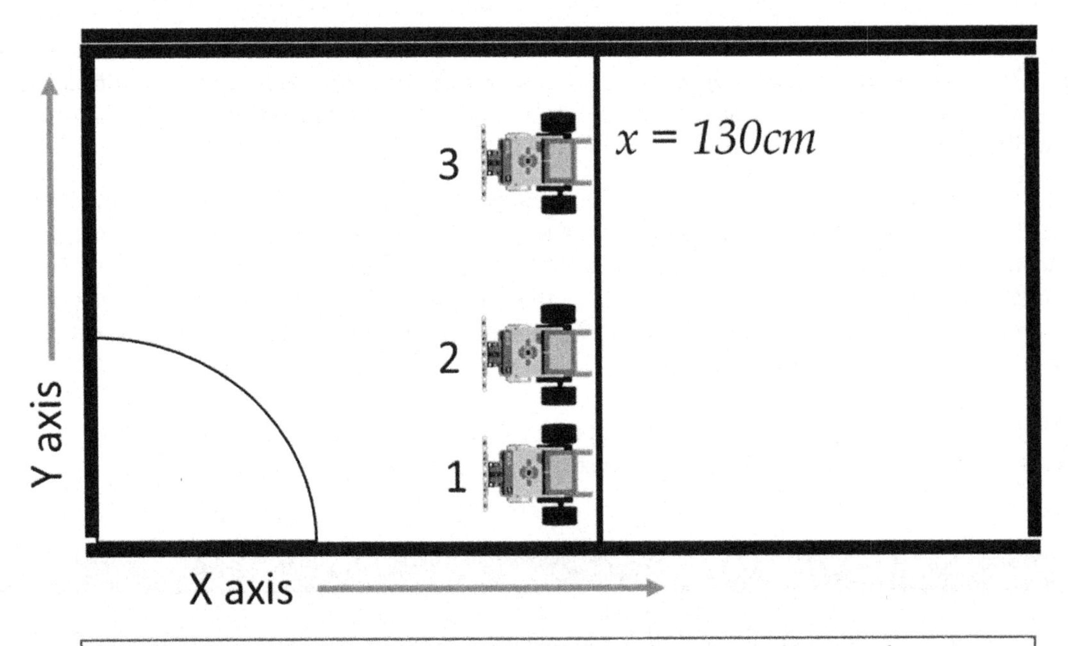

If the robot is certain that it is at *x=130cm,* then it can be anywhere on the red line i.e. the Y position of the robot is uncertain. Sample positions 1, 2 and 3 are shown to indicate this. Having an uncertain Y-position means that the robot needs another reference point on the Y-axis before it can be sure of its position and get ready to perform missions.

Figure 227: Ascertaining the robot's position requires at least two reference markers.

Once you understand what the term *Island of Certainty* means, we can rephrase our path planning strategy as an exercise in jumping from one Island of Certainty to another and solving high value problems that are close to an island of certainty.

We would argue that the FLL Robot Game is an exercise in jumping from one Island of Certainty to another and solving missions that are close to each Island of Certainty.

The Islands of Certainty are marked using green circles (solid edge) and red circles (dashed edge). The orange rectangle does not provide an Island of Certainty since in this region the robot can only infer that it is on the black line but cannot infer where it is present on the black line. The Orange section (rectangular, dash and dot edge) is still quite useful because the black line on it leads to an Island of Certainty.

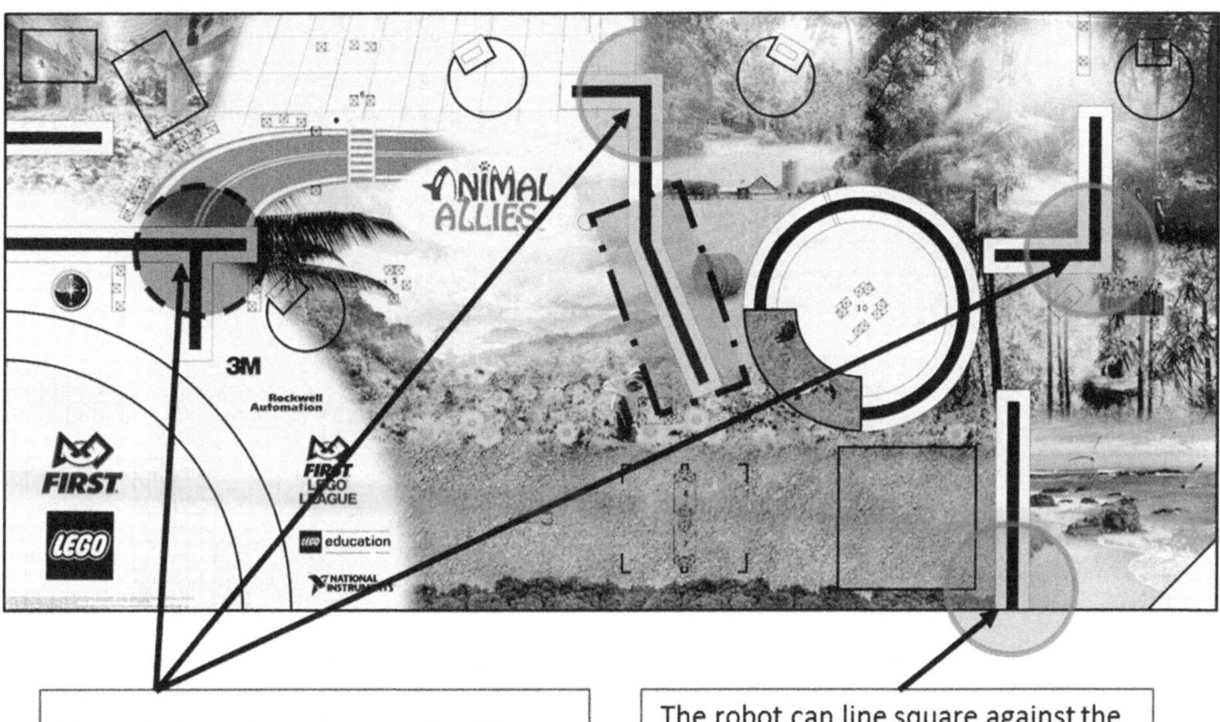

The robot can line square against the two perpendicular black lines using two color sensors, thus fixing both its *x* and *y* positions.

The robot can line square against the black line using color sensors and use bumper based wall alignment against the table wall to fix both its *x* and *y* positions.

Figure 228: Island of Certainty marked on the FLL 2016, Animal Allies mat. Although the red circle is an Island of Certainty, it is too close to the base to be of use. The robot can easily travel the distance between the base and the red Island of Certainty without drifting or otherwise losing its bearings significantly.

Overshoot your goal

The concept of overshooting your goal derives from a navigation paradigm in hiking. Let's say you are hiking in the woods and while coming back, you are a bit lost.

You need to get to the parking lot at the trail head. The parking lot happens to be at a dead end. You know that you are fairly close to the parking lot and that the parking lot is to your left. If you aim exactly for the parking lot, you may miss it as you are not a hundred percent certain of your current position and your guess of the parking lot's bearings may be incorrect. In this case, you aim much further to the left so that you will end up half a mile to a mile left of the parking lot. This means that when you finally get to the road leading to the parking lot, you can simply turn right and walk to the parking lot instead of being lost. You traded precision for certainty.

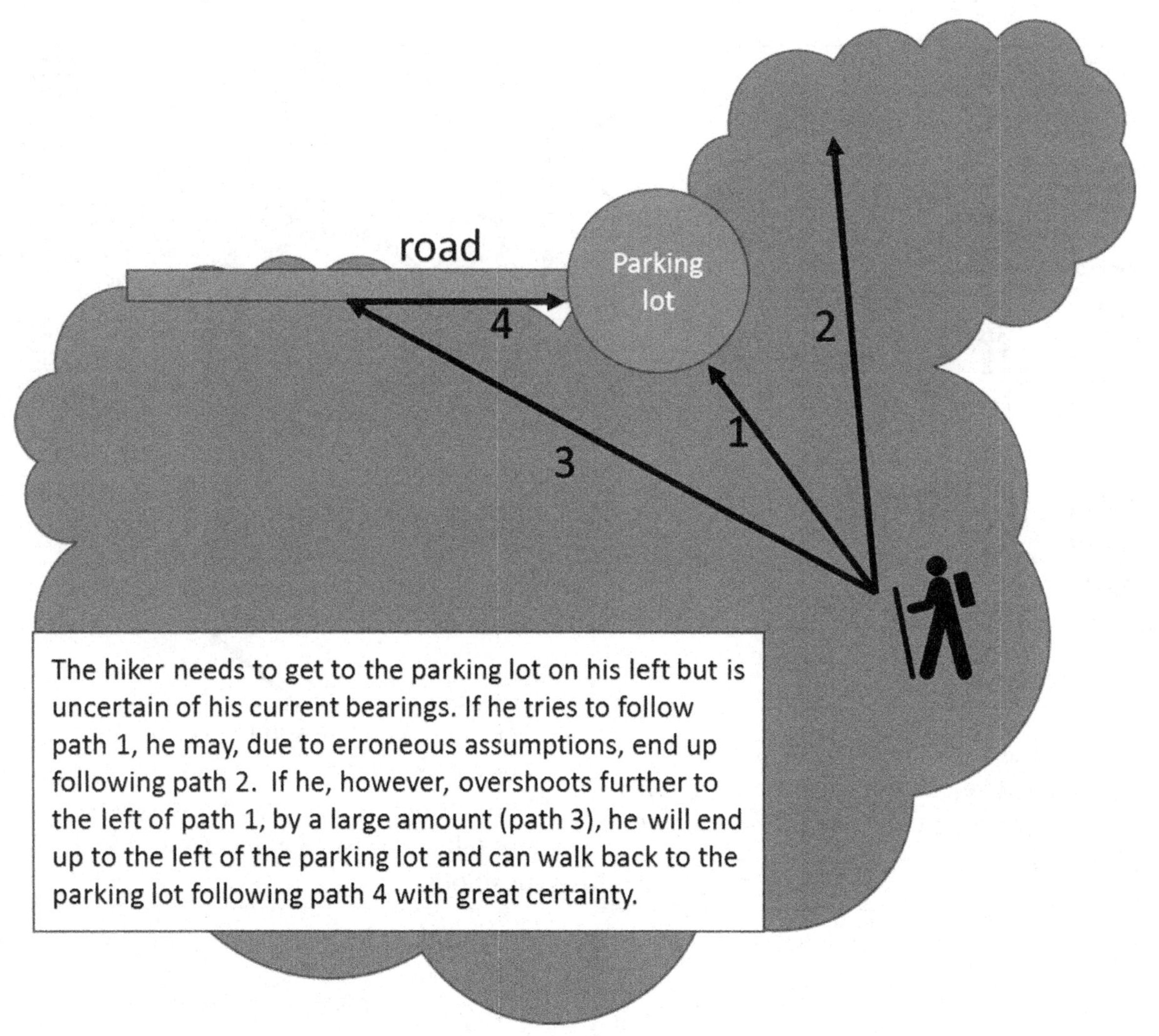

Figure 229: Overshooting your goal explained in terms of hiking navigation.

An example of overshooting your goal in FLL is shown in Figure 230 which illustrates the problem in context of FLL 2014, World Class.

Overshooting to compensate for errors. The desired robot path is shown with blue arrows (solid edges) to catch the highlighted line for line following. However, due to slippage, the robot might miss the line and end to the left of the green line (the line to be followed). Thus, we program the robot to overshoot its intended target to ensure that it falls someplace further on the green line and follows it to the intended target.

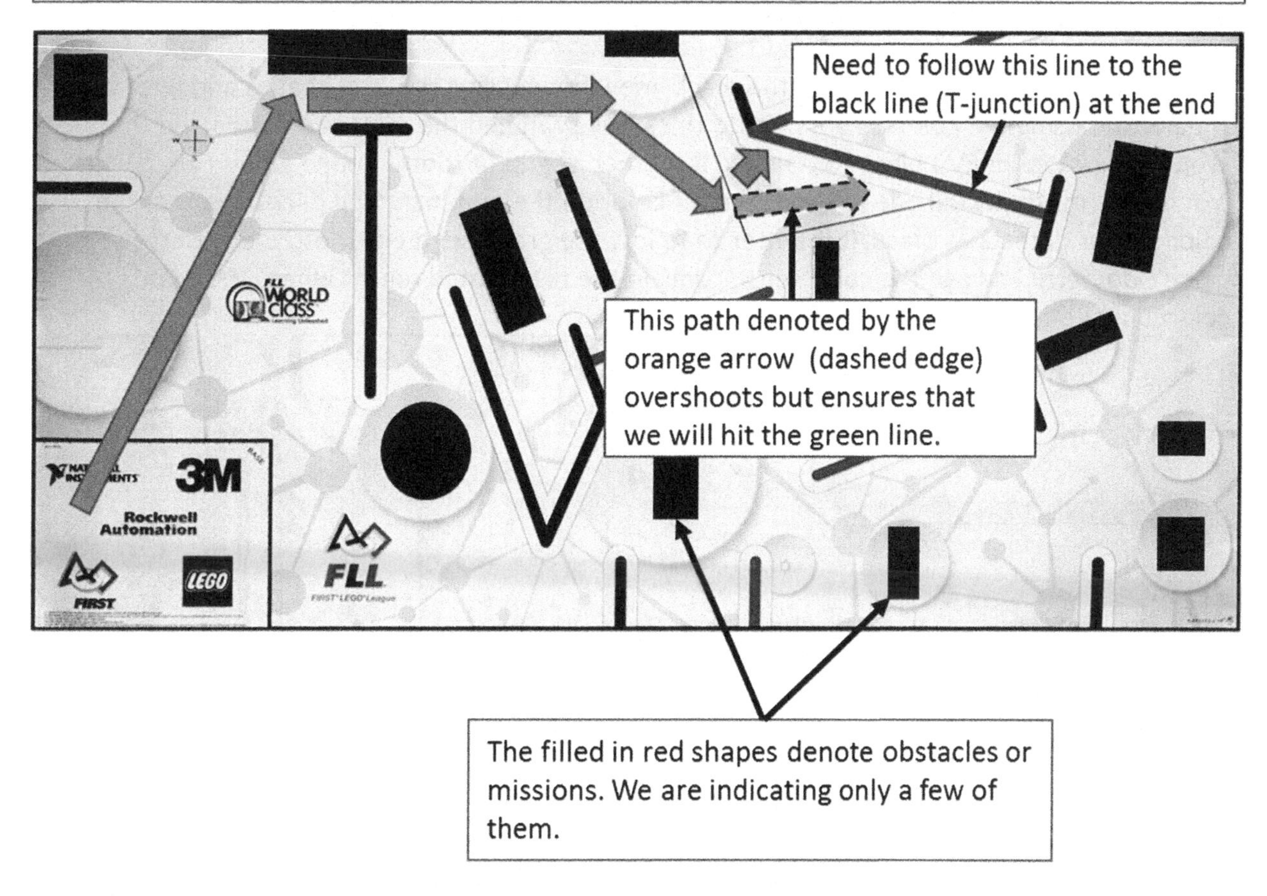

Figure 230: Overshooting the goal to ensure that the robot will for certain hit the intended goal.

Use techniques when you can be certain of them

Quite often, we see teams discounting techniques because they use them in an inappropriate fashion. Every technique, whether mechanical or programmatic needs a certain set of pre-conditions to be true to be most effective. Instead of discounting a technique because it is not applicable in the generic case, try to engineer circumstances where the technique becomes applicable.

As shown in Figure 231, to catch the black line for line following, the color sensor needs to detect when it has reached the black line. Participants would often switch the color sensor on for the entire duration of the segment indicated by the green arrow and wait for the black color to be detected by the color sensor. This would often fail because the mat is a full color graphic and has many hues that the sensor may misidentify for black color. Instead, it would be better in this case to use the more basic odometry based movement till the robot gets close to the black line and then use the color sensor to wait for the black line only when we are reasonably confident that the robot has reached the white color around the black line.

Many participants wanted to reach the black line (shown using the oval shape) and line square to it using the color sensors. To do this, they switched the motors on and used the *Wait for black* color EV3 block to traverse the green segment (dotted edge arrow) which would intermittently fail due to a plethora of color on the mat including dark hues that the color sensor detects as black. It is better to follow the green segment (dotted edge arrow) using *Odometry* and use the color sensor only in the highlighted region where the mat is colored white just before the black line.

Figure 231: Engineer circumstances so that a technique becomes usable in the specific circumstance.

Minimum number of segments on a path

During Mission path planning, if we have two paths to an objective, then we will likely pick the path that has minimum number of line segments in it. The reasoning behind this is that every time a robot turns or stops and then starts again, small errors in its position creep up due to slippage and drift. In absence of other reference points by means of which the robot can ascertain its position, the more number of segments, the more the chances of error. Apart from more chances of error in a path with more segments, there is the additional issue of errors in previous steps impacting the later steps strongly. What this means is that errors keep on accumulating and they have cascading effect on later segments where a later segment might diverge from the intended path by a large amount.

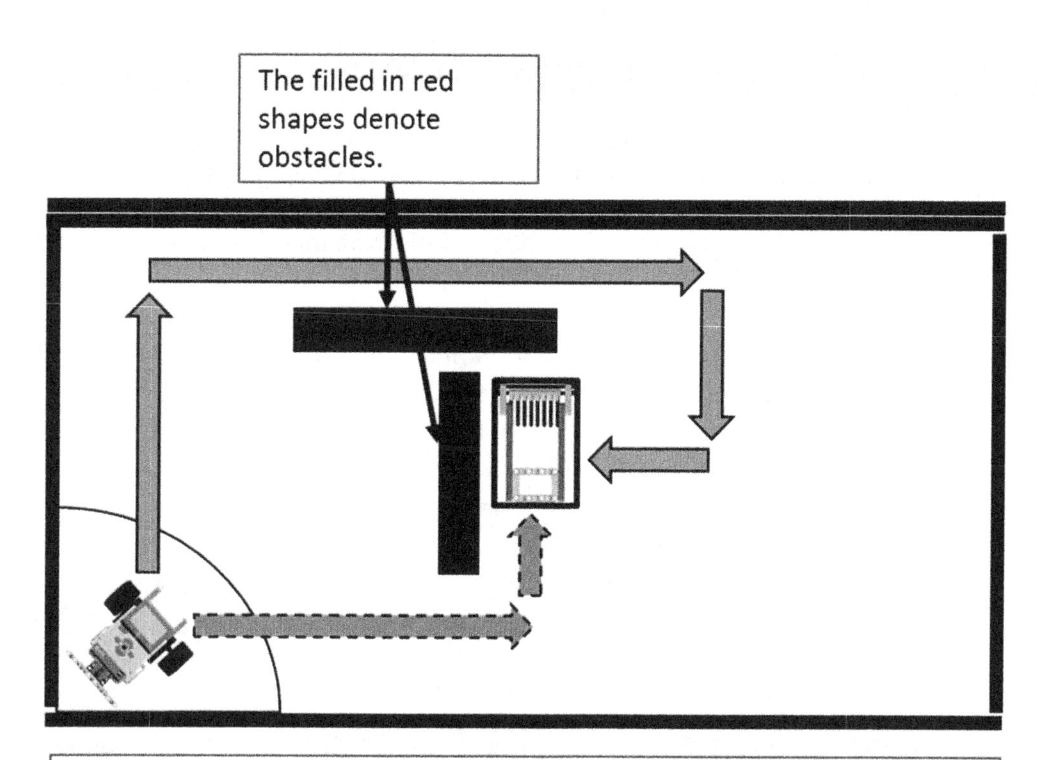

Figure 232: When planning a path, pick the one that has lesser number of linear segments and turns.

It is a great source of dismay to us when participants keep using only 90 degree turns to reach their objective when, travelling at an angle would introduce lesser number of segments and turns and hence less variability. Make sure that when path planning you are trying to find the shortest path with any angle, instead of the shortest path with only 90 degree turns.

Opportunistic Mission completion

Although we discussed this earlier, we believe that the topic of opportunistic mission completion merits another mention. If you have a high value mission that you are completing, then always look for other missions that may be accomplished while attempting to accomplish the original mission as long as it does not interfere with the high value mission and takes relatively little extra time. We show an example of this in Figure 233.

The robot had to travel the green path (solid edges) to finish a high value mission. Instead of coming back the same way, the robot takes the opportunistic view and tackles another mission close by (indicated by the orange arrow – dashed edges) before being picked up with a penalty.

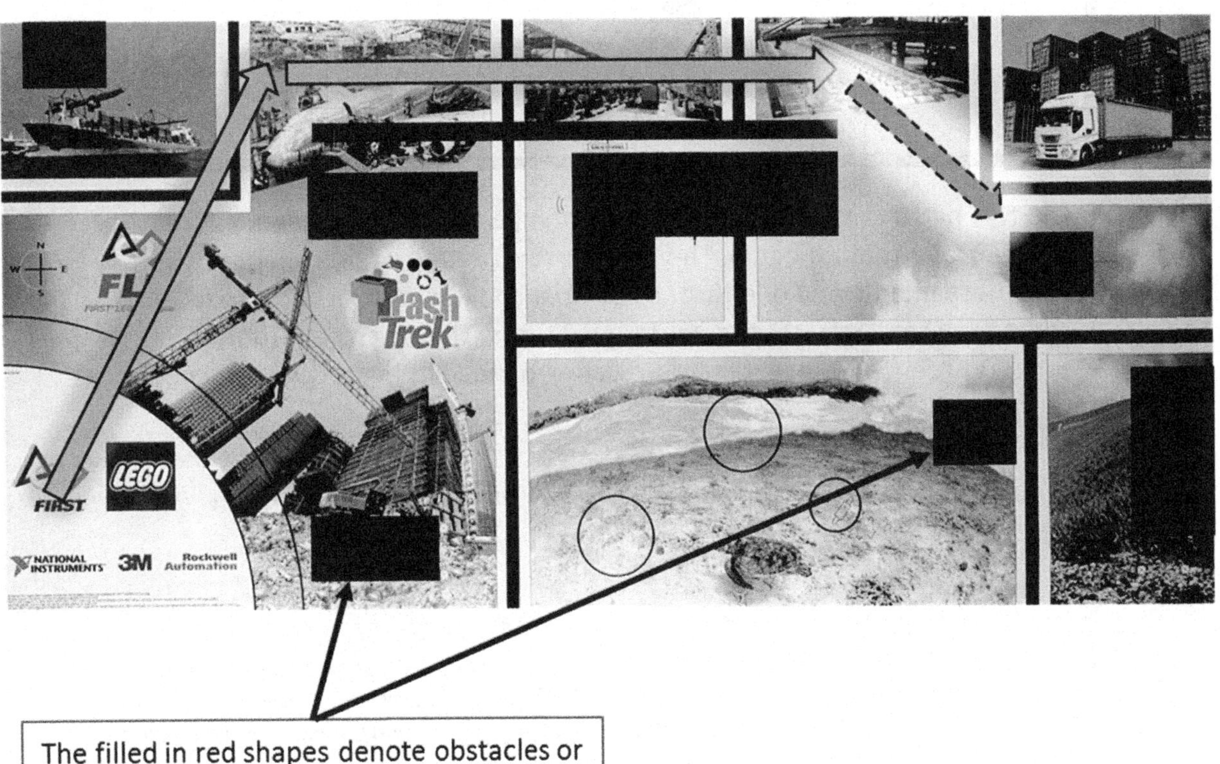

The filled in red shapes denote obstacles or missions. We are indicating only a few of them.

Figure 233: An opportunistic mission completion close to a desired high value mission path. FLL 2015, Trash Trek.

Taking penalties as a strategy

FLL Robot Games have very well-defined rules and they include penalty points if your robot misbehaves and the robot driver needs to stop the robot and pick it outside the robot base. However, stopping and picking the robot outside the base is a perfectly valid strategy in some cases. The reason you may want to do this is because usually the missions that count for a large number of points are quite far away from the base. Getting to these consumes quite a bit of time and it takes similar amount of time to get back to the base. If your planning and score counting reveals that you are better off taking a penalty by picking the robot and then scoring more points in the time you saved, you should consider making that part of your strategy (Figure 234.)

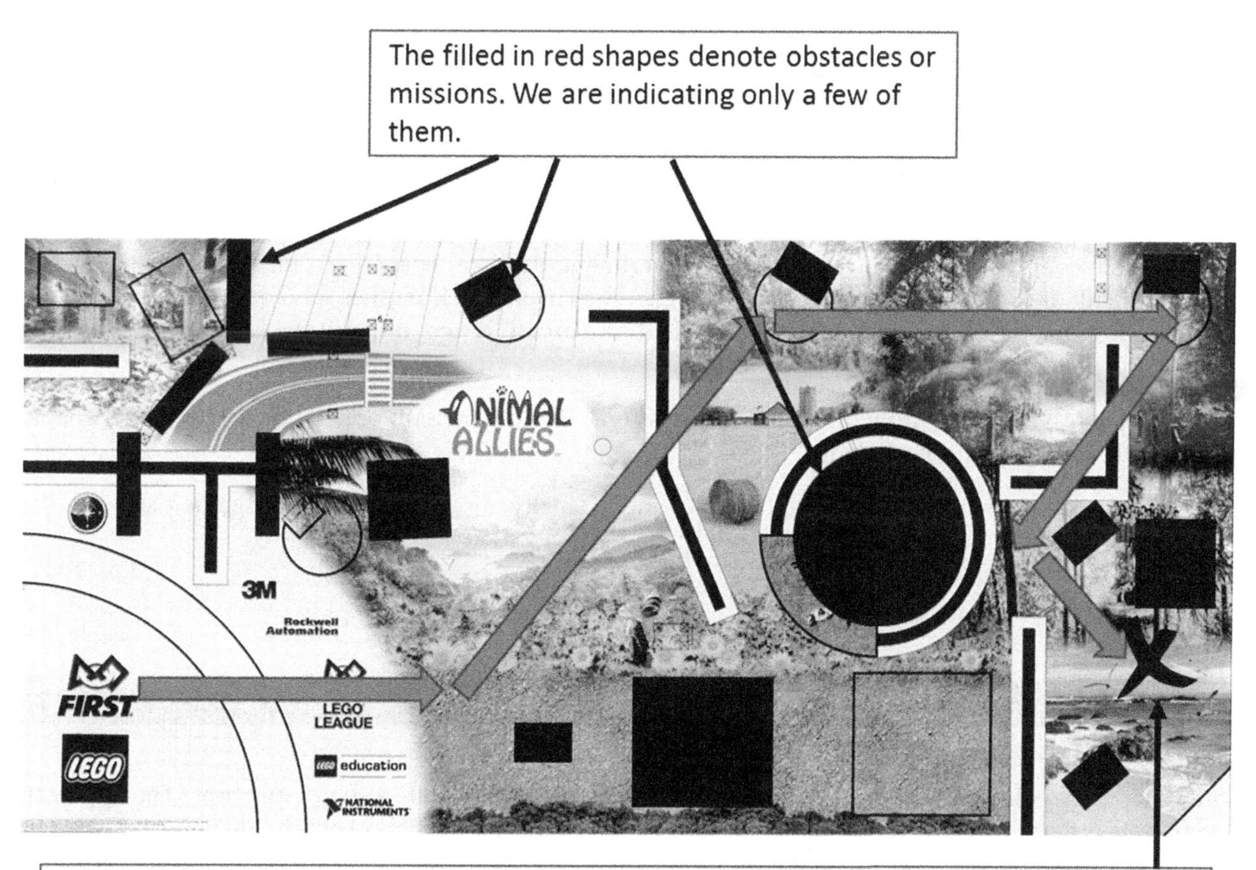

Figure 234: Sometimes, it simply makes sense to use a penalty as part of your robot strategy.

If you are incorporating taking a penalty as part of the game plan, train the participants to move closer to the robot where the penalty needs to be taken and be ready to grab the robot as soon as it is done with the last motion. This will save precious time.

Tying it all together

Once we understand all these tips and techniques to ensure that our robot can find Islands of Certainty for navigation and follow the best practices for reaching them, we can recap our Path and Opportunity based mission strategy as the following:

- Rank missions in order of the highest score.
- Re-rank the missions in order of the likelihood of success
- Find the Islands of Certainty on the FLL mat and hop from one Island of Certainty to another while solving the highest value missions first.
- Iterate over the paths again and again. If experimentation later bears that a mission is becoming too difficult or consuming too much time, change the plan and pick another mission. Alternately, if you realize that a high value mission you had discarded is too difficult to achieve successfully, incorporate it into the mission plan at a later stage.
- While traversing a path with high value missions, if a low mission opportunistically seems doable with relatively little cost, do it.
- If you can save time and score higher by taking a penalty, incorporate it in your game plan.
- Refine, refine, refine till the missions run like clockwork.

Summary

In this chapter, we discussed how to plan the paths for robots to optimally attempt missions. The approach we advocate is to prioritize all the missions in order of points and the likelihood of achieving them. Once you have the prioritized list, you can attempt to fit the highest value missions in as few robot outings as possible. You may not be able to finish every mission that you planned, but this provides a great starting point. Iterating over the missions again and again will eventually provide you a well working solution.

Appendix A: Winning Strategies at FLL - beyond Robotics

Here are some of the key learnings that we can impart based on our experiences. This is a short summary and only contains a list, but it is borne out of years of experience and might help you run a team.

- Make sure the team has no more than 4 kids ideally. Any more than that and you will be wasting time shepherding kids who are bored, not engaged and in general a nuisance. If the kids are well prepared and understand robotics fundamentals, for a team larger than 4 kids, there will not be enough robotics works for the kids.
- FLL is not a robotics competition. It is, instead a STEM competition with the robotics portion as an add on. Teams move from one level to another level based on all the non-robotics work such as Project Research and Teamwork. Even with a great robotics score, if your Project Research and Teamwork are lacking, your team will suffer greatly.
- No one wants to do the Research Project. One of the parents should be designated as the project lead and should be in charge of:
    - Having the team pick a meaningful research topic.
    - Organizing field trips and talking to experts (MUST)
    - In charge of sharing the proposed solution (MUST) - common vehicles are YouTube, website, presenting at school, local library etc.
    - Collecting and coalescing all the research and references
- It is highly recommended to create a booklet showing project research along with the project poster.
- For project presentation, many times team wants to go with a skit. An effective skit might help the kids but many times, the participants get too excited about the skit and forget about the fact that the skit is simply a means to present the idea and it is the idea that matters, not the skit. In fact, skits have been extremely overdone, so you might skip it altogether and find other ways of presentation.
- Create a robot design notebook with ideas, things that worked, did not work and take it with you to the Robot Design Judging. This is not a requirement but makes your team look well prepared. Things that go in a robot design notebook are: Pictures of your robot showing key things about your robot, design considerations for your robot, design philosophy of your attachment model, pictures of your attachments, screenshots of your (well commented) programs and a list of all your MyBlocks.
- Even though this is not required, and judges might not even look at it in qualifiers, create a core value poster and ensure that kids memorize the main core values. These are listed out in the coaches' handbook.
    - Gracious Professionalism
    - Coopertition
    - Inclusion

- It is ok to look on YouTube for inspiration although we do not recommend using it as a crutch. FLL offers a great opportunity for kids to apply a very scientific approach to hypothesizing, prototyping, testing and the perfecting an idea into real solution. Simply looking at a solution on YouTube may limit the team's creativity and learning opportunities.

- It would be wise of the participants and coaches to remember that FLL is NOT A ROBOTICS COMPETITION! Many teams never seem to realize this even after going through multiple seasons. FLL rewards Core Values the most and then in the order, it rewards the project research and robot design (Figure 235). All the work the team does on attaining the high score is used only as a validation point. Essentially, if a team stands out in Core Values, Project or Robot Design and the judges have decided to award them a prize or want to move them from one level to another, then in those cases, the Score is used ONLY for validation that the team is in the top 40% of the highest scoring teams. This is not anecdotal, rather, if you search for it carefully, you will find this in official FLL literature.

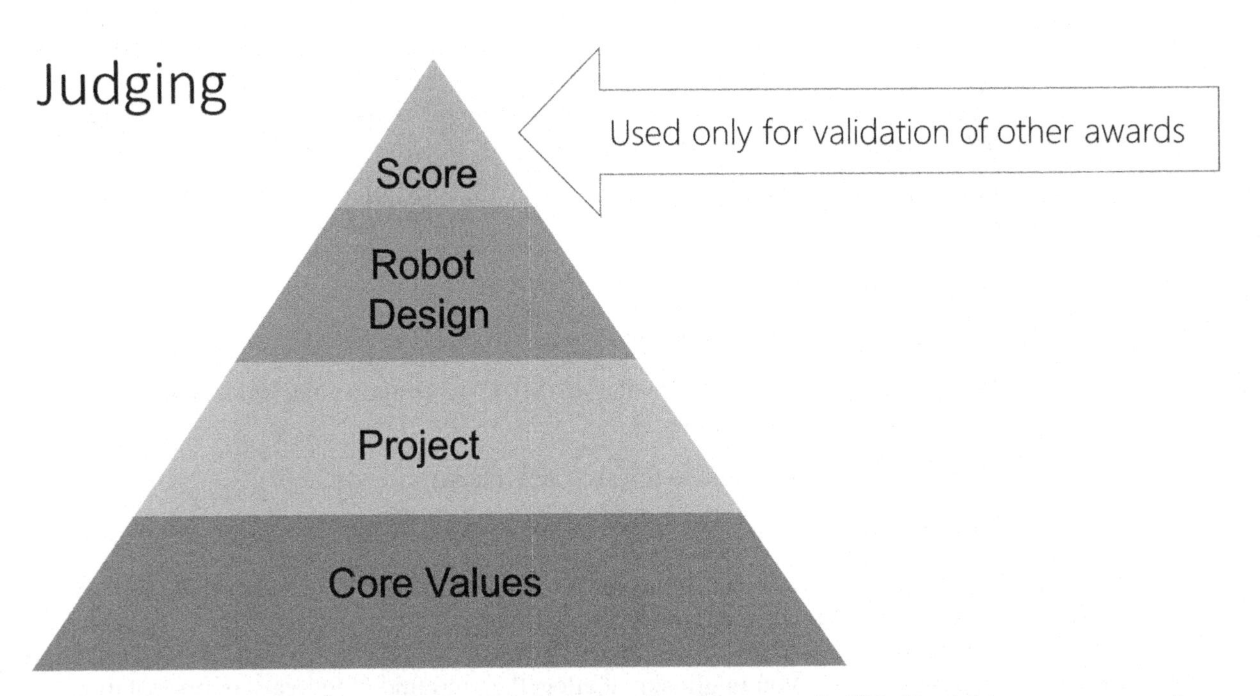

Figure 235: The order of importance of various parts in FLL. It might come as a surprise to you, but the Robot Game Score is the least important portion of FLL.

- Participate in Scrimmages - It is key for you to remember that practice makes perfect. At the very beginning, the participants move a bit slower and do not fully understand how they stand with respect to other teams. The trick to getting kids motivated is to participate in one or more scrimmages. A scrimmage is a gathering of teams where each team runs through the Robot Games exactly like they would in the real competition and find their strengths and state of readiness. If there are no scrimmages around you, find a few other teams and setup a scrimmage at one of the tables. Scrimmages, in our experience have been the single-most important factor leading up to the readiness of the teams. Additionally, for rookie teams, they also quickly align them to how experienced team approach FLL.

- Have the kids practice the final sequence of runs. Once you have finalized the number and sequence of missions for your Robot Game, have the robot drivers execute the runs in a flawless way. In FLL, only two team members may stay at the base and handle the robot. Everyone else must stay at least 2ft away from the table. Thus, if you have multiple drivers for the robot for various missions, then they need to quickly switch in-between missions. To avoid errors and confusion, at home have the participants practice switching between the robot drivers.
- Plan for contingencies - You should have multiple contingency plans for the various runs. In FLL, you are likely going to be attempting a variety of missions and every so often, a run will fail because of user error, robot error or even because the table has some flaws. In these cases, the participants should have multiple plans that they have practiced that they can move to. Without planning and practicing for contingencies, the participants will likely panic and make errors. With contingency plans on the other hand, the team will know that they simply have to enact another plan that they have already practiced for. Here are a few examples:
    - Team planned Missions A, B, C and D. If any mission fails, the team has decided that they will abort that mission, the participants in charge of that mission will stop the robot and move on to the next mission.
    - Team planned Missions A, B, C and D. Mission B is the lion's share of the Robot Game score and on the first attempt, it failed. The team has agreed beforehand that they will attempt mission B 3 times in case of failure before moving to the other missions.
    - Team planned missions A, B, C and D in that order. However, only missions A, B and C fit within the allotted time. If one of A, B or C is failing consistently, the team may decide to run the spare mission D to try and salvage the run.

Appendix B: Things to remember for the competition day

Having gone through many FLL seasons, we have learned a few tips and tricks that make sure that your team has a productive and pleasant FLL competition. In no particular order, we are listing them out. Again, this is not a MUST, but rather things we have found that helped our team.

- *Teach the participants "the Kitty hold" for the robot. It* is very tempting to pick the robot up by holding the brick. Unfortunately, in a majority of robots, the brick is connected to the rest of the robot by simply a few pegs. If the pegs come loose or if the brick slips from someone's hand, you will have your robot shattered in pieces. This has happened in front of us multiple times. To avoid this, we teach participants the *Kitty Hold* i.e. you hold the robot like you would hold a cat with one hand below the robot and the other on top of the robot. This ensures you have a good grip on the robot and will not let it go.

- *Port view is your friend when inexplicable things are happening.* Travelling with a robot is always a bit precarious task. They have tons of parts and many electrical cables that come loose as they travel to the competition site. When weird things start happening that just simply do not make sense, use the port view to make sure that the EV3 brick is detecting all the motors and sensors and that they are showing correct readings. Some of the inexplicable things that you may note are:
  - Robot seems to be going in circles - likely the cable in one of the driving motors has come loose.
  - Robot seems to be going in circles when following a line - the color sensor cable has come loose, or the color sensor needs calibration.
  - Robot seems to have a bias i.e. it is drifting more than usual - likely your chassis has become a bit loose while moving the robot. Tighten the pegs all over the robot.

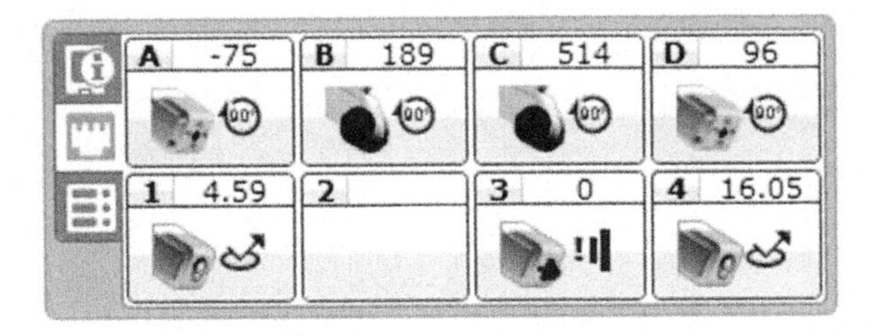

Figure 236: The Port View is your friend when inexplicable things happen.

- *Clear programs and load just the programs you need.* Before you head over to the competition, make sure to delete all the programs you accumulated on the brick during the season and reload only the ones you need (Figure 237.) This will avoid confusion and save time.

- *Carry the robot charger and an extension cable with a power strip.* In many locations in FLL, the organizers will supply each team a table, a few chairs and a socket for electrical power. Given that there are multiple participants, coaches, mentors etc, there is a fight for that socket. Not only that, it may so happen that for some reason, your table did not get an electrical socket. Thus, carrying an extension cable and a power strip will solve this problem and might even help other teams.
- *Carry a spare fully charged battery.* This applies especially to the Robots that have been used over years. After long usage the capacity of battery to hold the charge diminishes and becomes uncertain. The Motors and sensors are power hungry. Having a spare fully charged battery is very reassuring.
- *Carry spare Lego parts.* This goes without saying, but we have had this happen where team needed to tweak the robot just a tiny bit because of variances in the table at the competition. Not having spares will put the team at a disadvantage.
- *Document the Robot and attachment design.* For any crucial robot parts or attachments, use LDD (Lego Digital Designer) or take multiple pictures to record its structure. Participants HAVE broken or lost attachments, right up to the competition. One time, this happened after the qualifiers and before the semi-finals. It is no fun trying to remember and rebuild an attachment from vague memory. The kid who loses it is immediately treated with some hostility with other team members which will affect your core values.
- *Carry backups of your programs.* Carry a laptop and a USB stick with all the final versions of the programs. Robots have malfunctioned and completely wiped all the programs on a robot, including JUST AS THE TEAM GOT TO THE ROBOT GAME TABLE!! The judges were nice enough that time to give the team 2 minutes to run and reload the programs, but you might not be so lucky. Be warned!!

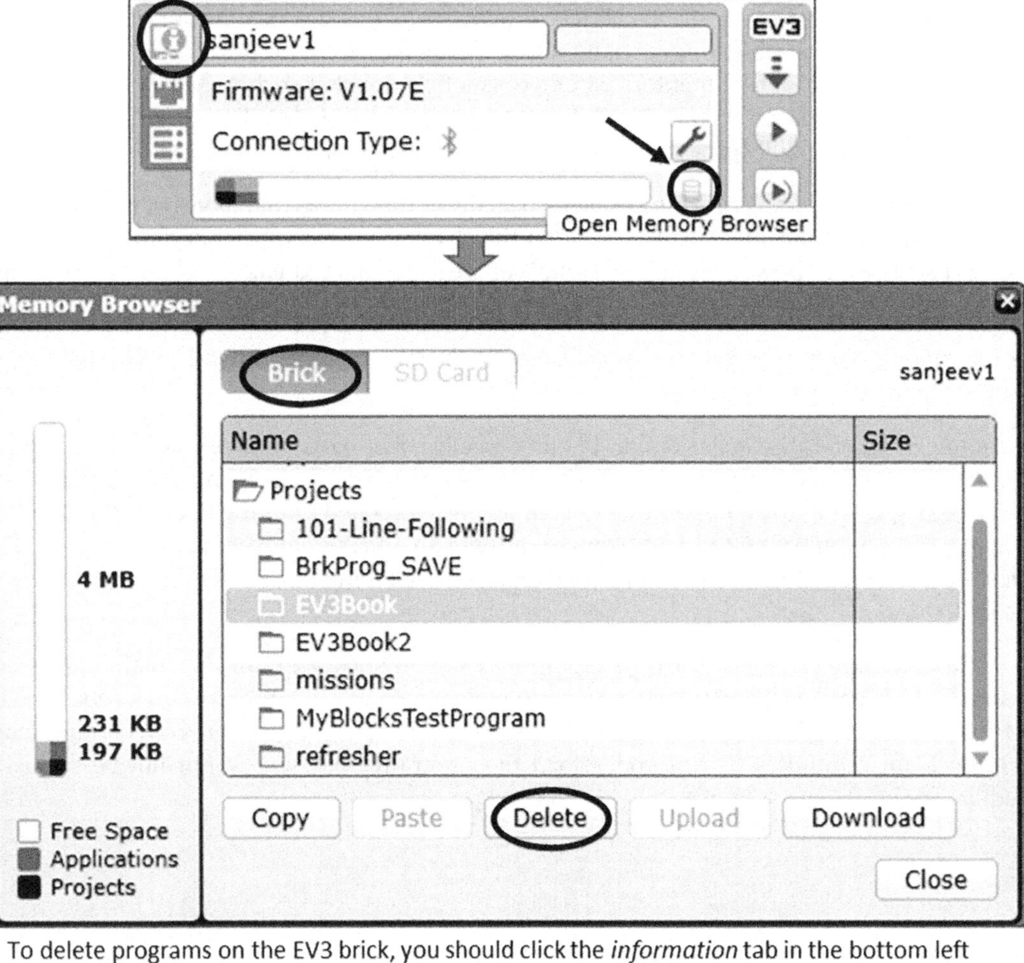

To delete programs on the EV3 brick, you should click the *information* tab in the bottom left corner and click the *Open Memory Browser* button. From there, you will get a list of the programs on the brick that you can delete. You cannot delete the BrkProg_Save program which comes built in into the EV3 brick.

Figure 237: Clear all programs on the brick using the memory browser.

- ***COACHES and PARENTS*** *DO NOT TOUCH THE ROBOT OR TEAM COMPUTER.* Absolutely no touching of team's stuff by coaches or parents, including the team computer at the venue as this can only hurt you. The higher the level that you are competing at, the higher the chance you will hurt your team's chances. We have been in the judging room with teams being denied prizes in spite of being spectacular because judges think that the adults did the work. An adult touching the Robot, robot parts or the computer including just holding boxes for kids is an absolute taboo that no one talks about. At the competition the team MUST demonstrate total independence. You have been forewarned!!

- *Make sure that your team behaves nicely at the venue.* There are volunteers and judges and judge helpers roaming around and they pass notes back to the judging room. A single bad note for your team can mean a total disaster for your team. On the other hand, a positive note may boost your team and increase the chances of winning. Make sure the kids stay at their table and explain the project to anyone who comes around. It is common to see kids engrossed in their phones and ignoring folks who are asking questions or visiting their booths which certainly ends up hurting their chances.

- *Make sure all kids participate in every event.* Make sure that during each of the judgings, EVERY KID gets to speak at least a few times and that the kids are smiling and behaving like they are enjoying being at the competition. A kid that is aloof or does not speak out will likely dent an otherwise stellar performance.

- *Adults should not enter the judging room.* As coaches, mentors, parents, you did everything you could do to help the team BEFORE you got to the venue. At the venue, being in the room where the judging is happening can ONLY HURT your team as you are not allowed to speak, hint or in any other way assist the team. Your being there means a kid might inadvertently look back at you for validation or kids might get stressed because of your presence. Taking pictures, videos etc. can distract kids and will likely work against your team.

- *Bring food and snacks from home.* If you don't know where the competition is being held, bring food from home. Sometimes, the competition is held in a place where no nearby eateries will be available. The competition takes place from 9-5pm with teams starting to check in from 8am. It is a full workday for adults and an exhausting journey for the team. It is simply a great idea to bring extra food and snacks from home for the team and yourself to keep energy levels up.

- *Bring extra warm clothing if it is winter season.* The competition is usually held in schools during December and the heating might be low/down. In state of Washington, the finals are held in a stadium where the floor sits directly on top of a "frozen" skating rink!! Participants that are cold or are in the process of falling sick will dampen your chances of success.

- *Prepare for the next mission while the robot is running.* Train the participants so that as soon as the robot is off the base for a mission, the participants are getting ready for the next mission i.e. preparing the attachment, removing unnecessary pieces from the gameboard, aligning the starting jig etc. This will save precious time that may allow you to accommodate failed runs or yet another mission.

Appendix C: The Essential kit for a successful FLL season

Here we list out the parts that all teams should procure before season start so the kids are not limited by parts. Note that this list is a minimal requirement for kids to be able to experiment freely and you should get more pieces. The EV3 kits i.e. the Core Kit and the Expansion kit do not contain sufficient parts to be able to build many mechanisms or robots. Also note that the recommended number of parts here is AFTER you have finished building the robot. Finally, to purchase extra Lego parts, use **http://bricklink.com** or **http://brickowl.com**. Although Lego now sells bricks directly from their site, sometimes they don't have parts in stock or the parts are more expensive than on brick link or brick owl. Additionally, if you don't know the number for a part (also known as Lego Design ID)to look at, you can use **http://brickset.com** to find a Lego set and then visually look at all the parts to identify the one that you need. Once you have the Lego Design ID, you can use it a BrickLink or BrickOwl to order the parts. In the rest of this appendix, we are listing out the requisite parts you will need along with part numbers and in some cases, illustrations. Here is the essential list of parts (excluding the ones required to build the robot)

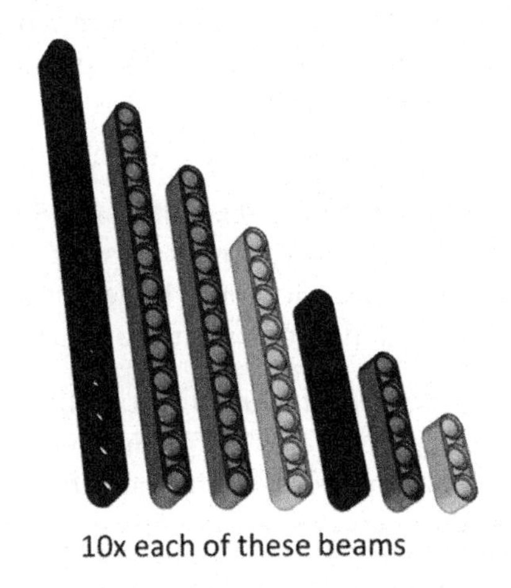

Figure 238: Assorted beams

- 20 beams of each kind lengths 3-15 (Figure 238.) Note that Lego beams are available in only odd lengths.
- 20 axles of each kind from length 2-8 and 10 axles of length 8-10(Figure 239.) Note that Lego axles are available in all lengths. Axles longer than 10 are hard to find and expensive.

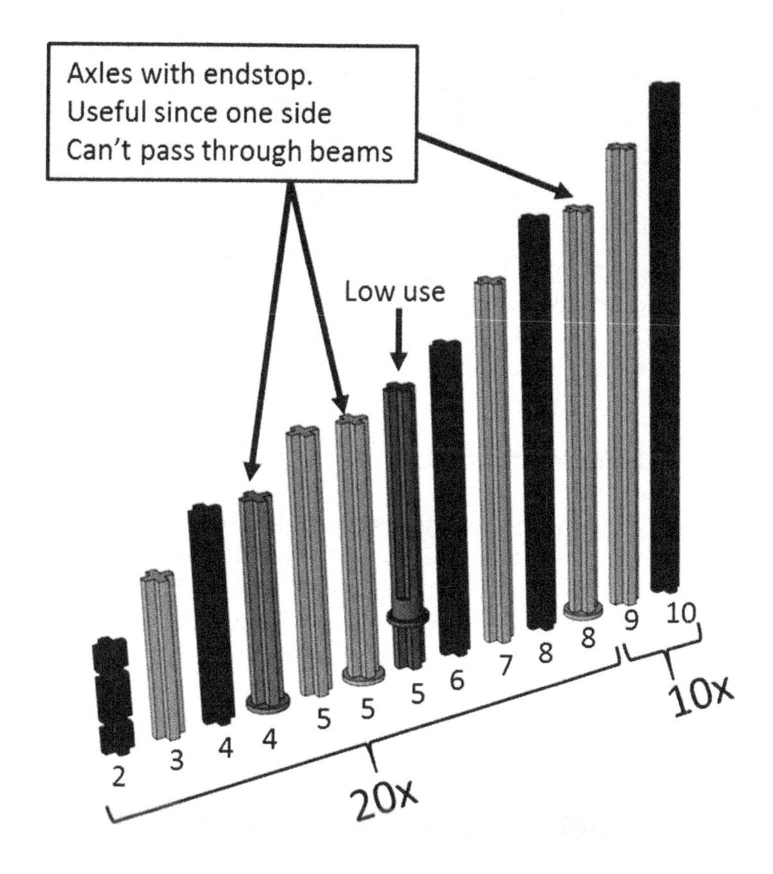

Figure 239: Assorted axles

- 20x L-beams - 3x5 and 2x4 kind (#32526 and #32140), 10x 90-degree snap connectors (#55615), 10x H connectors (#48989) (Figure 240.)

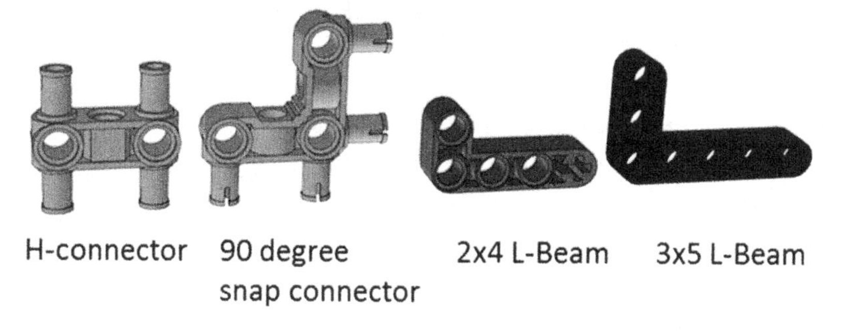

Figure 240: Structural connecting parts

- 10x h-frames (#64178) and 10x rectangular frames (#64179) - these are used as structural elements and as frames for dog-clutch/gear arrangements (Figure 241.)

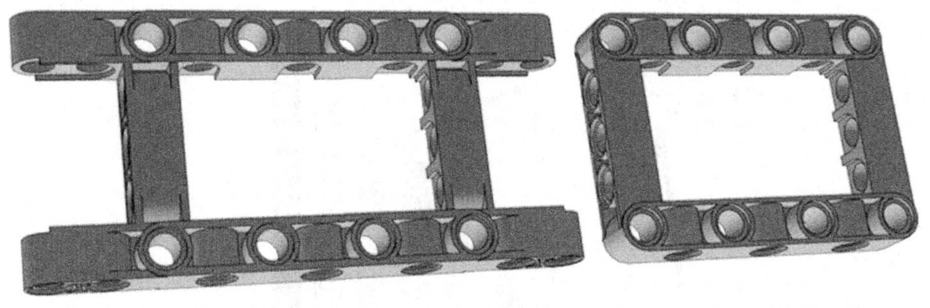

Figure 241: Frames are used to create dog gear mechanism and attachments

- 20x - Axle connectors, straight and 90 degrees (#32014, #59443, #32013, #32039) - the axle connectors allow you to connect two axles to make them longer, change direction and create rectangular frames, 5x - Axle Connectors angled - (#32016, #32015). Angled axle connectors are somewhat infrequently used but at times they are invaluable. One of our teams used the angled axle connectors to fulfill 4 missions using a single motor in a single robot run in 2016 Animal Allies season (Figure 242.)

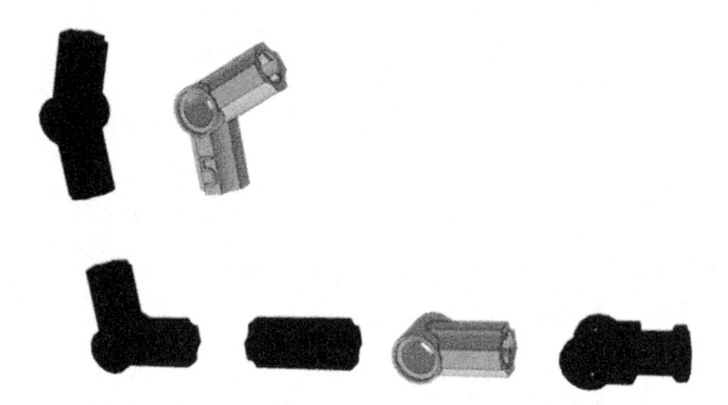

Various axle connectors that connect two axles at 90, 112 and 135 degrees, extend the length of axles and allow you to connect components with axle and pegs.

Figure 242: Various axle connectors

- 10x beam and axle connectors (#32184), 10x 2L-beam with axle hole (#74695), 2L-block with axle and peg hole (#6536), 5x - Worm Gear (#32905) and 5x - 24 teeth gear (#24505) and 10x pulleys (#4185) and 5x Y-connectors (#57585) (Figure 243) These are used to create dog gears and often change the direction that elements need to be attached.

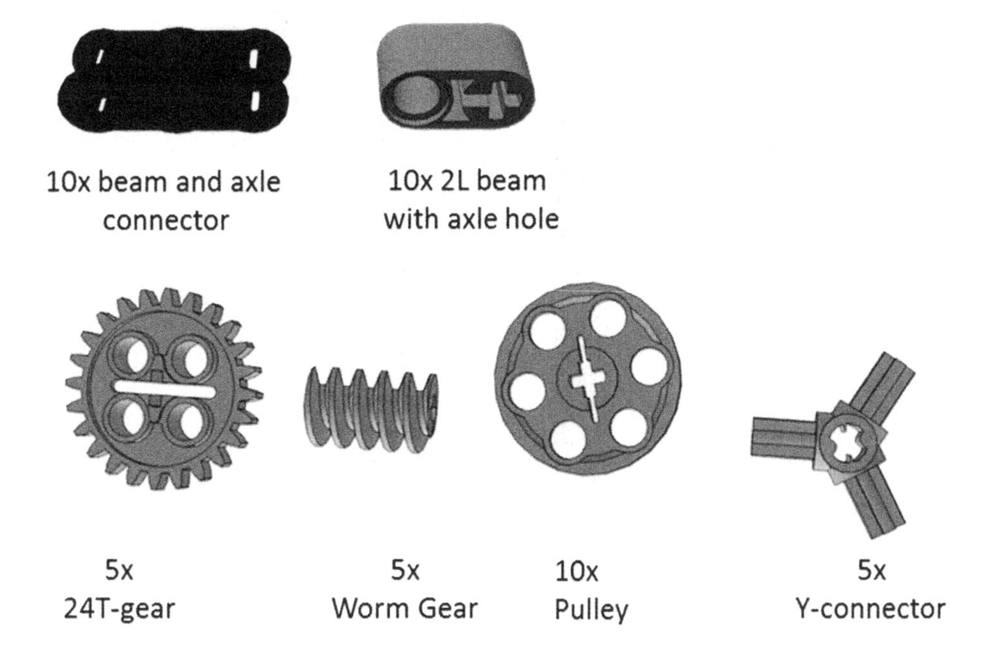

Figure 243: Parts for dog gear and wall following

- 300-500 black friction pegs (#2780), 50x blue pegs with axle (#43093), 100x 3L blue pegs (#6558), 10x tan frictionless pegs with axle (#6562), gray frictionless pegs (#3673), 20x half bushes (#32123), 20x full bushes (#6590) - these are used to constrain axles and as spacers on axles (Figure 244.)

Figure 244: Bushes and pegs

- 6x - Angled beams (#32348, #6629, #32271) and 10x J-Beams (#32009) - these are often used in grippers, as hooks to pick objects and as structural elements in self-aligning features on the robot (Figure 245.)

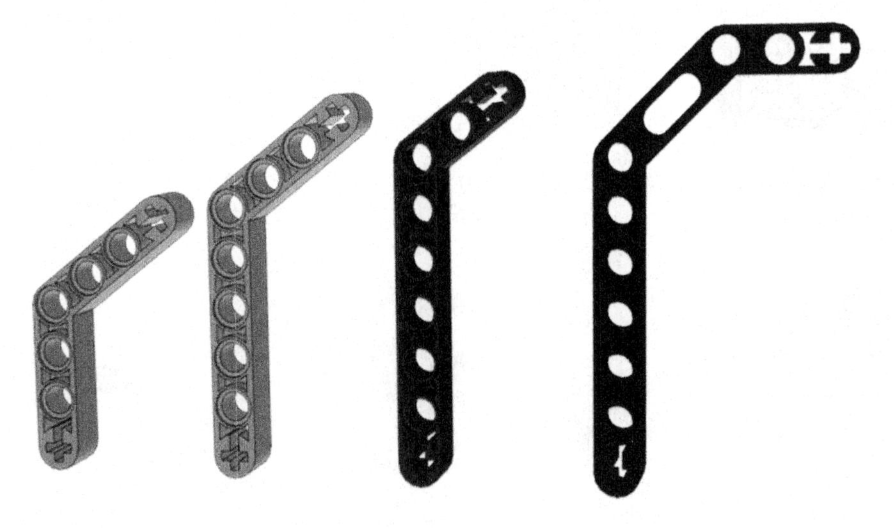

Figure 245: Angled beams

- 2nd color sensor and medium motor apart from the one that comes in the box.

Appendix D: A FLL Robot

Here we are providing a design for a robot that conforms to the guiding principles listed out in this book. The robot has only three points of support, thus providing consistency to the chassis in every run. The robot has a strong rectangular frame around it which provides rigidity and structural integrity to the robot. Additionally, the frame supports the wheels from outside. This makes the axles driving the wheels supported on both sides i.e. supported using the motors from the inside and supported using the frame from the outside. This ensures that the axles supporting the wheels don't bend and thus provide a consistent movement through the season. This is important since unlikely though this may seem, we have seen plenty of robots with axles that bent over time causing inconsistency as the season progressed. This robot uses the ZNAP wheels (Lego Design ID# 32247) which perform a lot better than the stock wheel in the kit in terms of Moving straight and turning. The robot has low height and thus has low centre of gravity. This ensures that even with a bit uneven load, the robot will not tip easily.

This robot is equipped with two medium motors balanced on the two sides of the robot as well as the dog clutch (also known as dog-gear) mechanism in front for attachments. The dog-clutch (gear) attachment also allows you to connect and disconnect various attachments very easily and quickly. This robot uses two color sensors that have adequate shielding from the robot chassis and it also contains an Ultrasonic sensor in the front, mounted upside down and low to the ground for recognizing obstacles. Although we have not mounted touch sensors or a gyro sensor on the robot, the rectangular frame provides plenty of places on the robot to attach these sensors.

Finally, our teams have used this robot design to great success in FLL. Many times, participants and coaches hit a bit of a mental block thinking of a robot design, so we have provided this battle tested robot design as an inspiration to design your own robots.

For the following robot instructions, the ones on white background are sequentially numbered. The instructions on yellow background indicate assemblies that need to be built separately and then used in the next page with a white background.

1

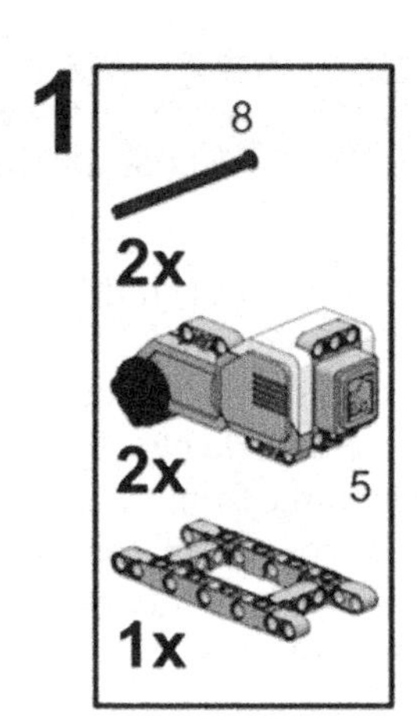

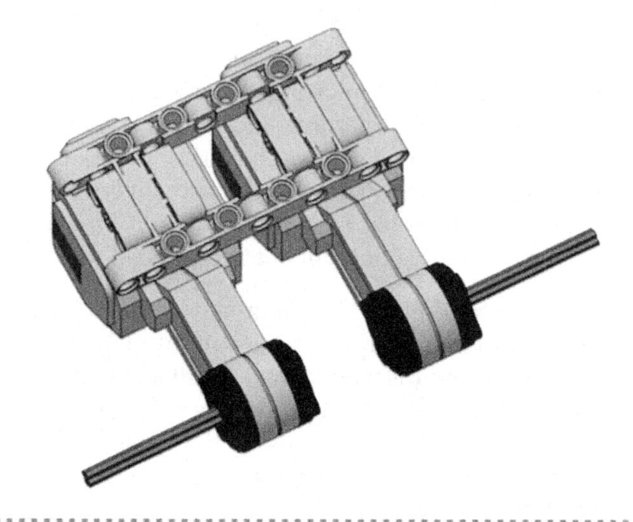

2

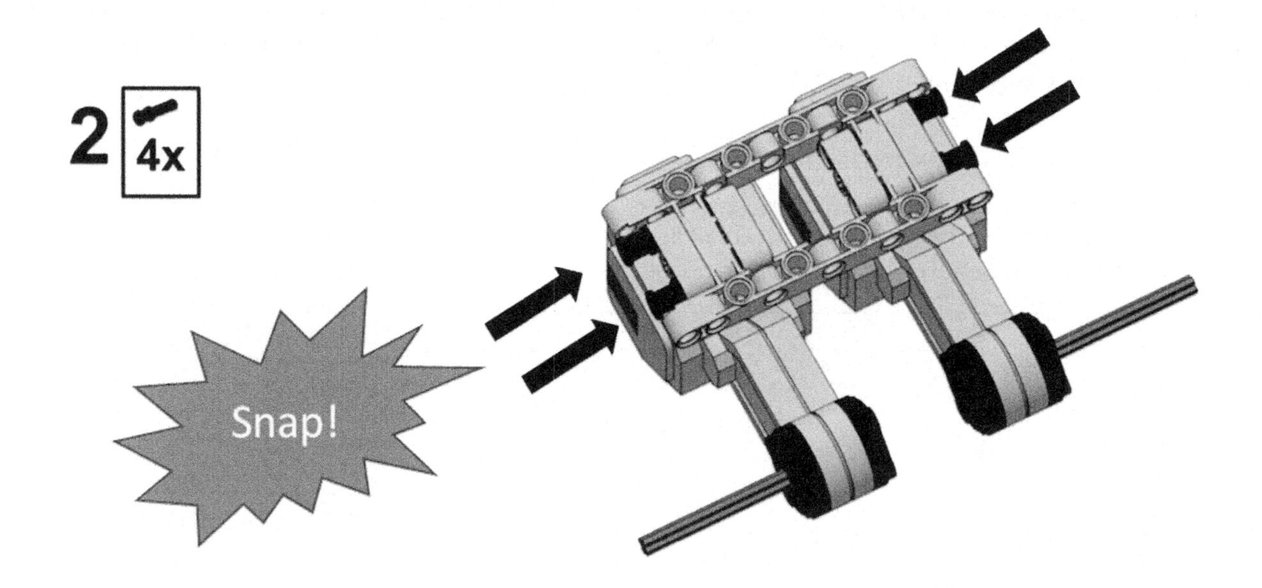

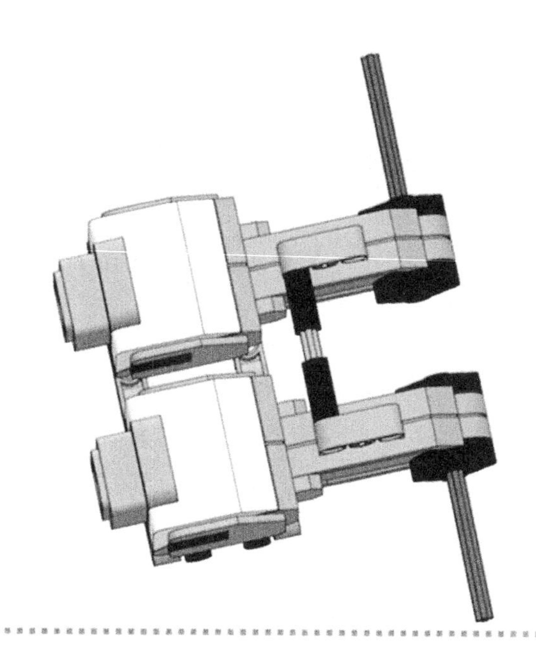

4

2x

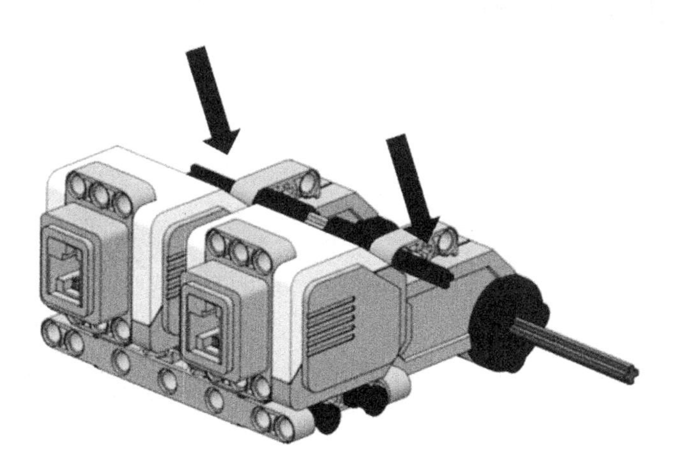

5
2x

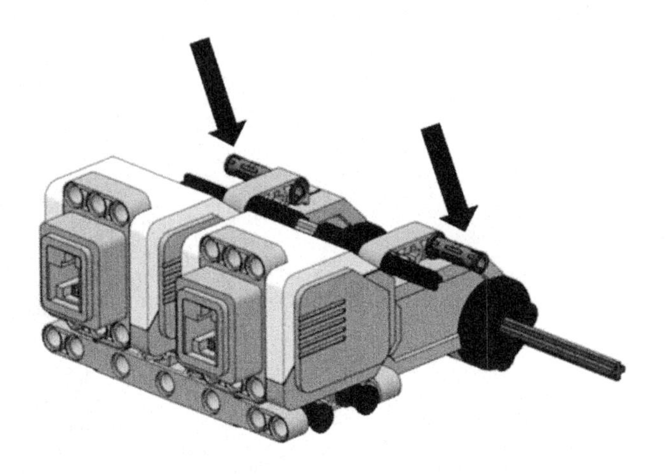

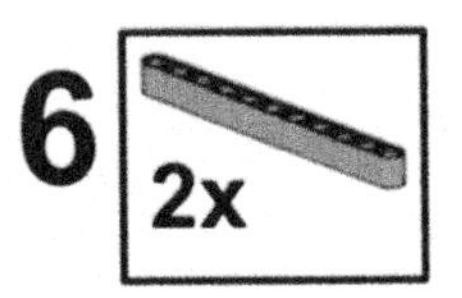
6
2x

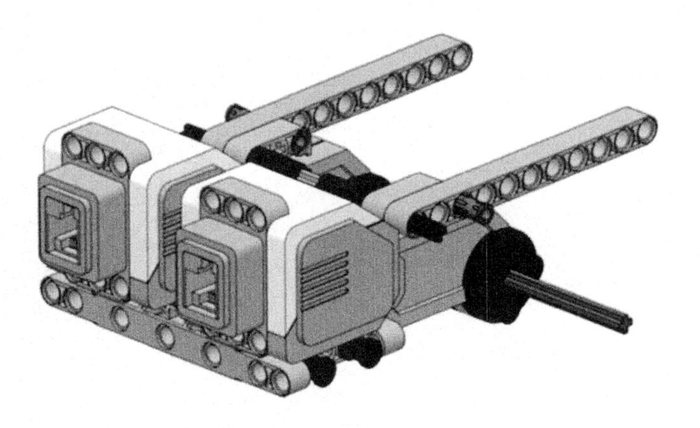

7

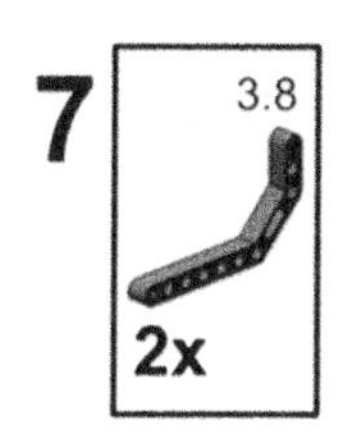

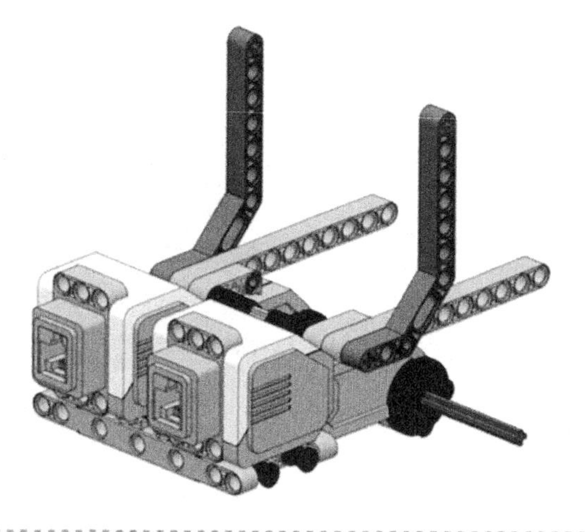

8

Only push the pegs partially inside the yellow beams in this step.

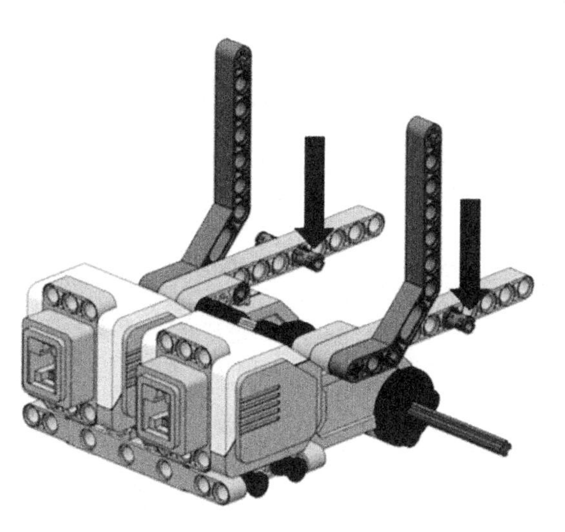

9

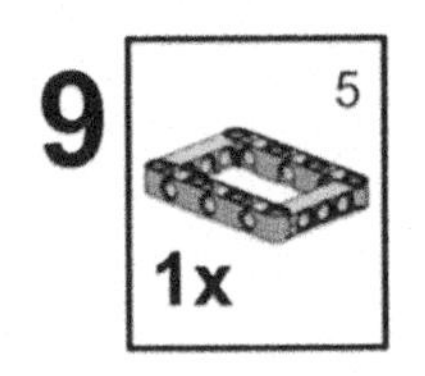

Place the frame and then push the blue pegs in, all the way.

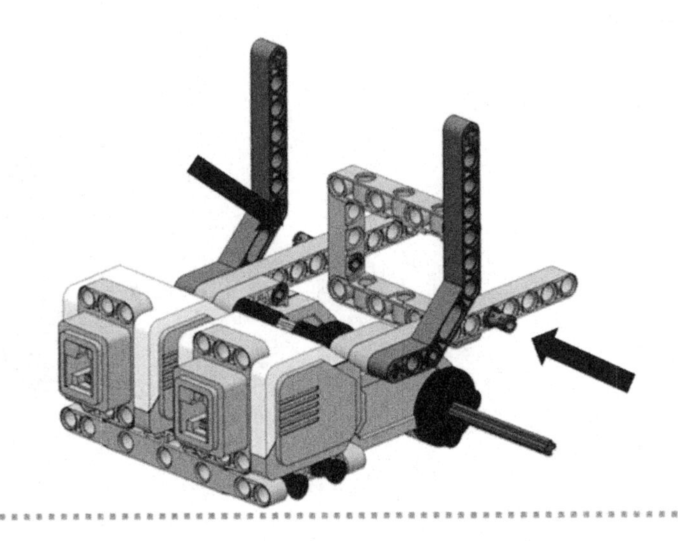

10

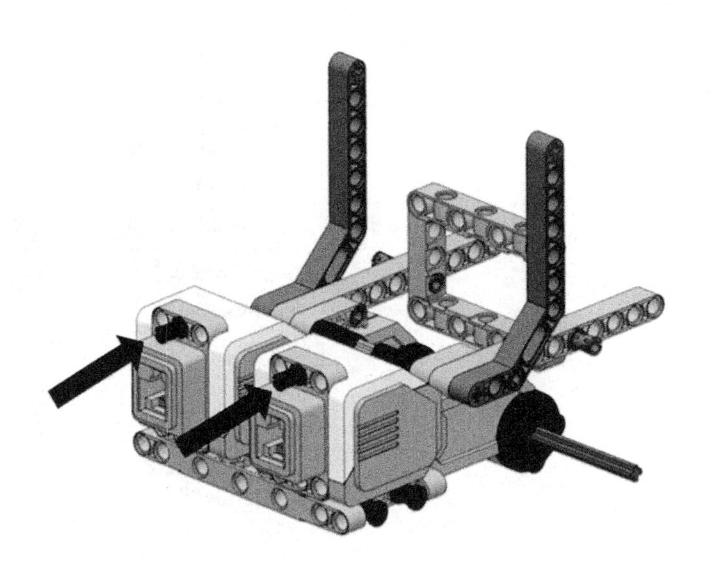

11 2x2 2x

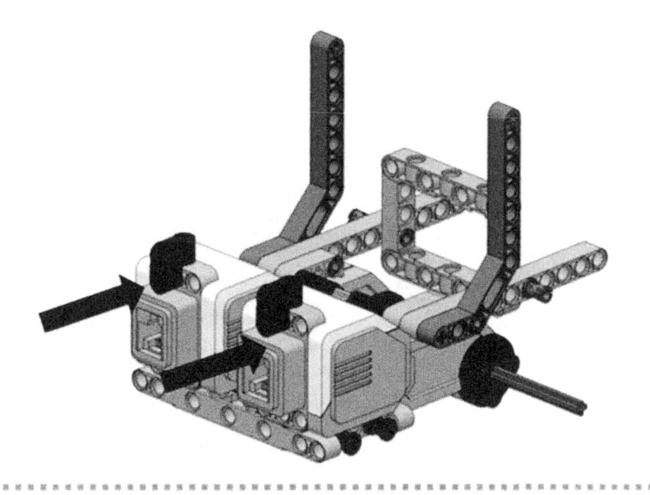

12 4x

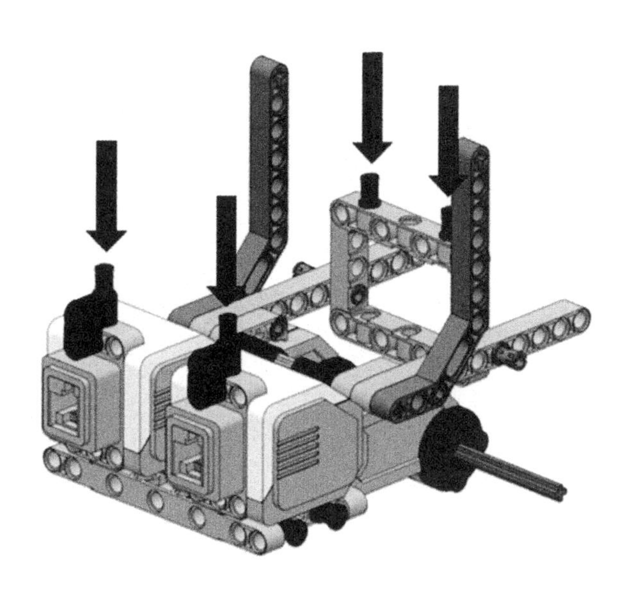

13 2x

Symmetrical placement on both sides

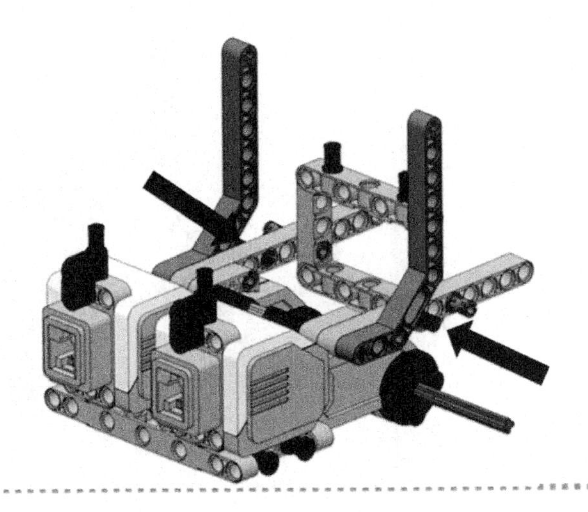

14 5 2x

Symmetrical placement on both sides

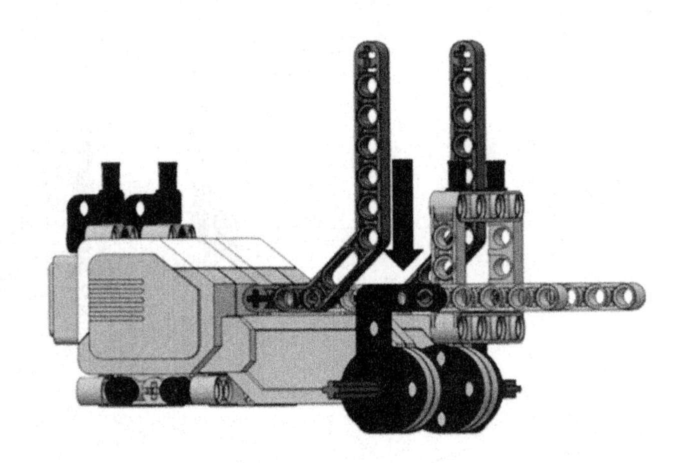

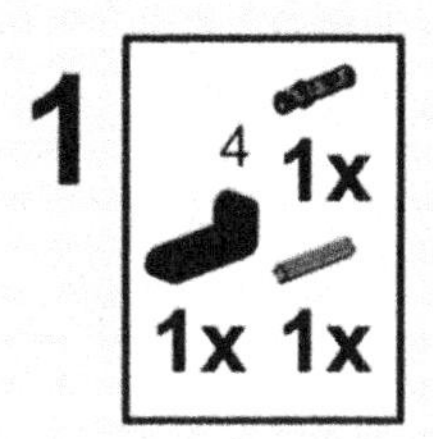

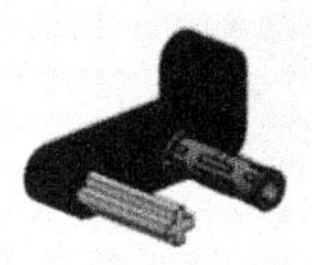

2

1x

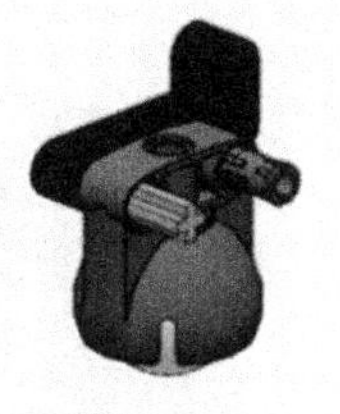

3

4

2x

5 

Symmetrical placement on both sides

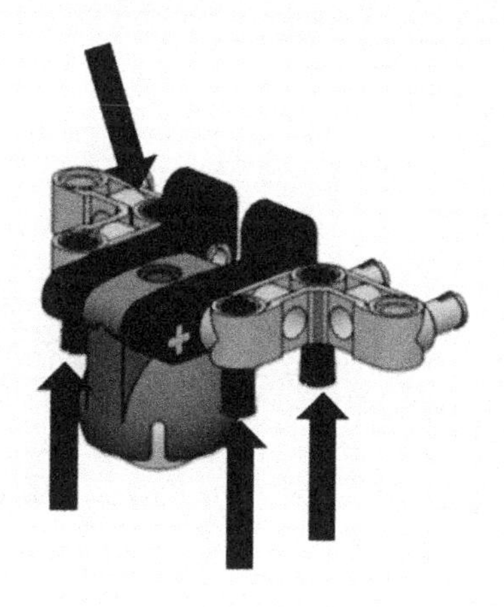

6 

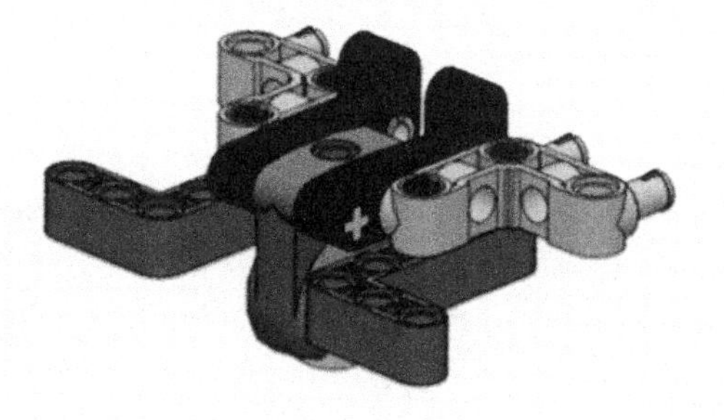

7

4x

8

1x

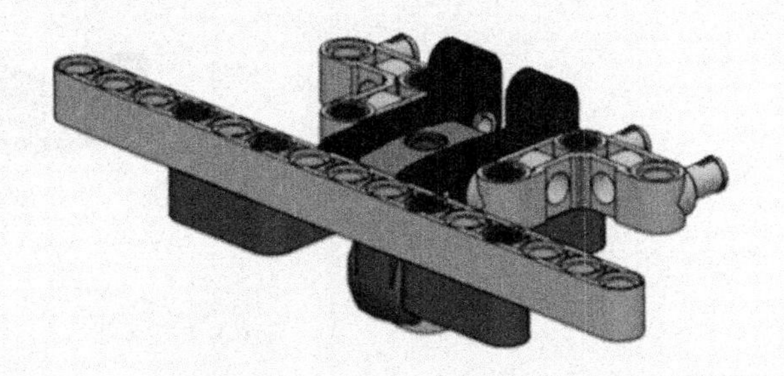

15

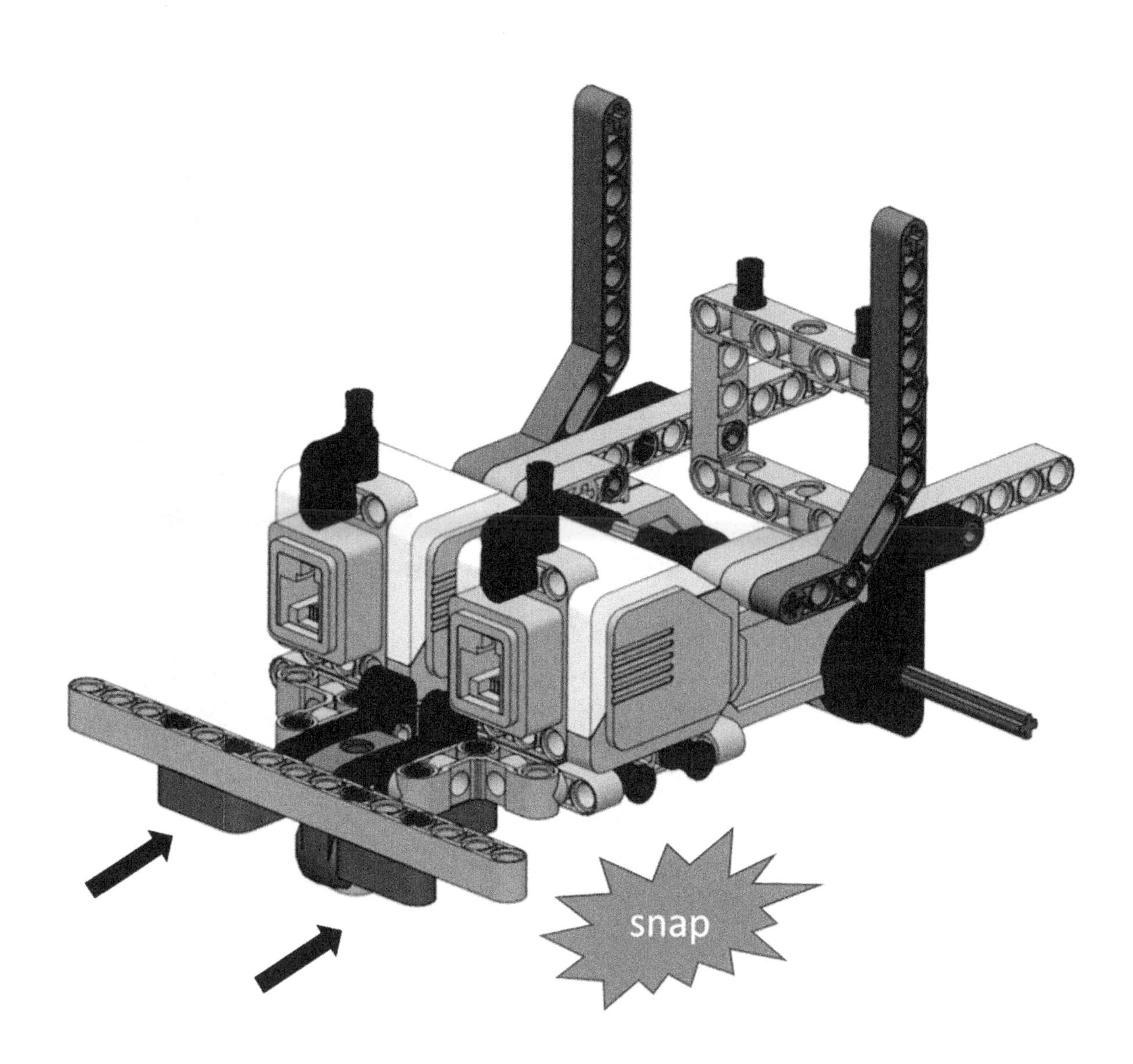

Seat the brick so it connects with the 4 black pegs on the chassis.

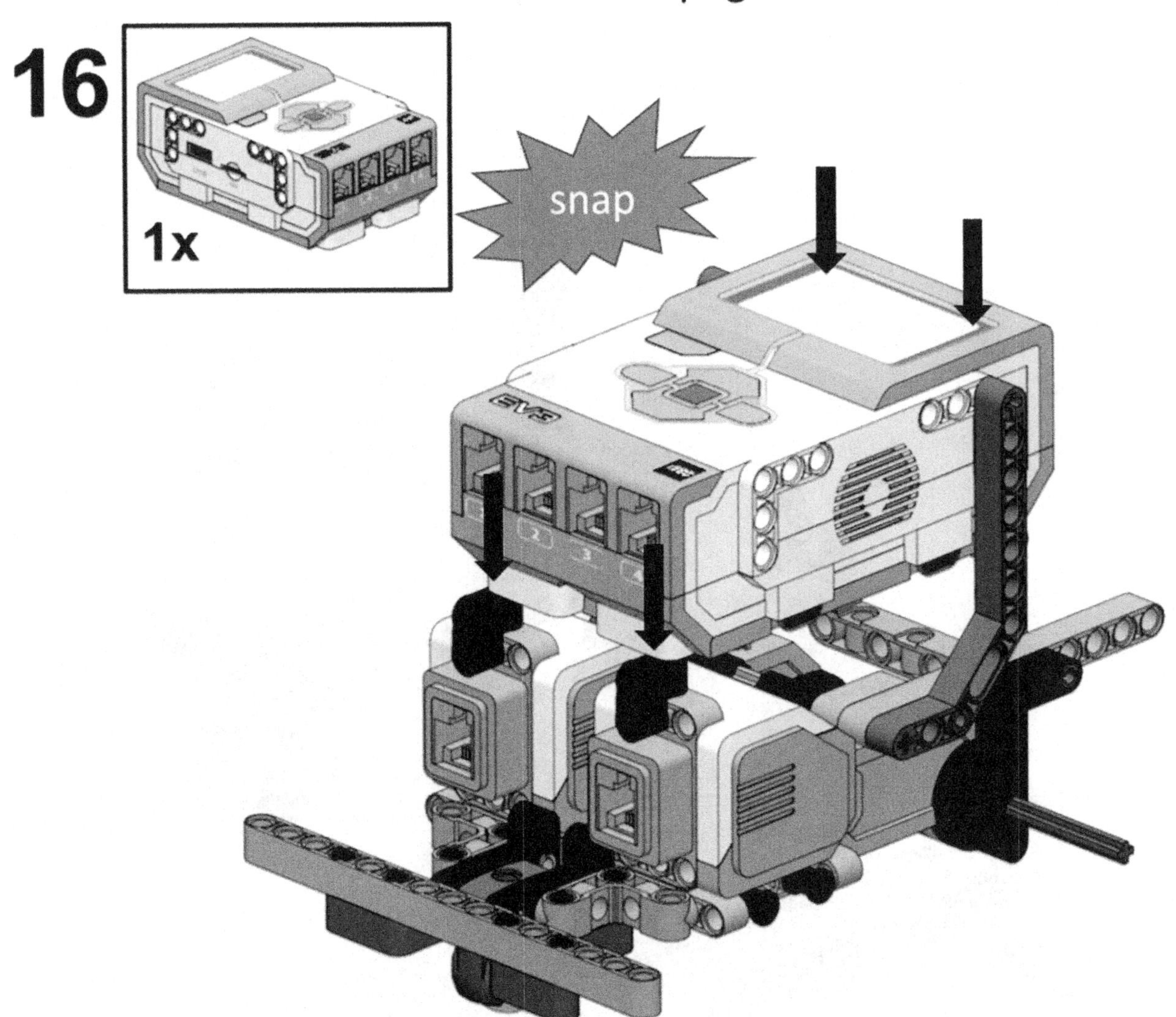

17 4x

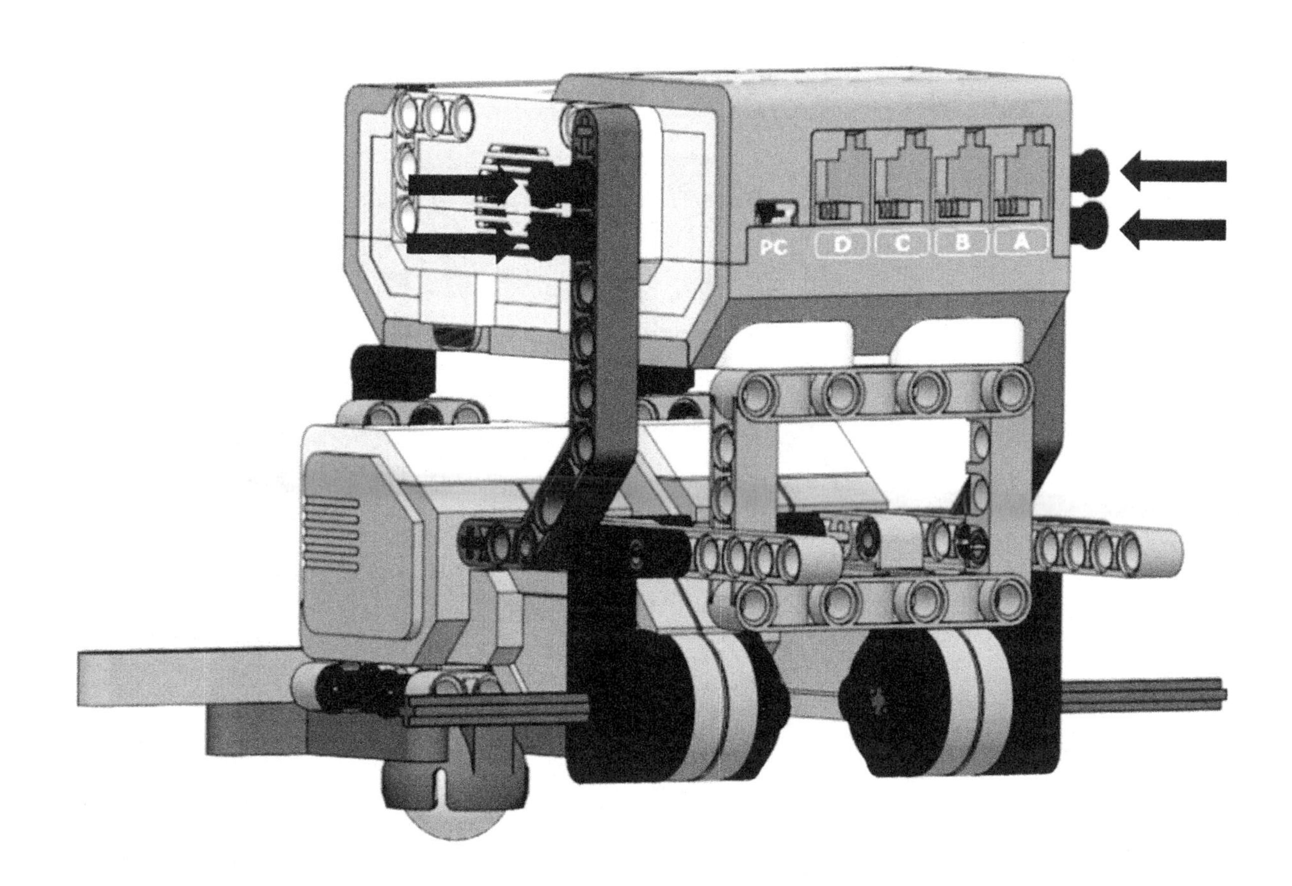

18

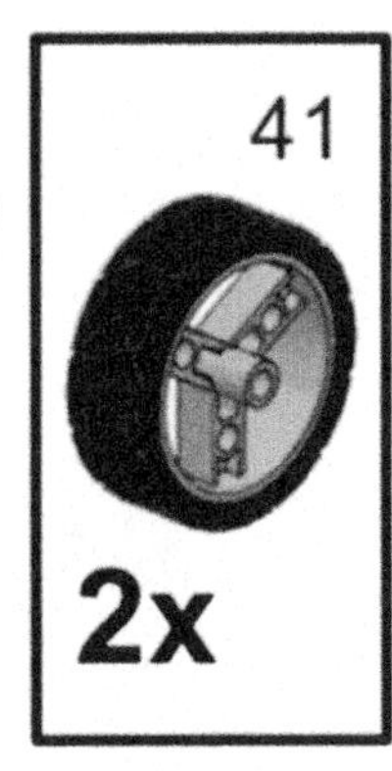

The wheels shown here are Znap wheels (Lego Design ID# 32247) We have found them to be very reliable wheels that go fairly straight.

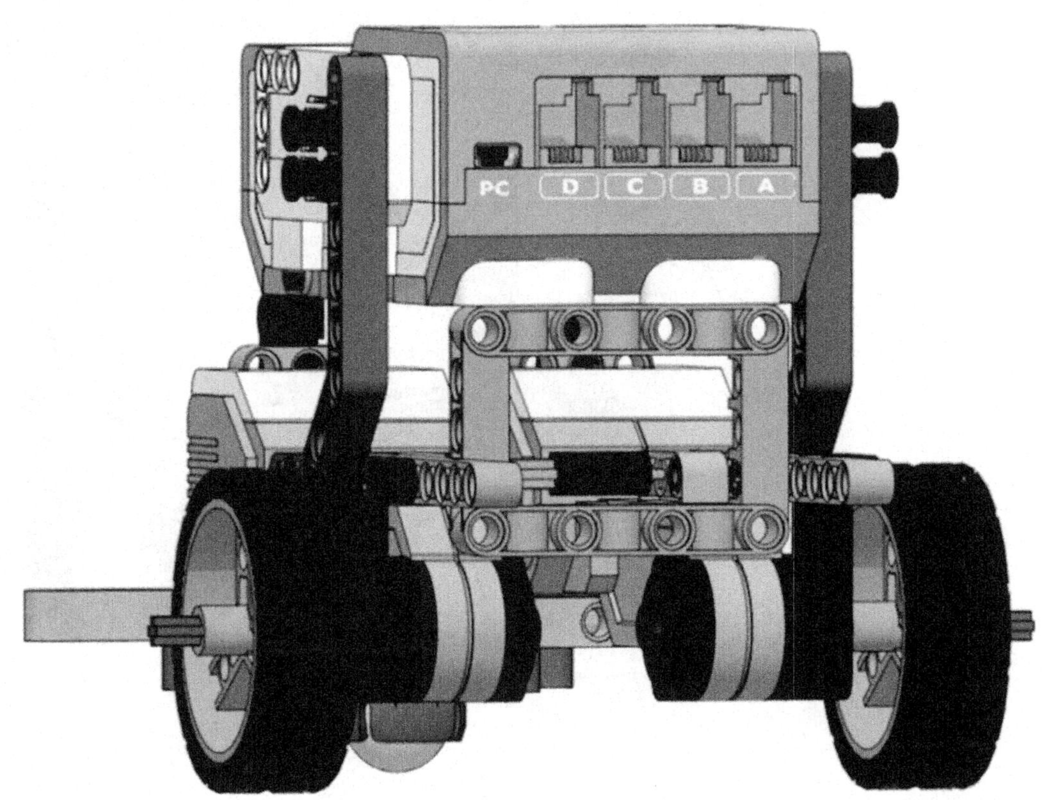

19 4x

Symmetrically attach blue pegs on both sides.

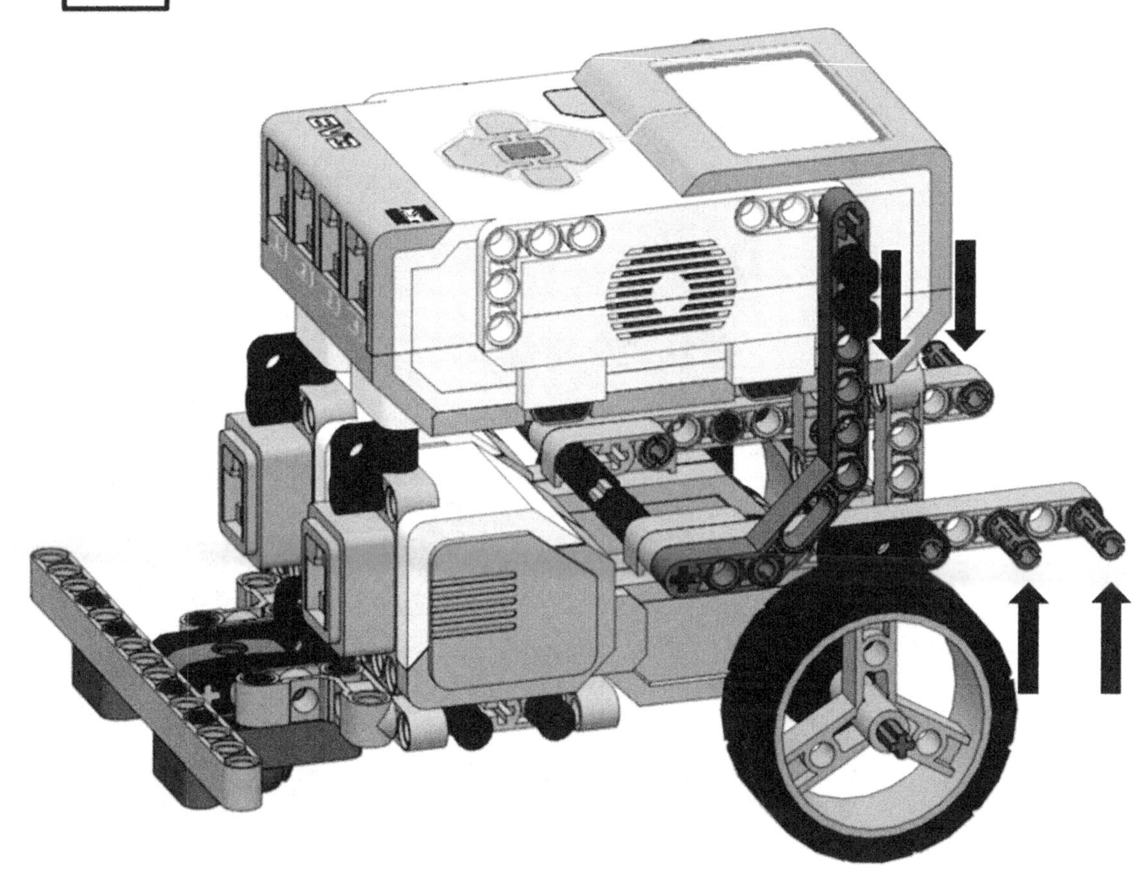

20

Symmetrically attach 2x5 L-beams on both sides.

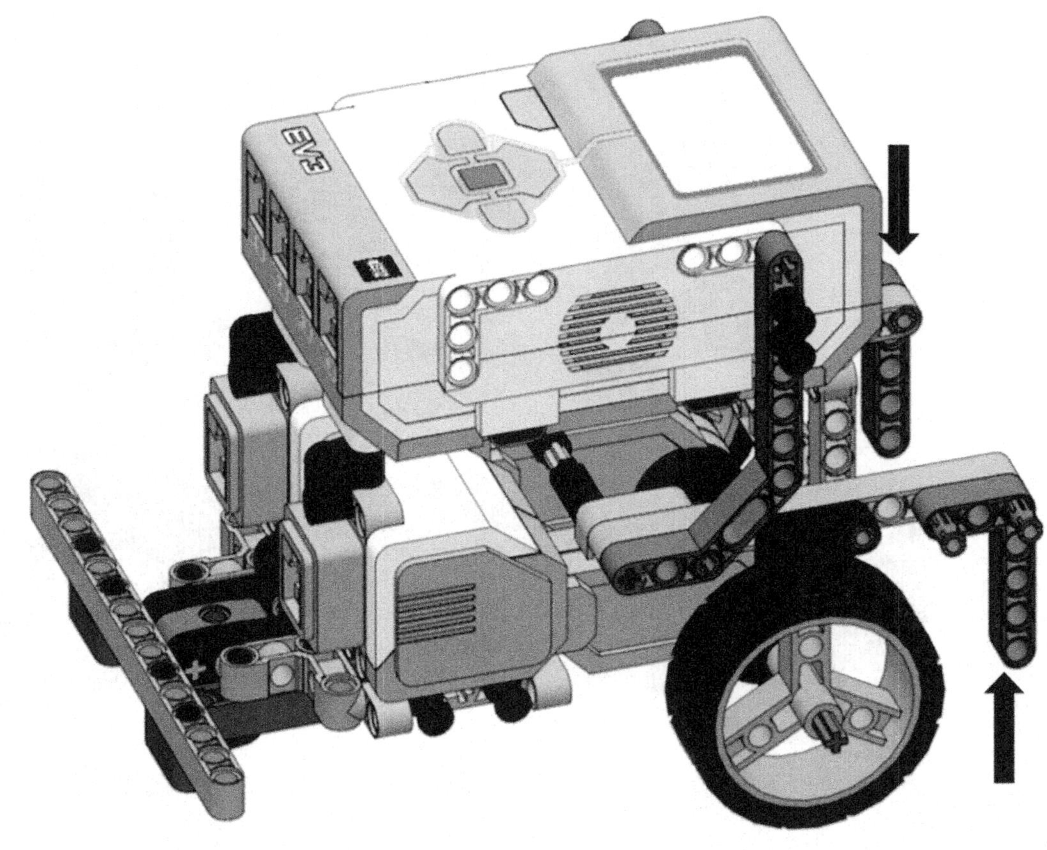

21

4x

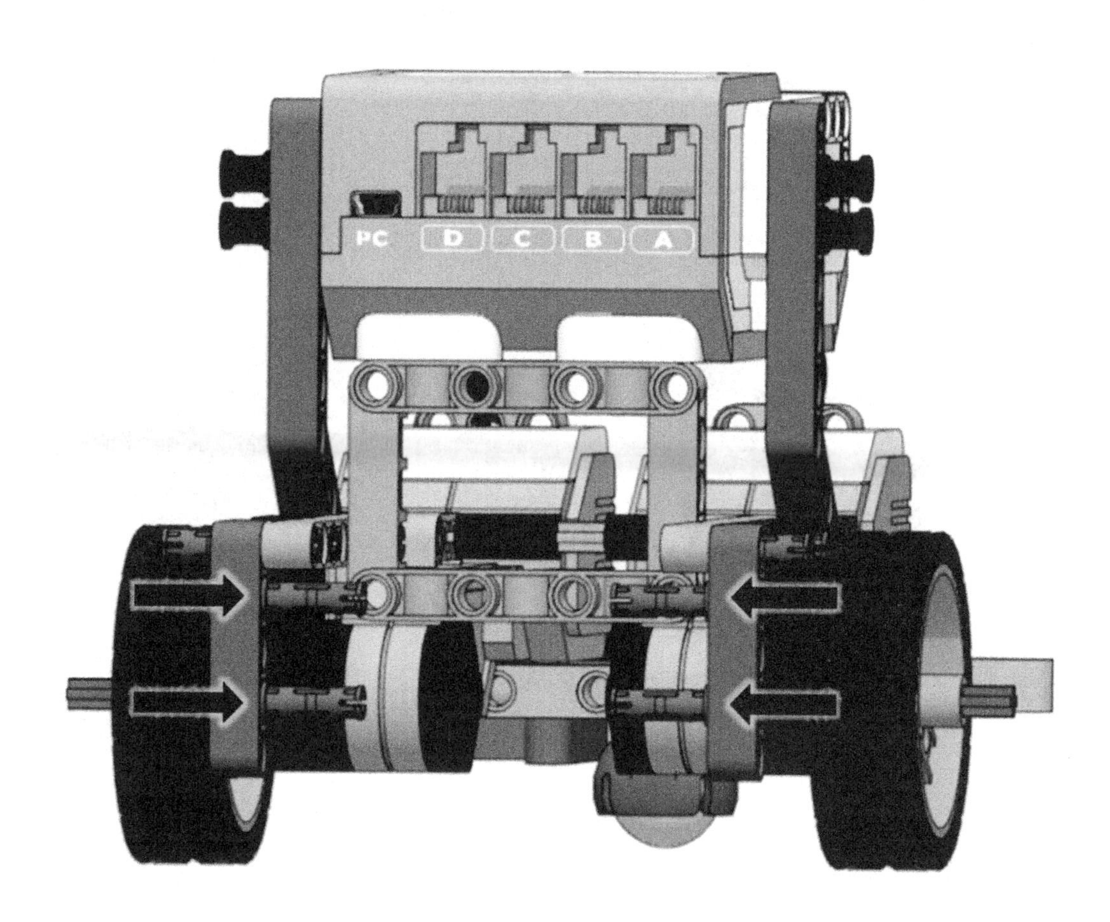

22 2x

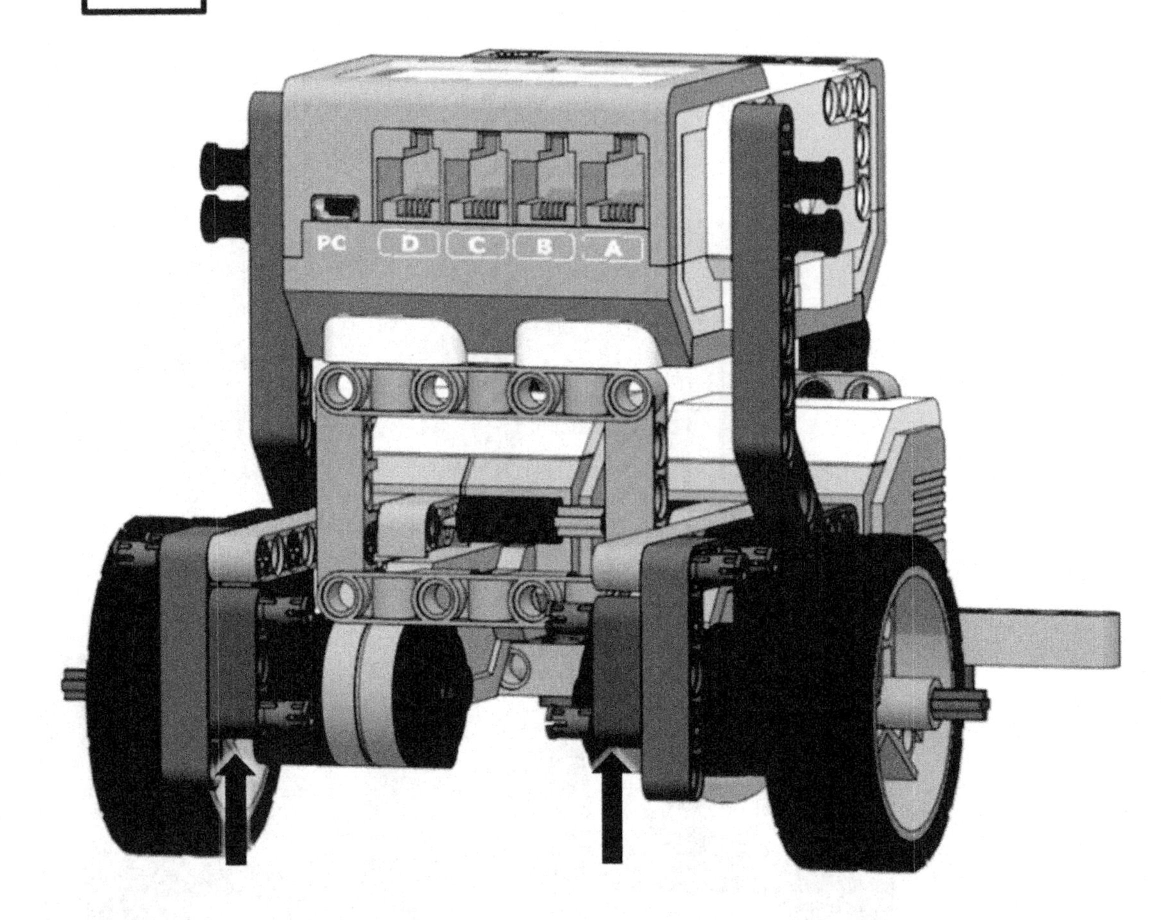

23

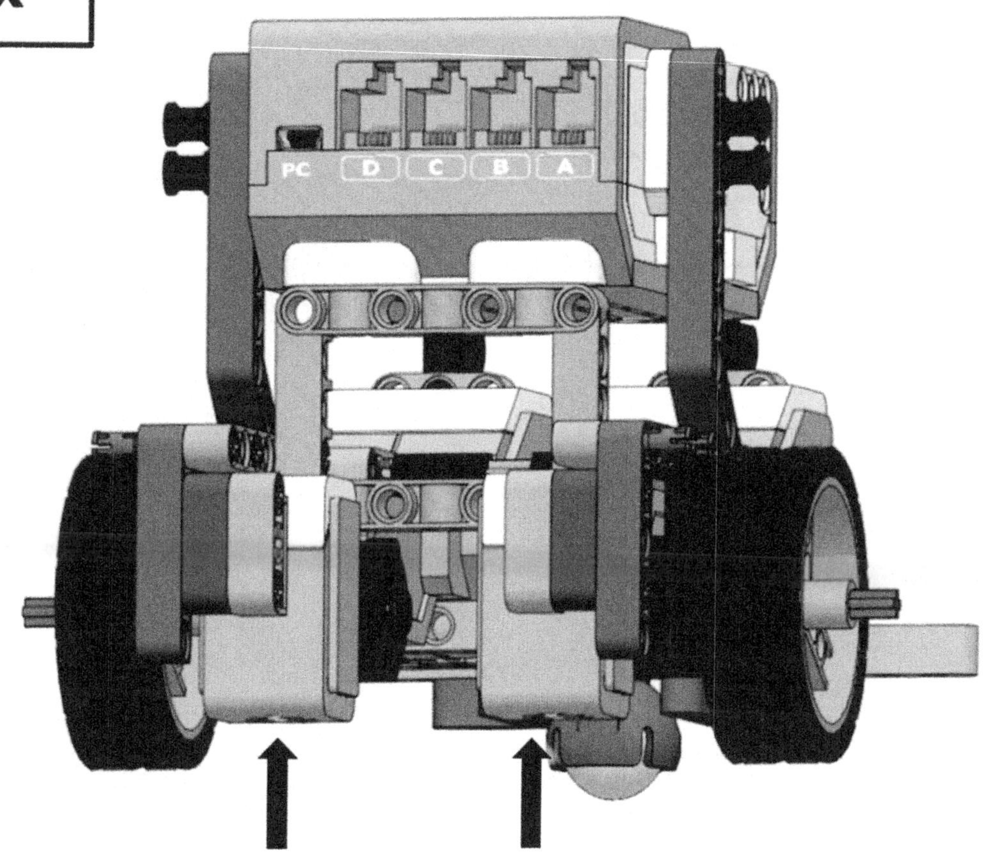

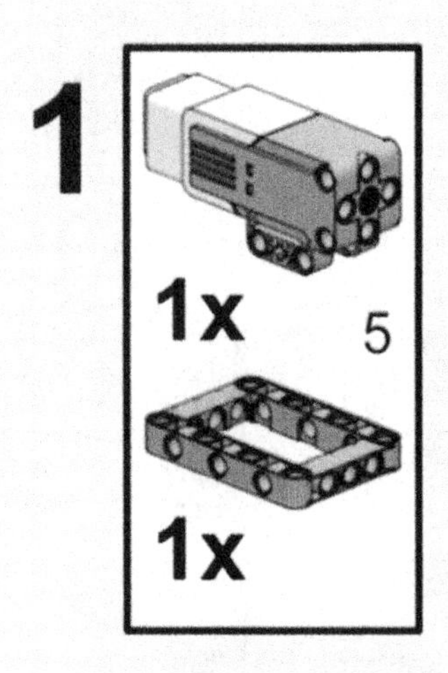

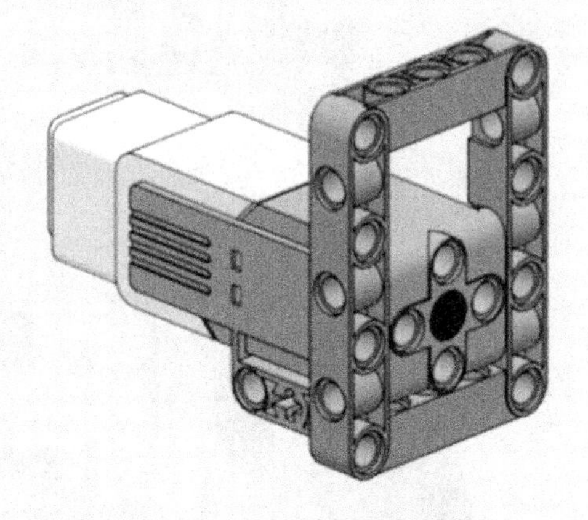

2

4x

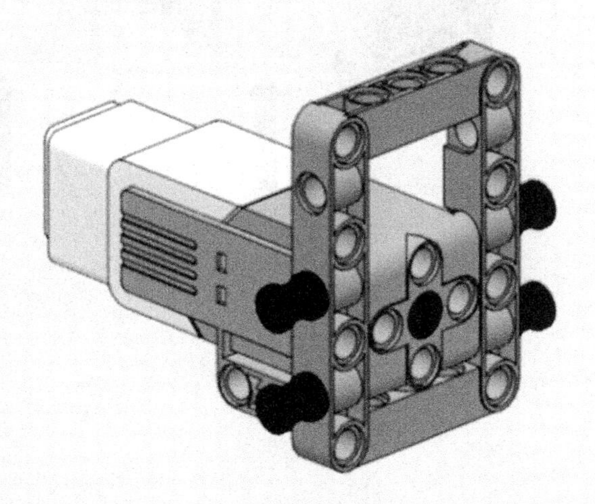

3

2x 4x

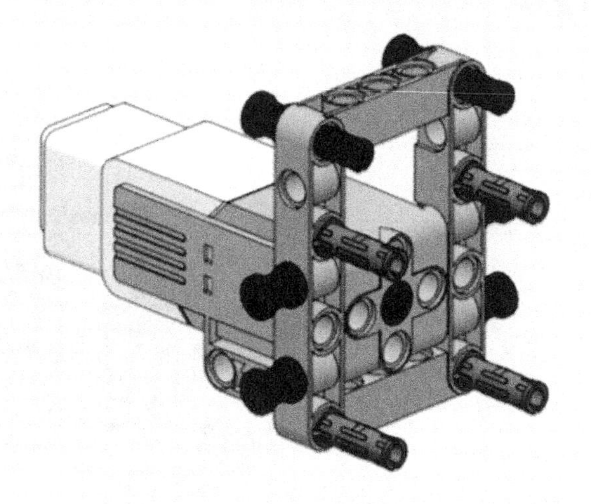

4

2x

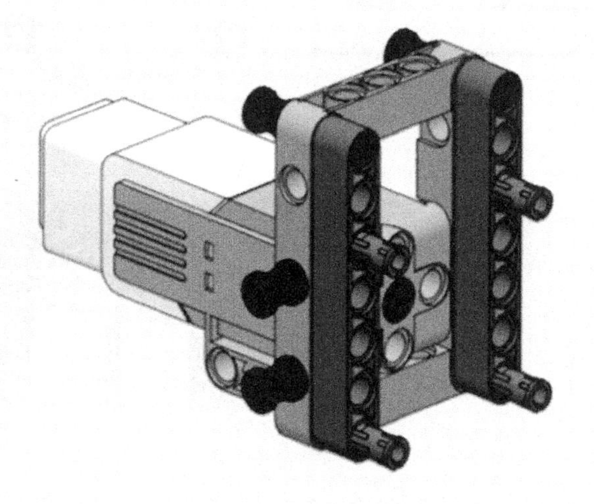

5
2x

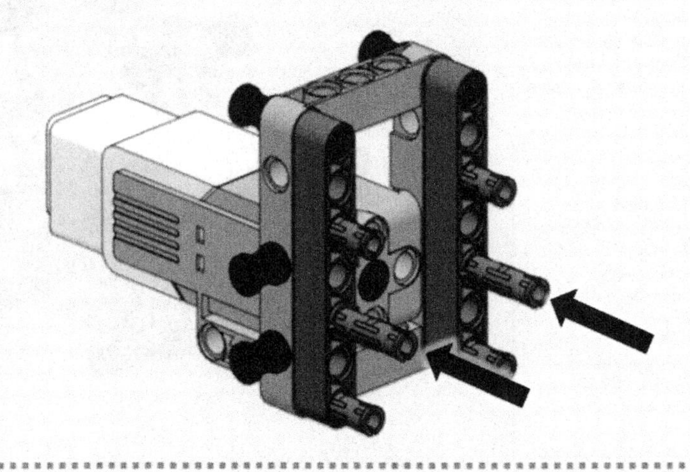

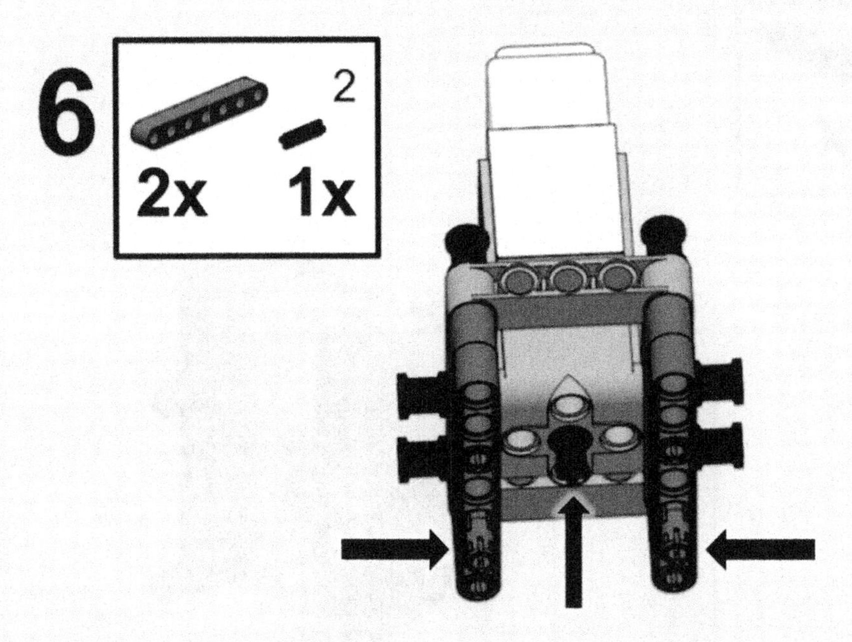
6
2
2x
1x

7

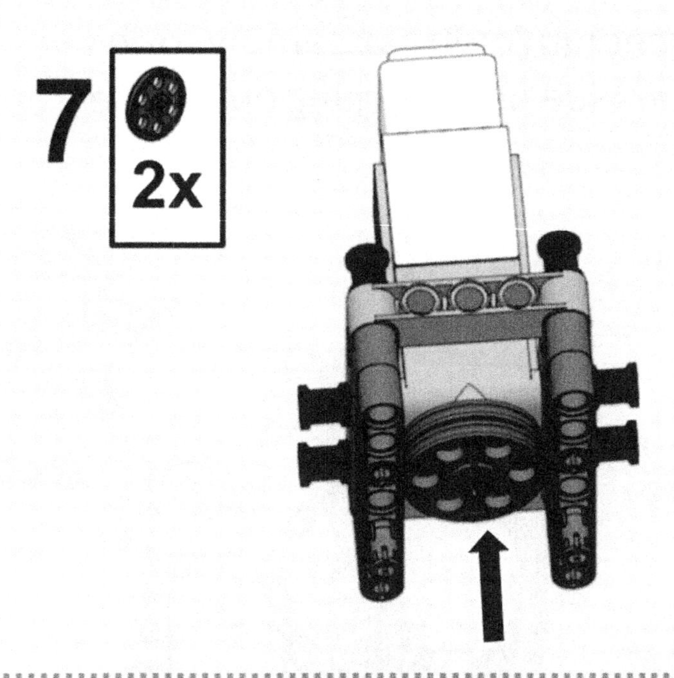

8

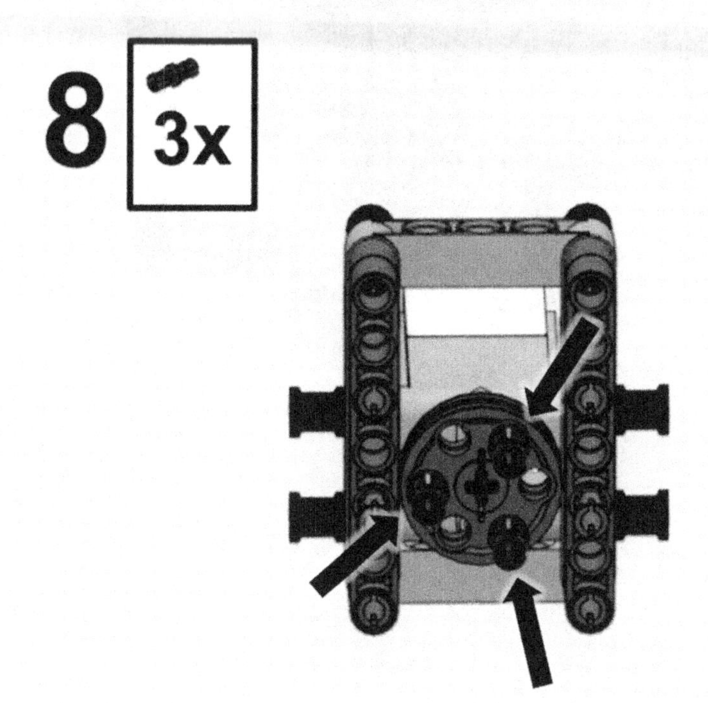

24

Connect the medium motor assembly to the robot using the blue pegs sticking out of the robot.

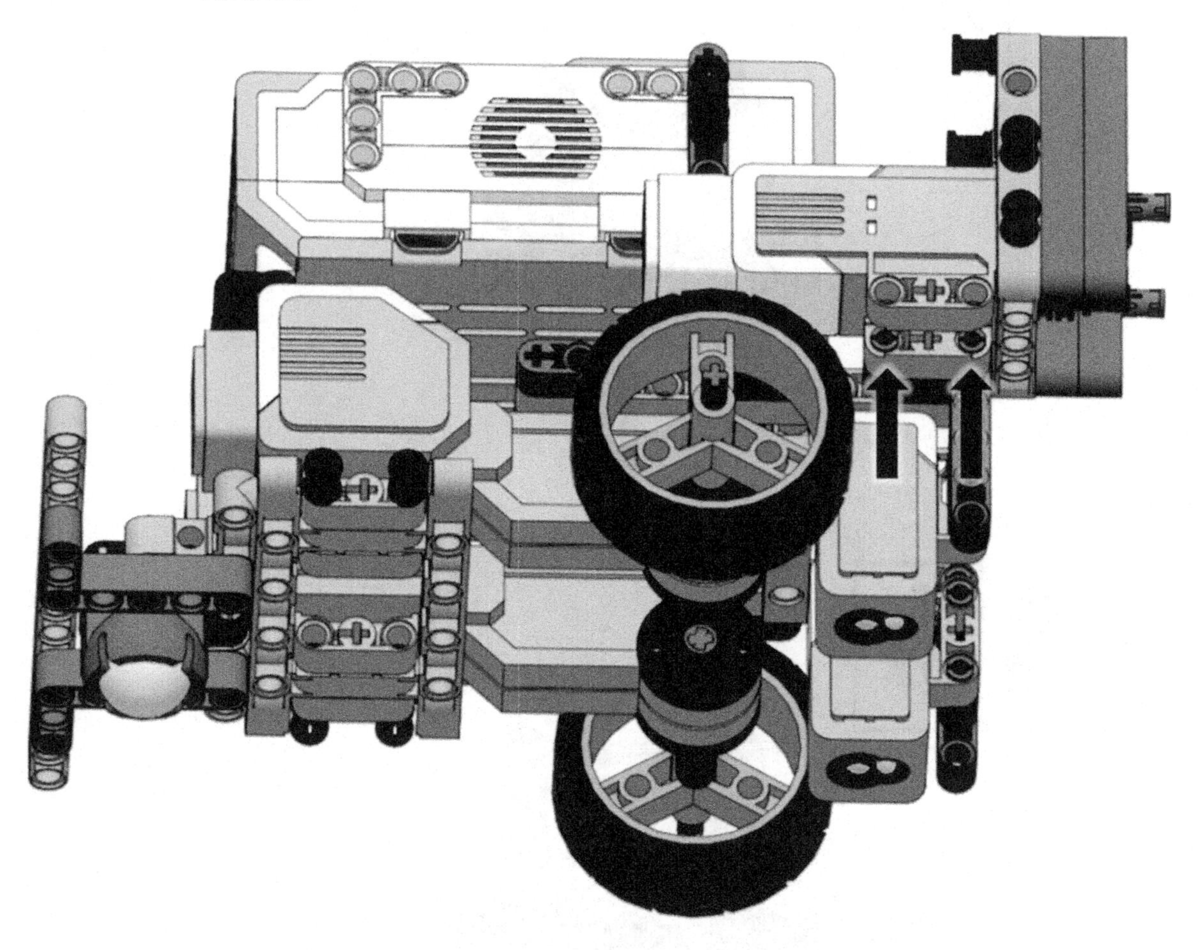

25

Connect the medium motor assembly to the robot using the blue pegs sticking out of the robot.

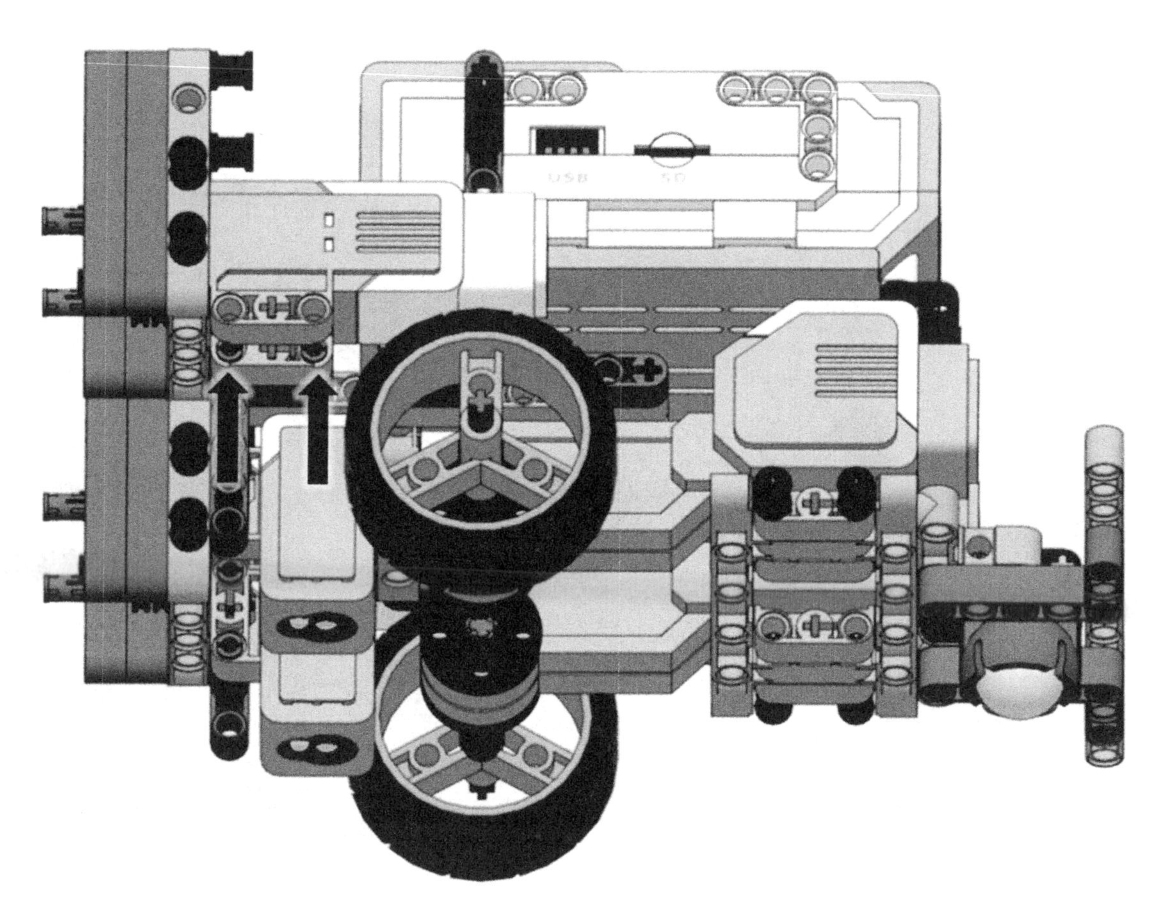

26 4x

Symmetrically add black pegs on the other side of the robot as well.

27

Symmetrically add the 2x5 L-beam to the other side of the robot as well.

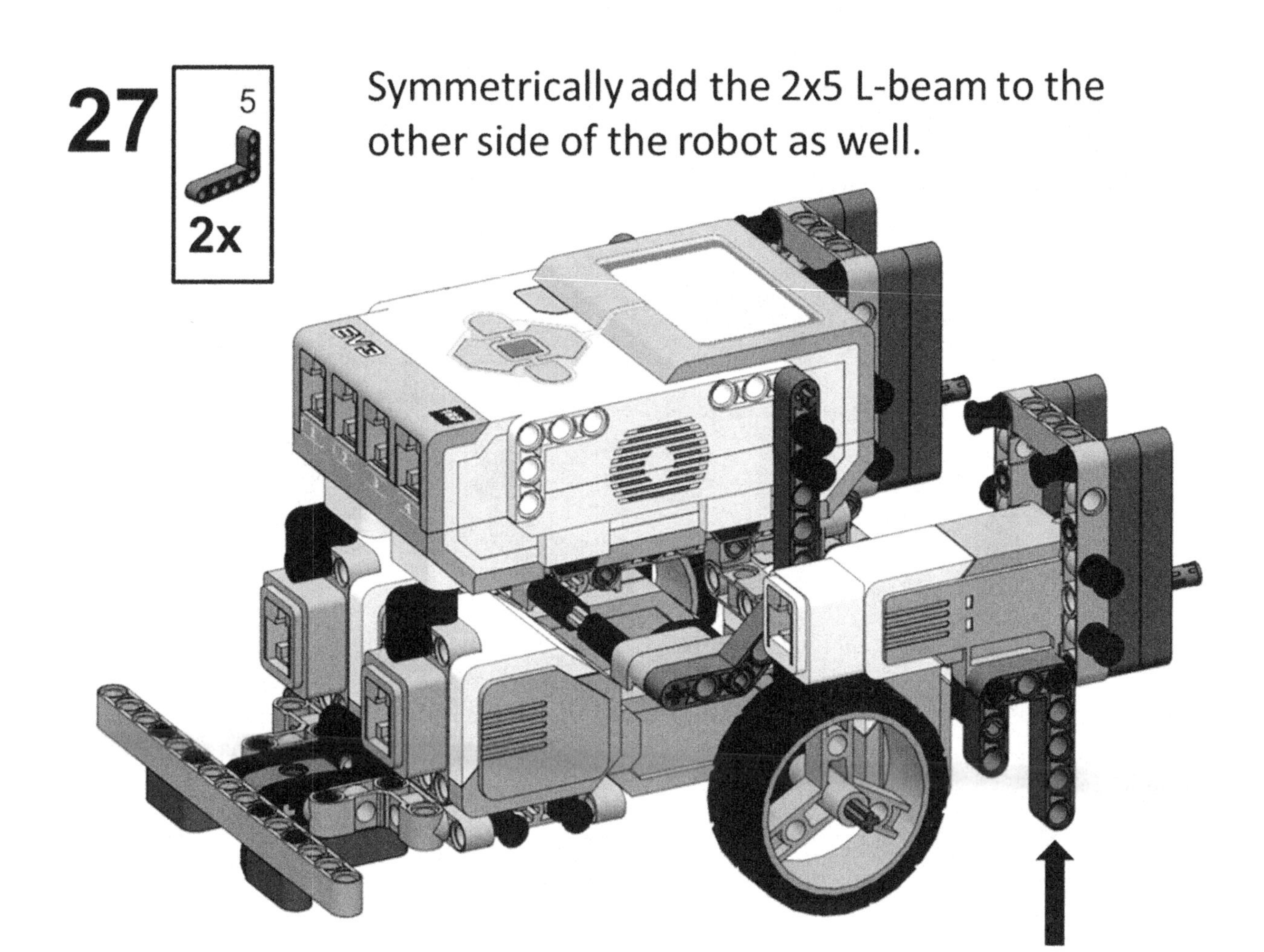

28 4x

Symmetrically attach the black pegs on the other side of the robot as well.

29 4x

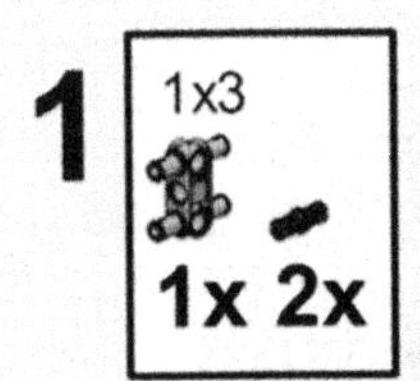
1
1x3
1x 2x

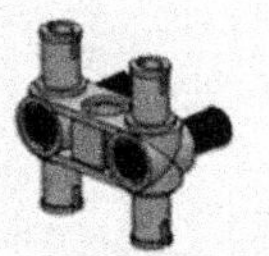

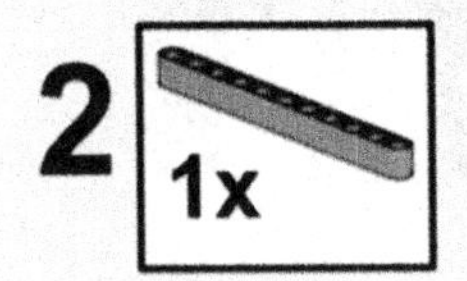
2
1x

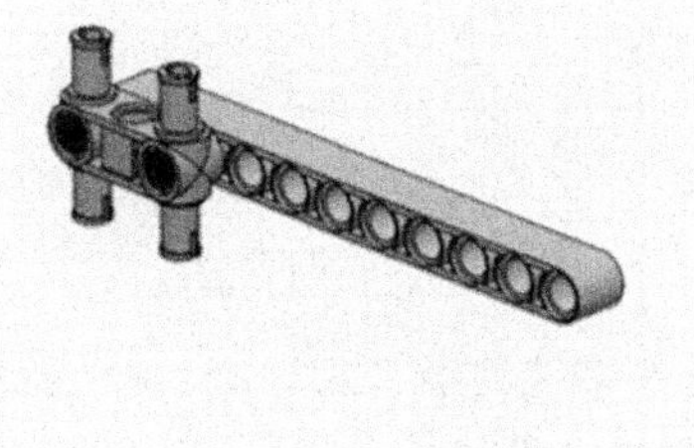

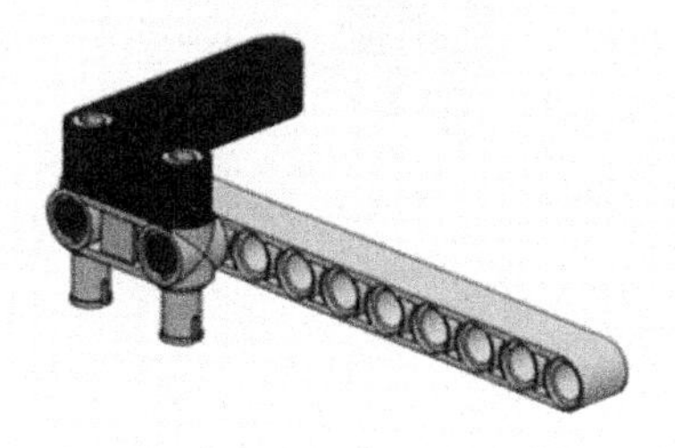

4 2x

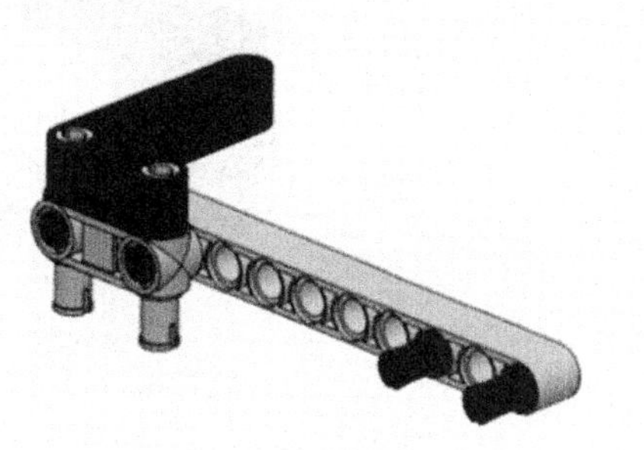

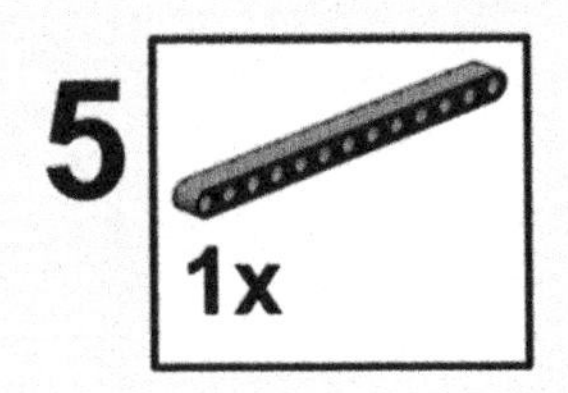

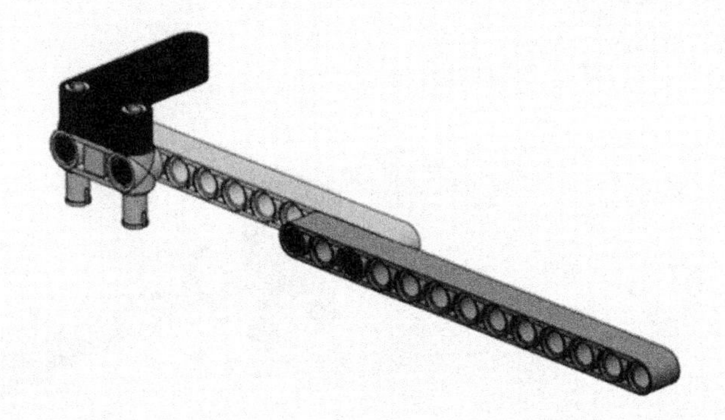

6

2x

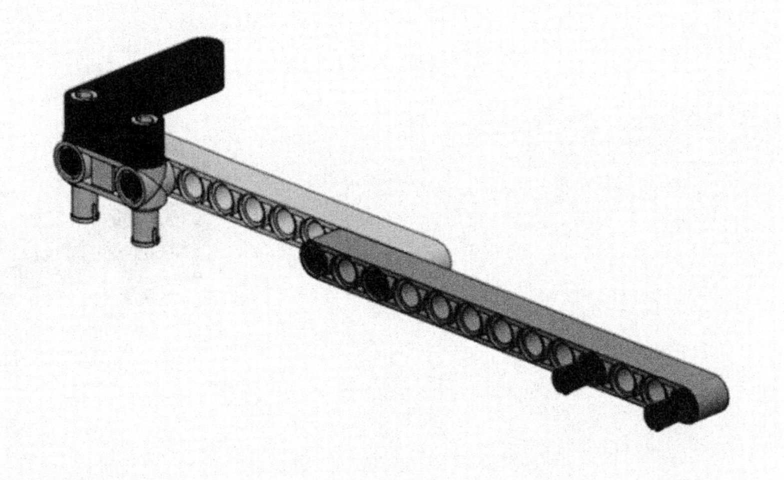

7

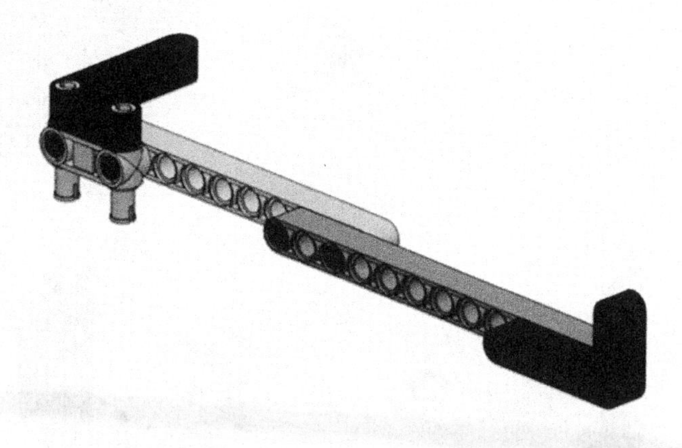

30

Connect the external frame assembly to the robot. Make sure that the axle extending out of the wheel enters the hole in the beam.

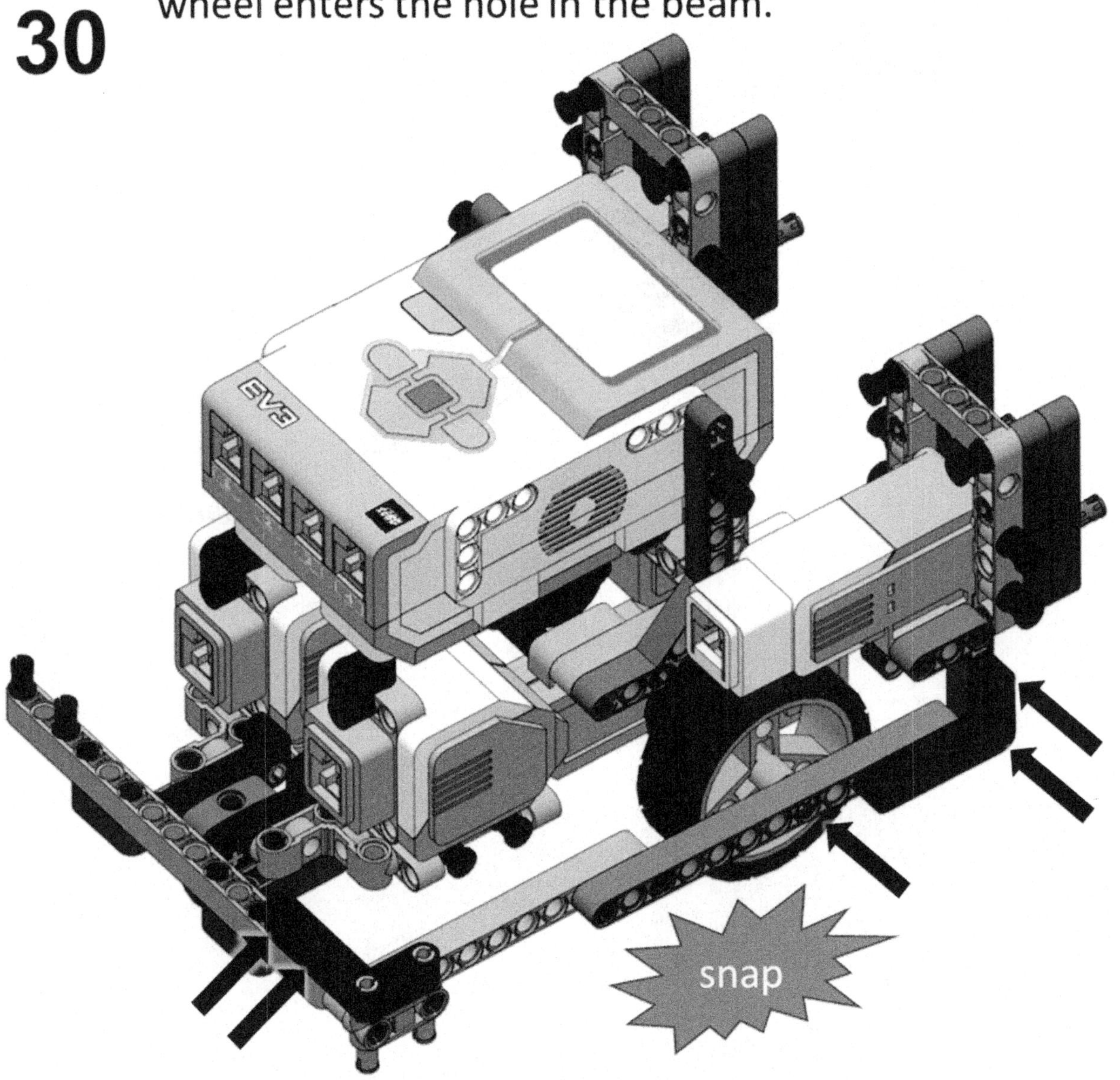

1

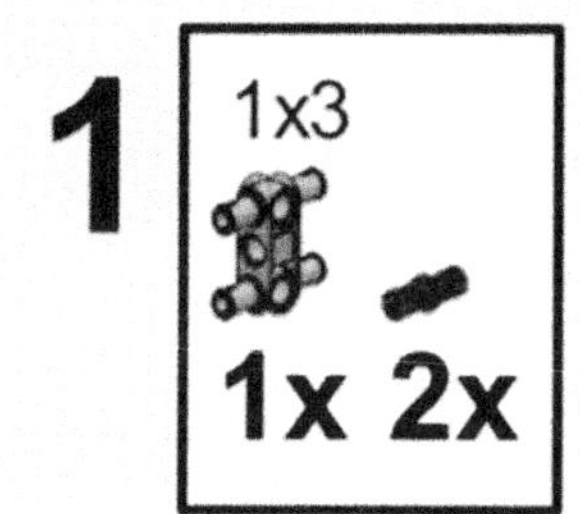

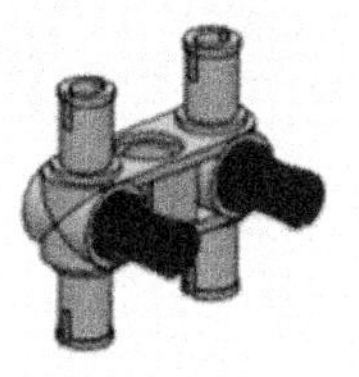

2

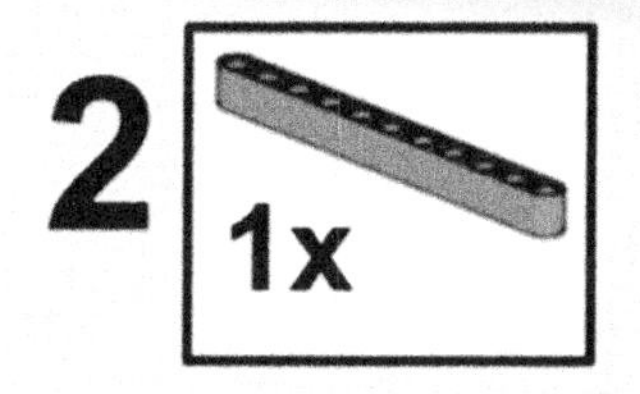

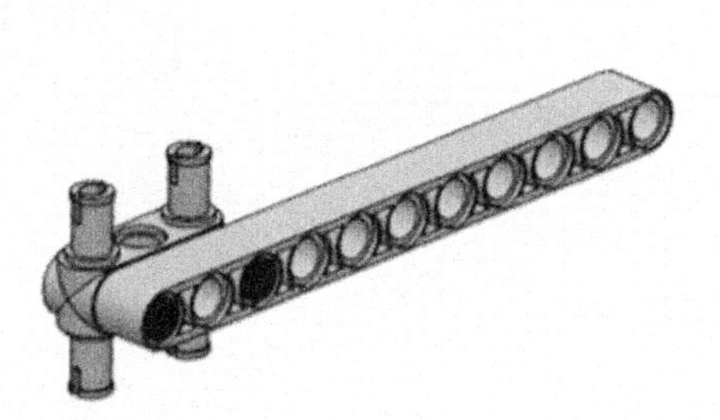

3

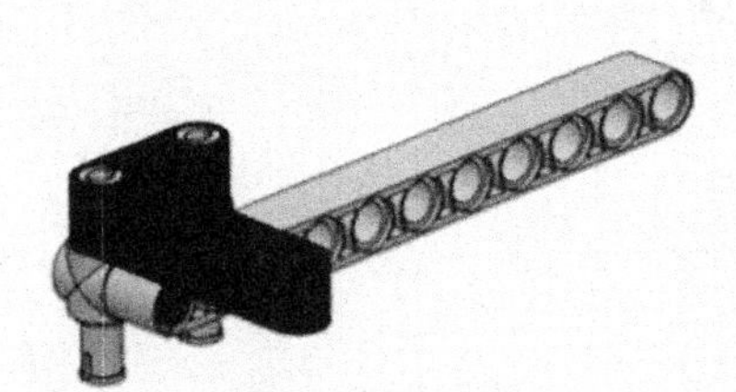

4

2x

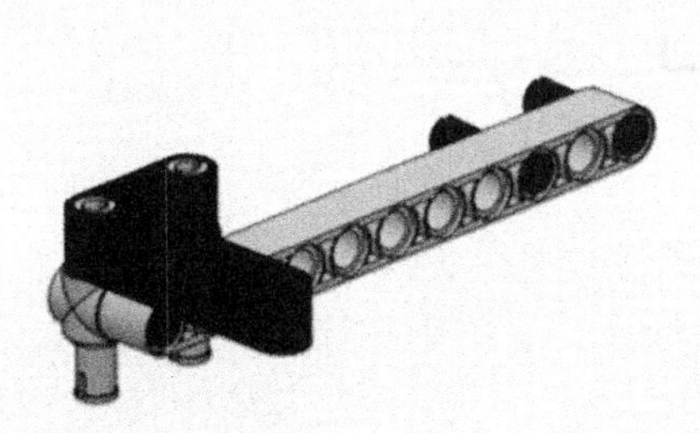

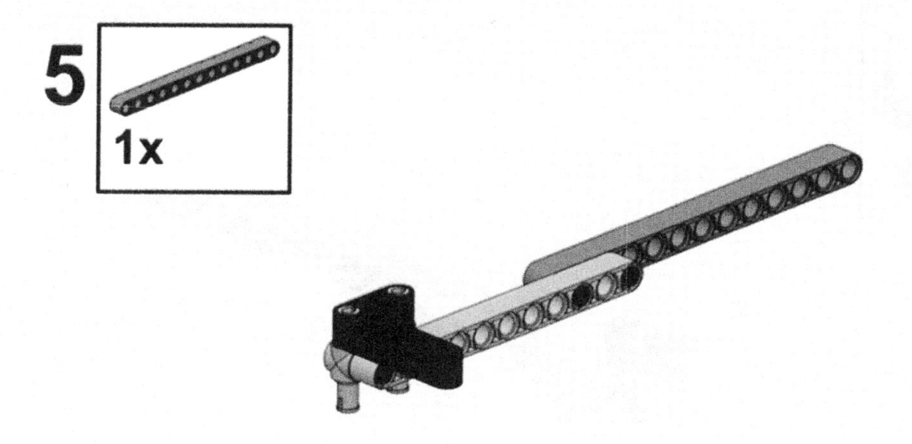
5
1x

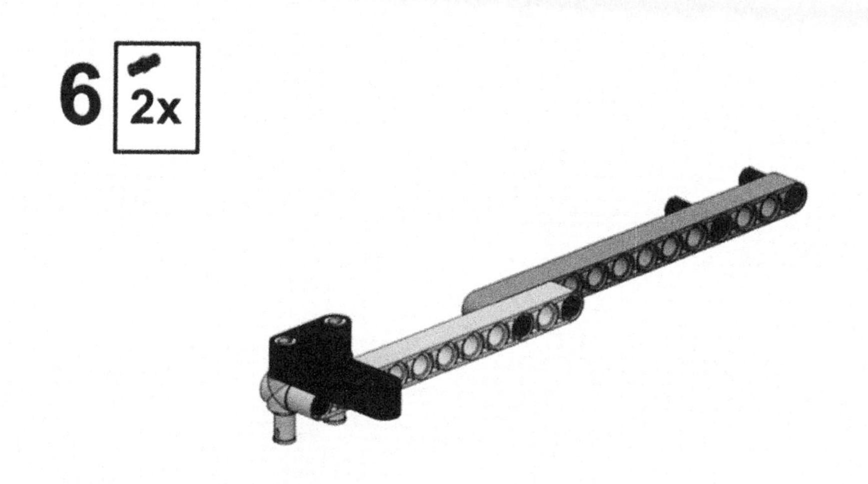
6
2x

7
5
1x

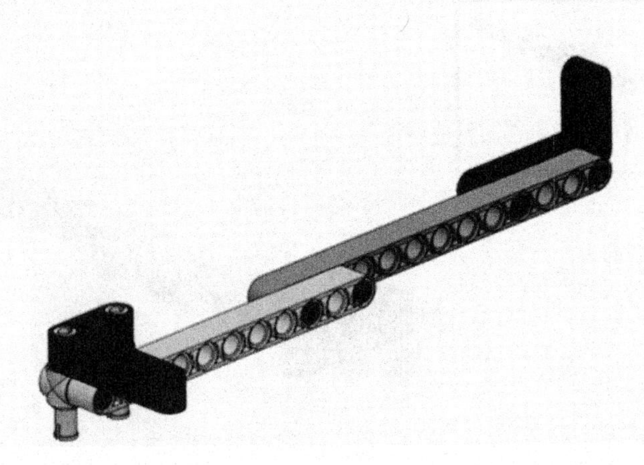

31

Connect the external frame assembly to the robot. Make sure that the axle extending out of the wheel enters the hole in the beam.

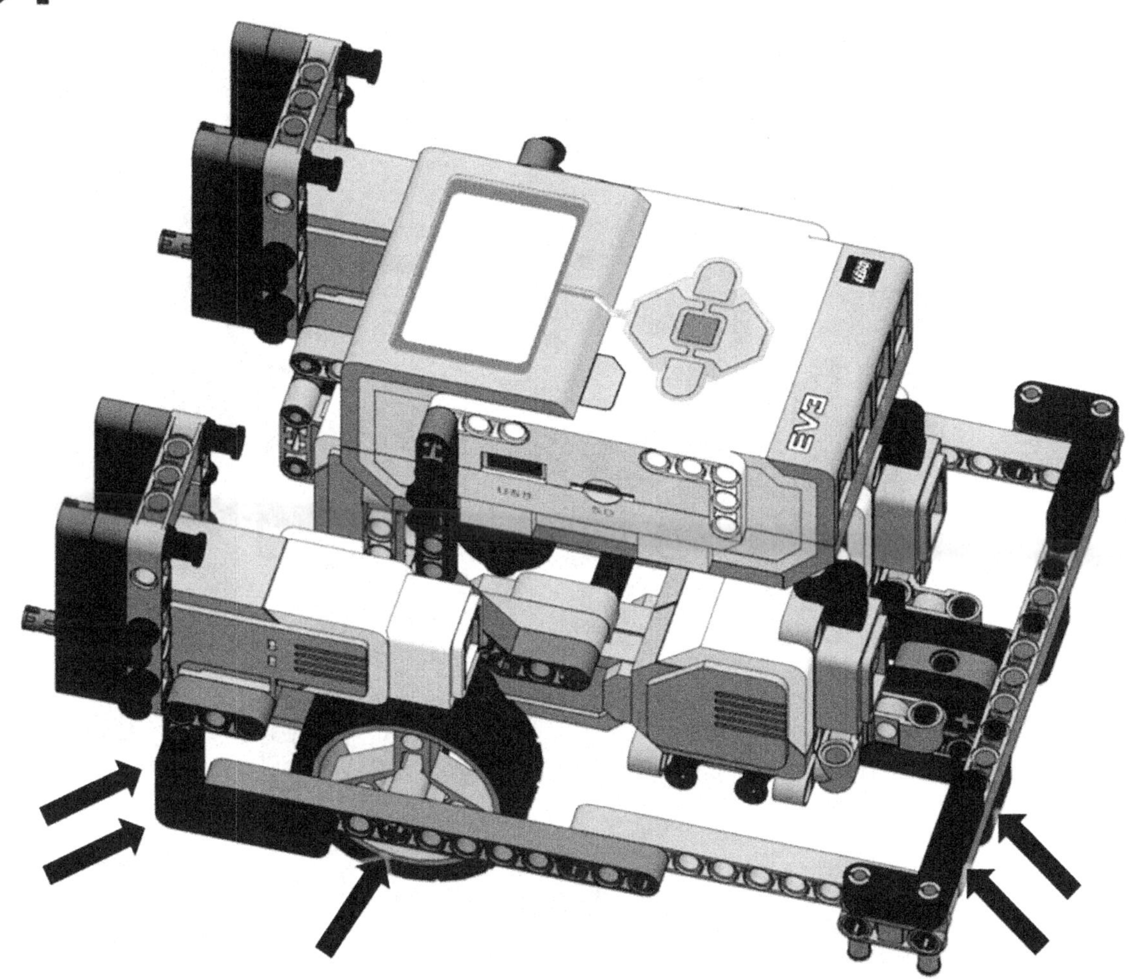

32

4x

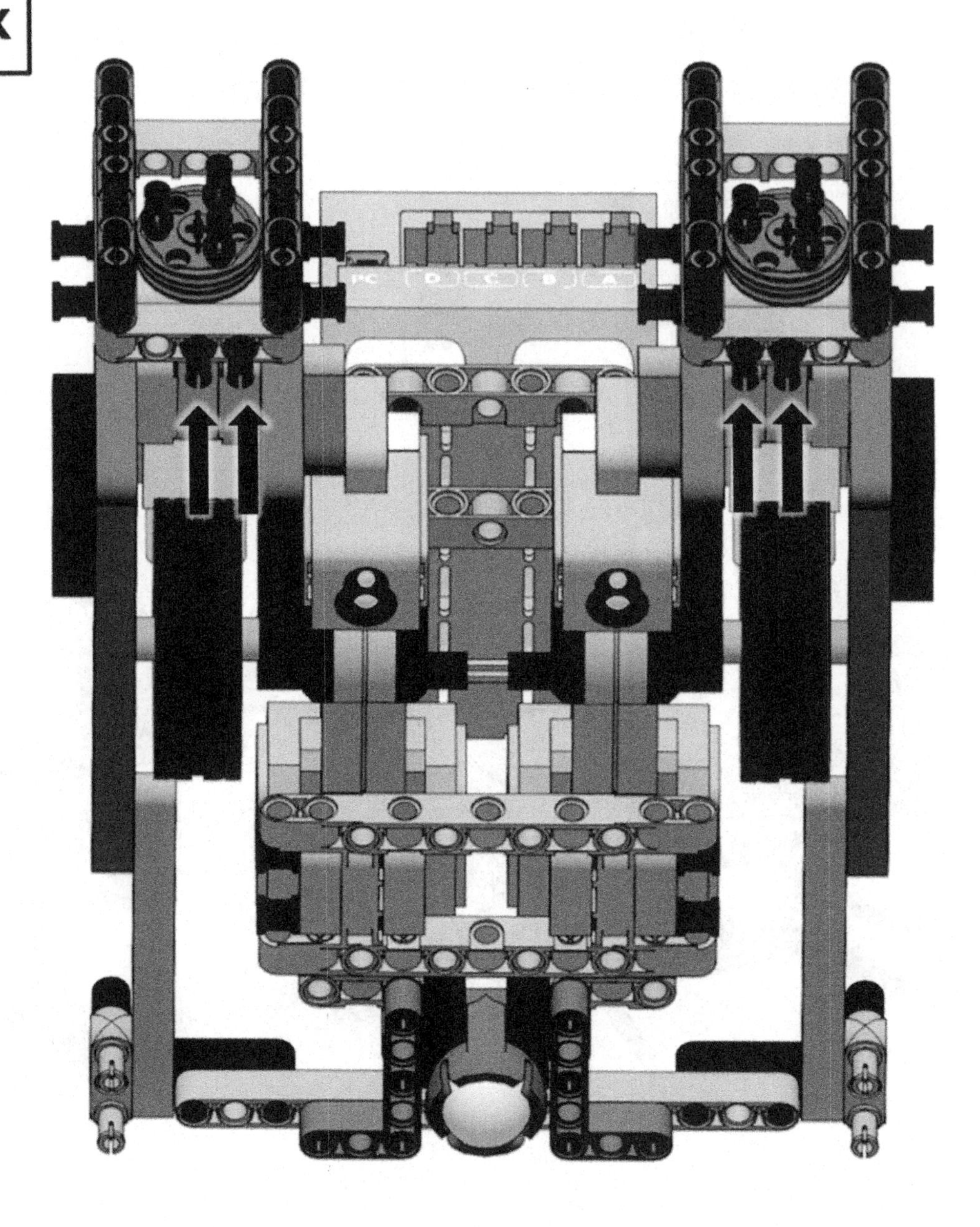

33

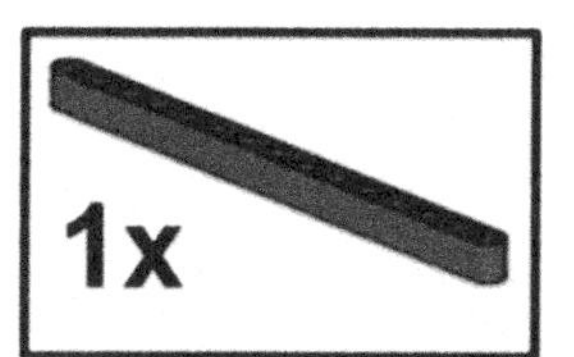

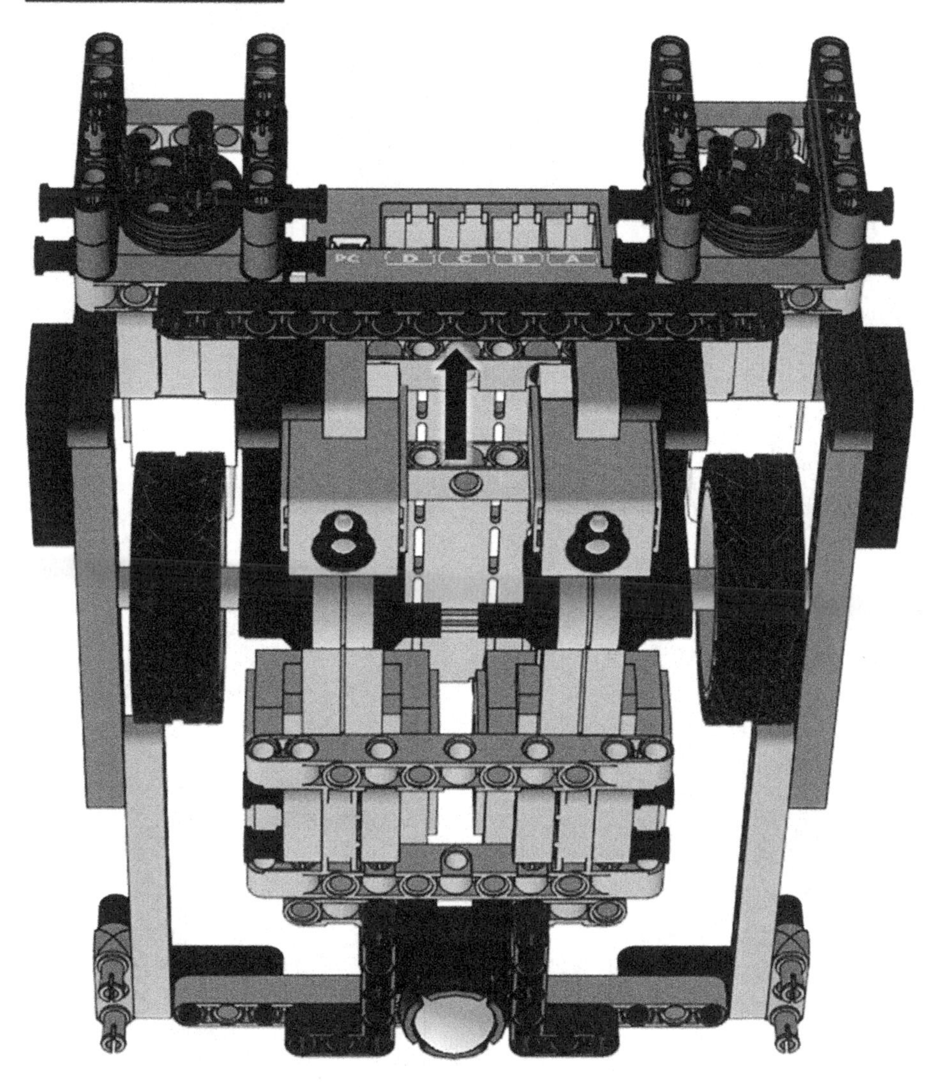

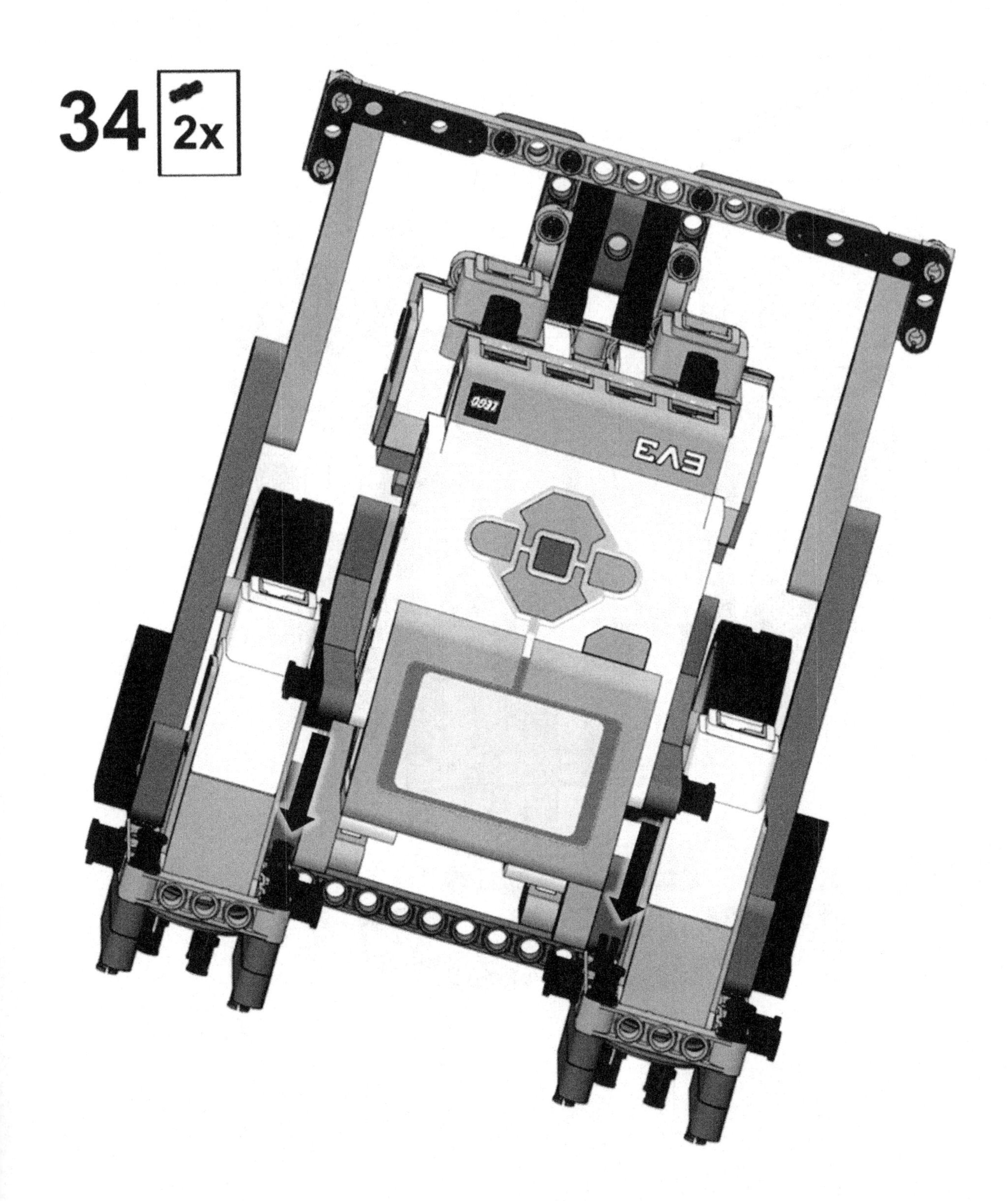
34
2x

35 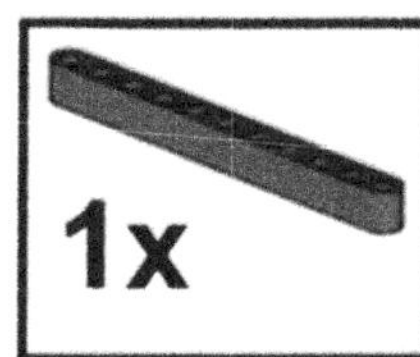

Connect the 13 beam to the pegs connected in the previous step. It provides further structural integrity to the robot body.

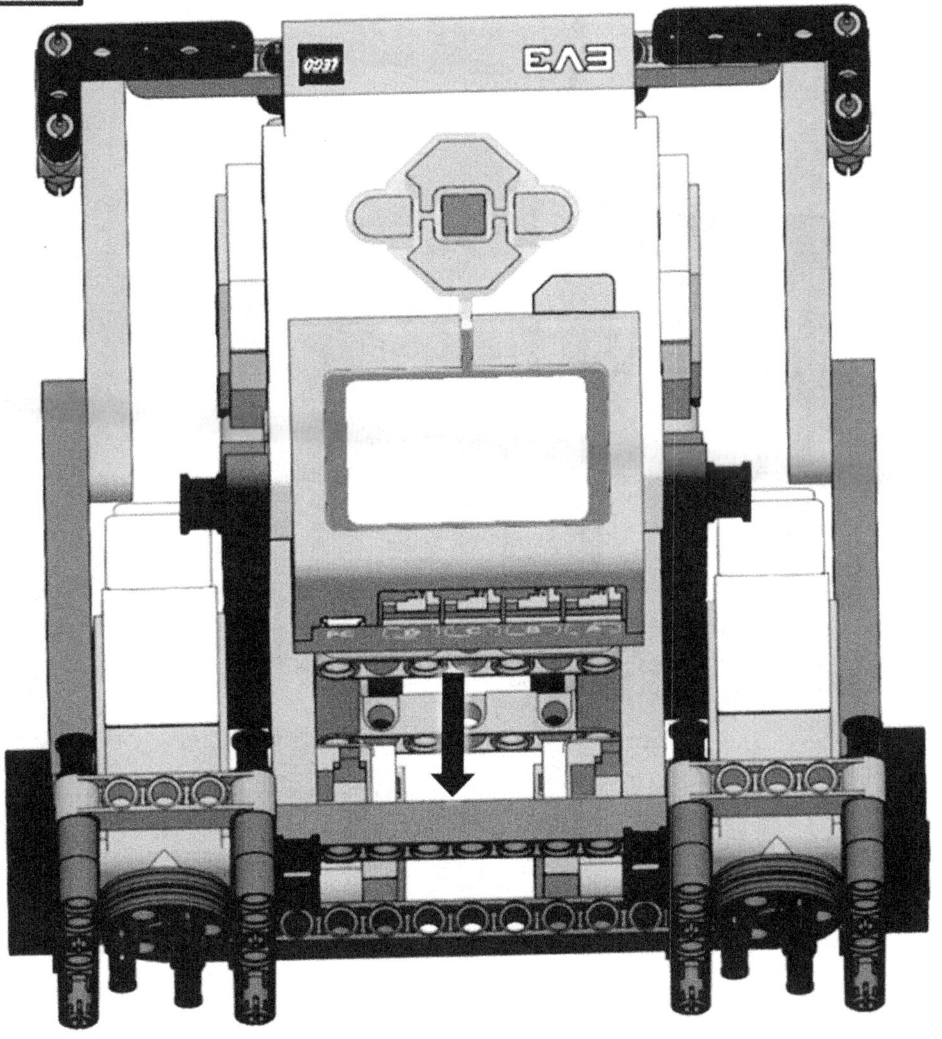

36

2x

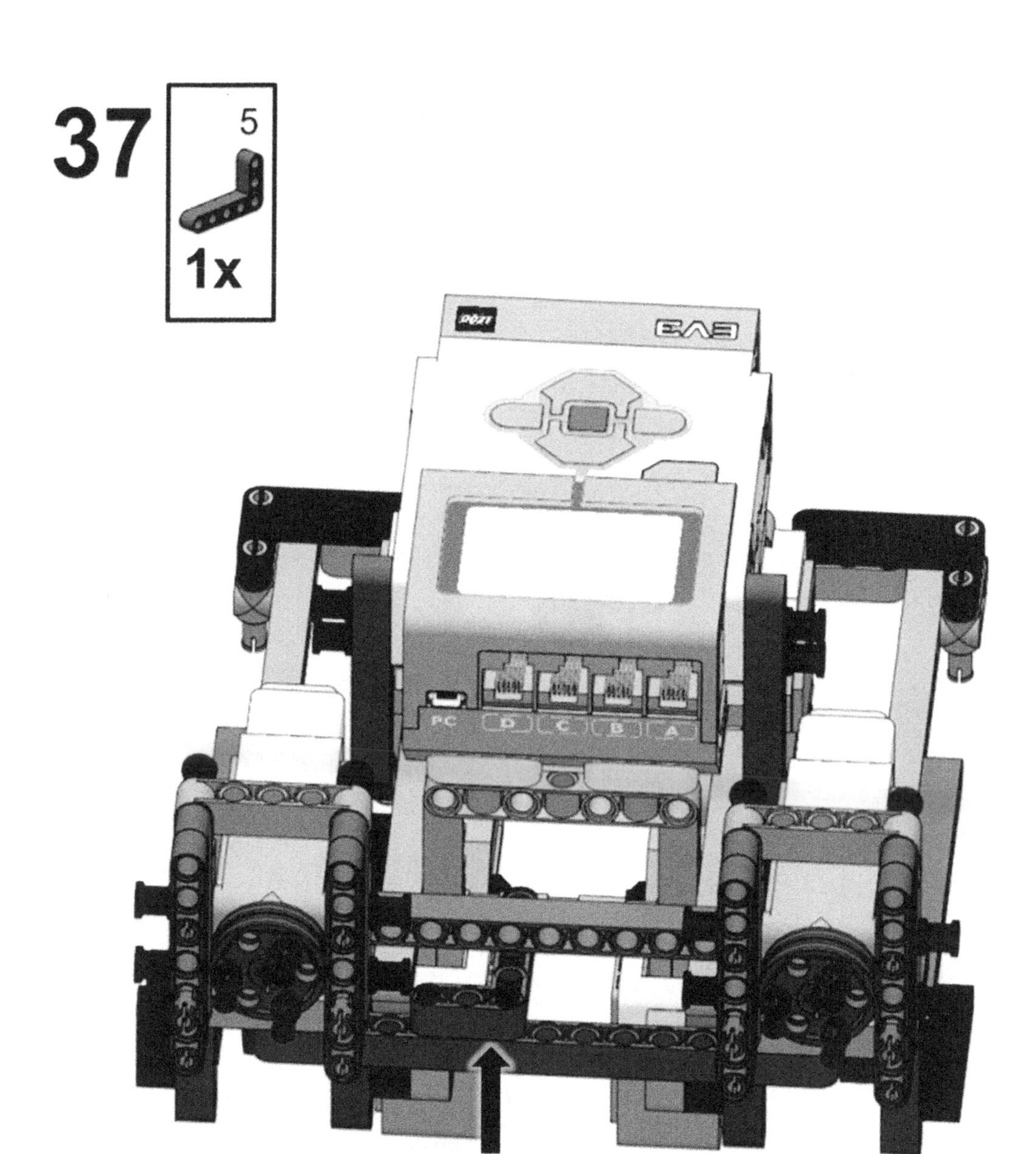
37
5
1x
EV3
PC
D
C
B
A

1

1x2

2x 2x

2

2x

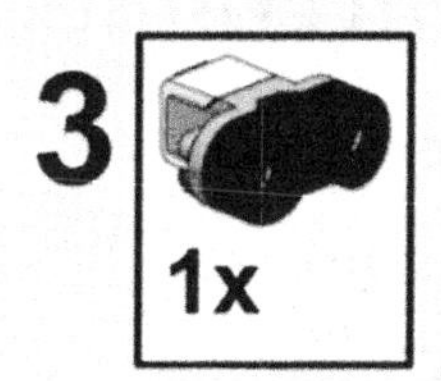
3
1x

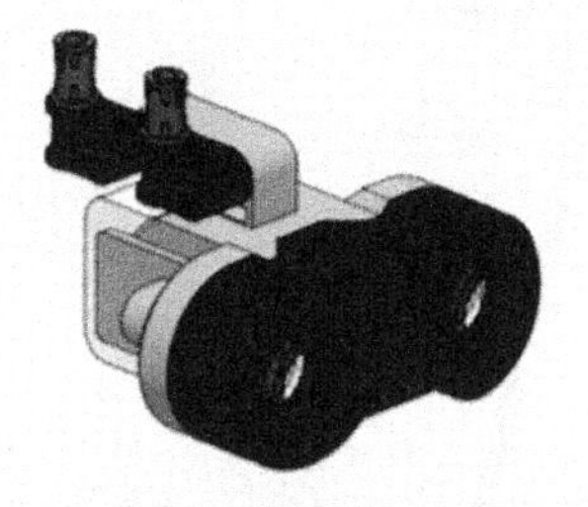

38 Although a bit hard to see, the Ultrasonic assembly attaches to the L-beam connected in Step 37.

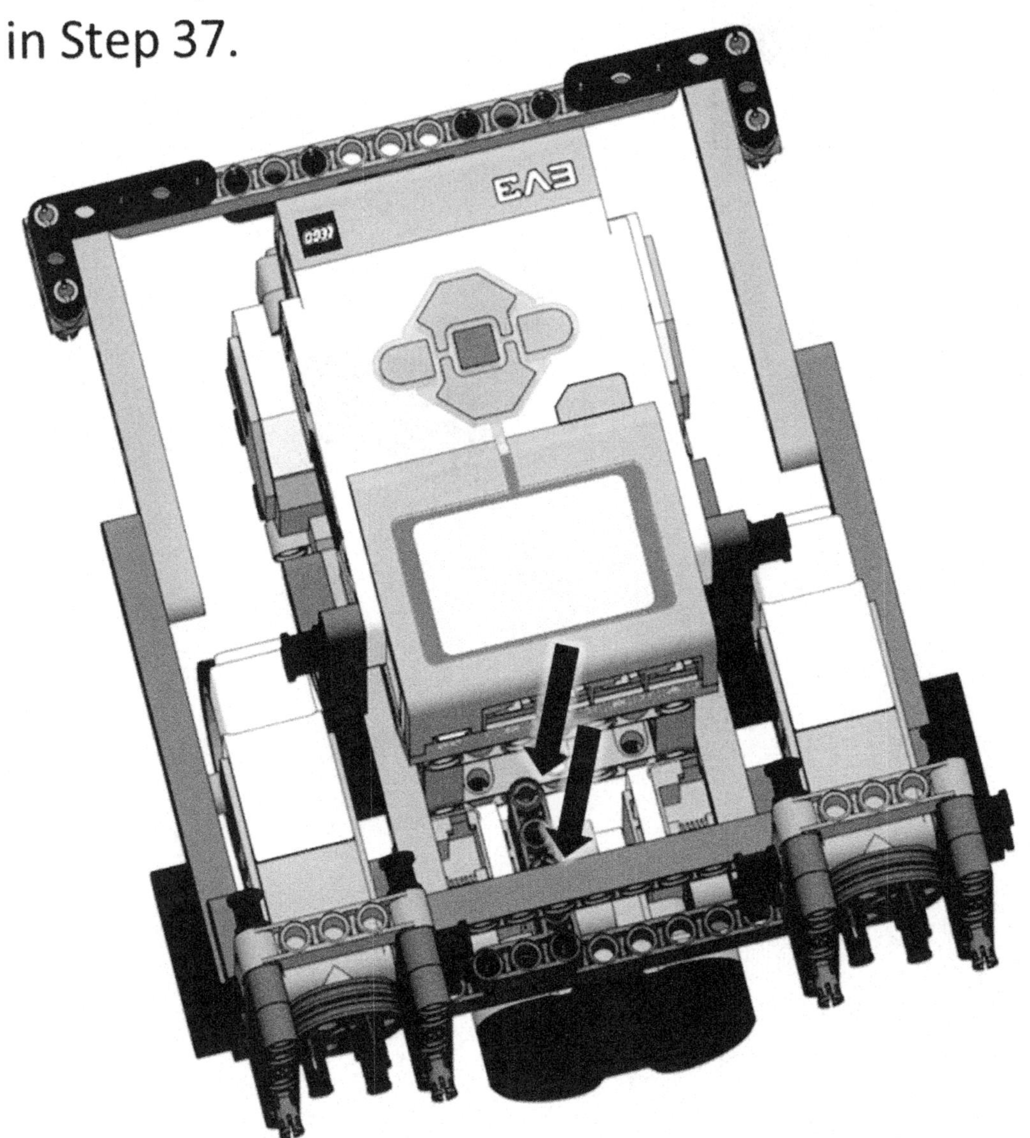

1

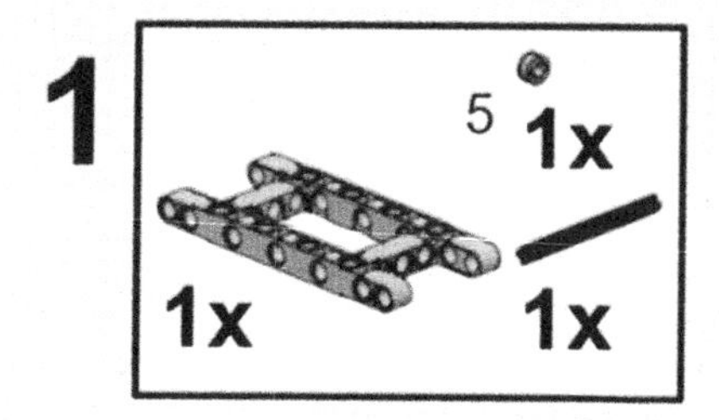

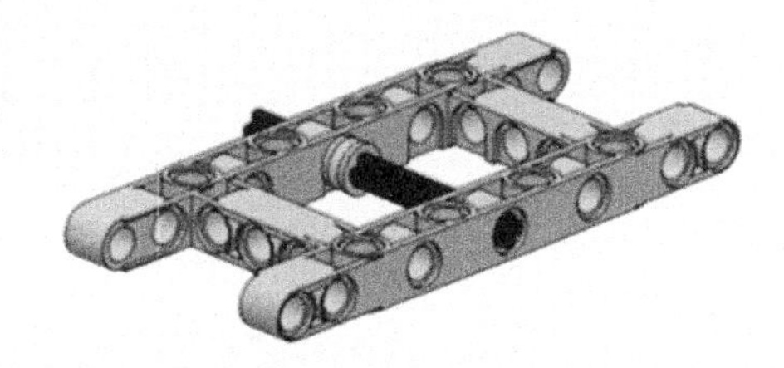

2

1x 1x

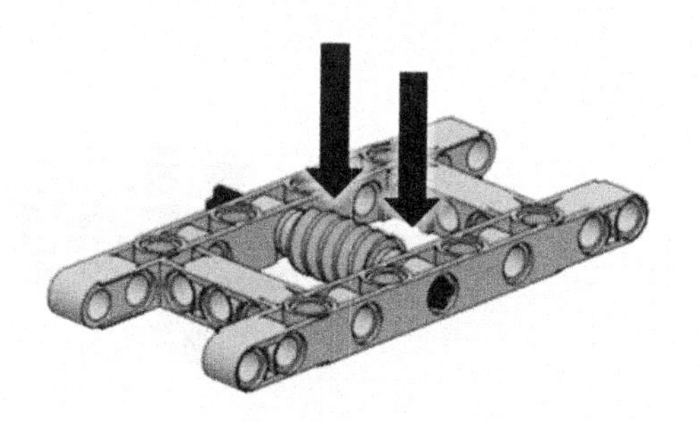

3 1x

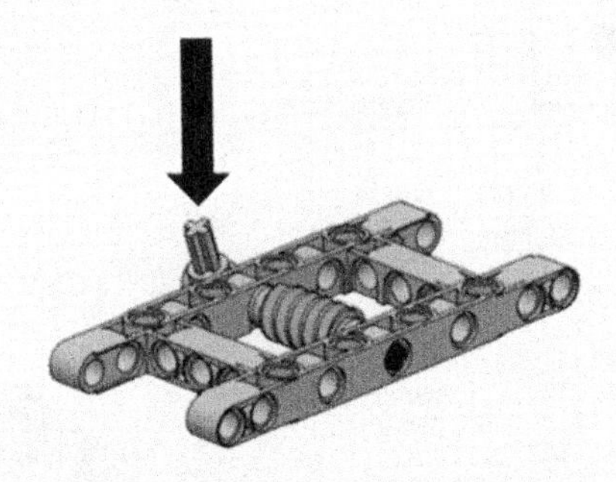

4 4x 2x

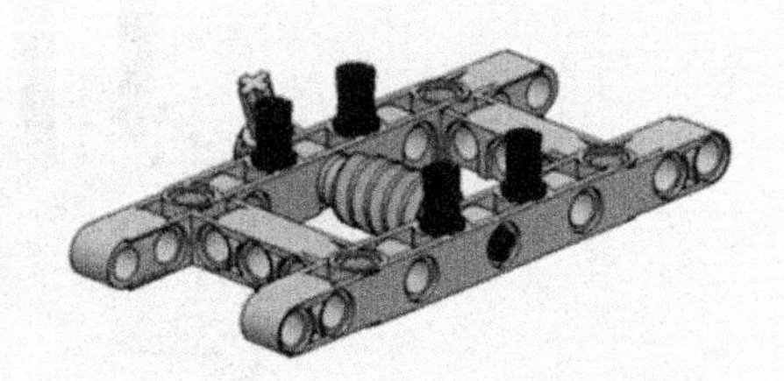

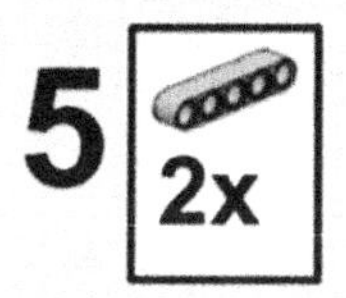

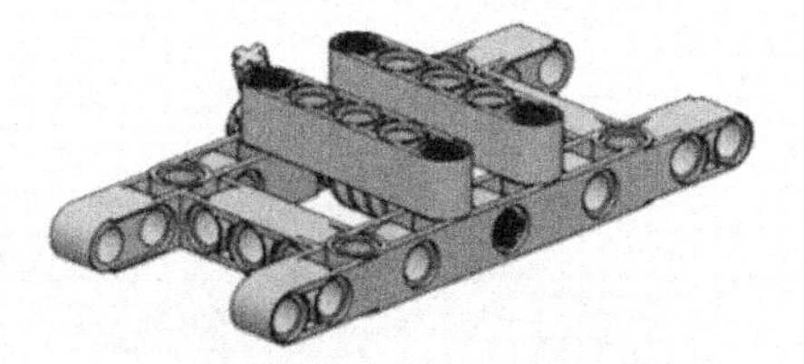

6

1x3

2x 4x

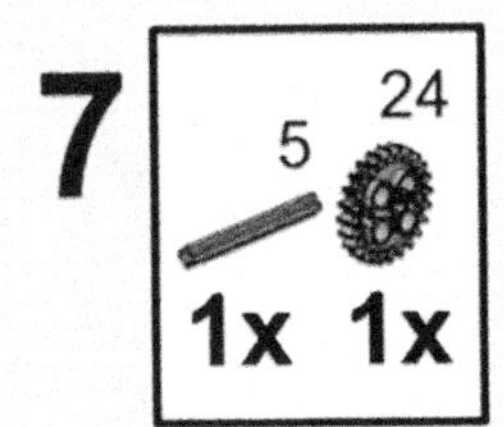
7
5
24
1x 1x

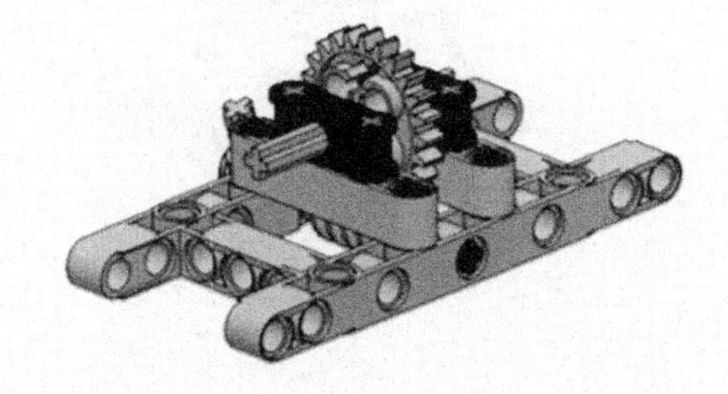

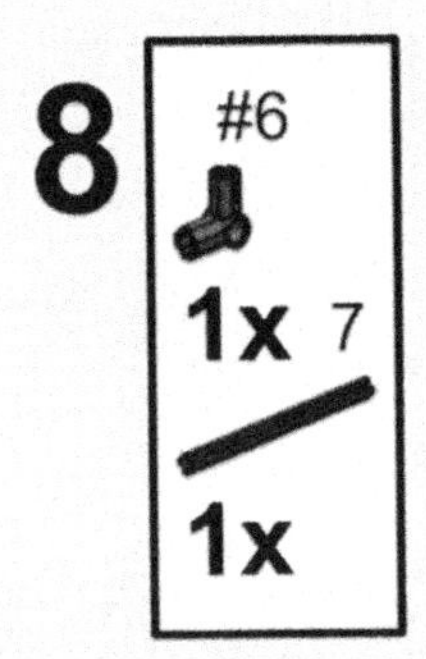
8
#6
1x
7
1x

You can create two of the dog-clutch (also called dog-gear assemblies and connect them to the pegs in front of the motors. This allows you to have a quick attach-detach uniform attachment method.

39

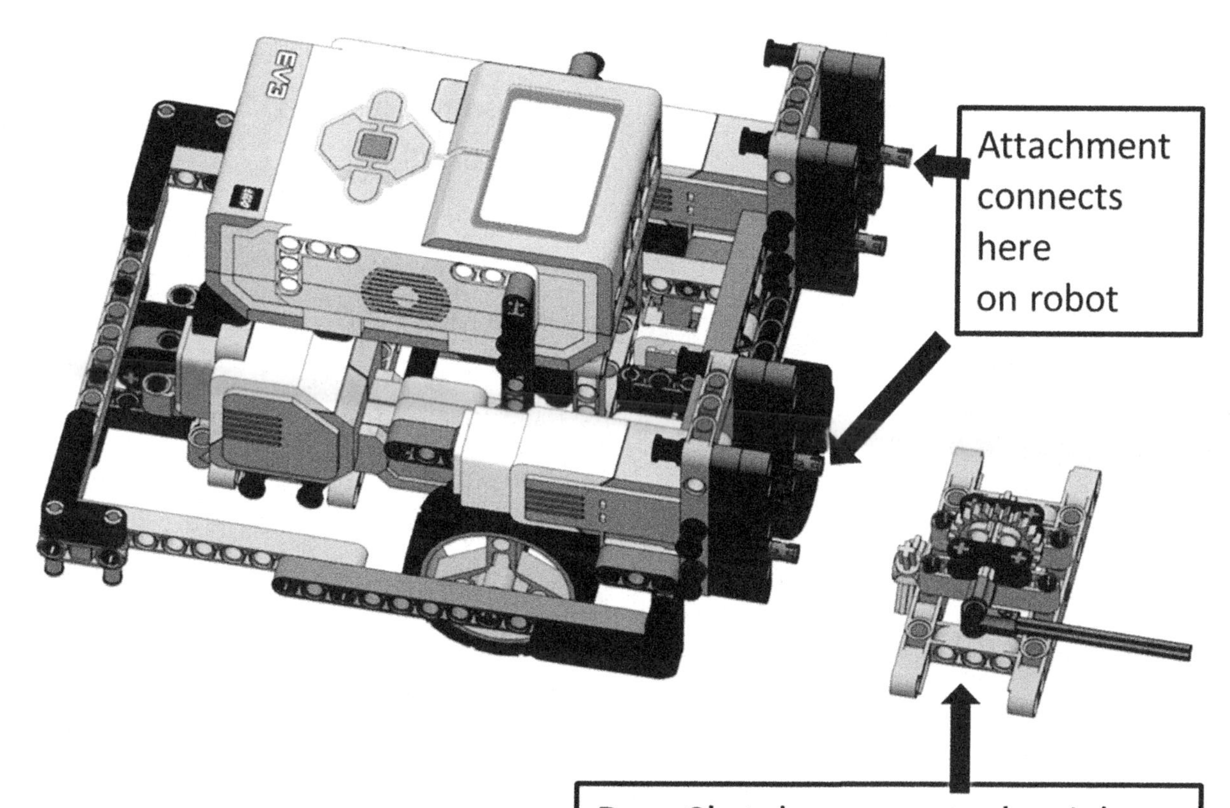

Index

Analogy, 24

Animal Allies, 11, 12, 16, 113, 188, 202, 203, 253, 254, 257, 274

autonomous, 12, 14

Autonomy, 12

Bumper, 91

calibration, 98

caster, 28, 31, 32, 89, 91

Caster, 91

center of gravity, 26, 27, 31, 92

Center of gravity, 25

Chassis, 17, 92

Chebyshev lambda, 57

Color Sensor, 83, 111, 142, 182, 183, 184, 193, 211

Composite drives, 61

configuration, 17, 25, 26, 27, 50, 51, 52, 59, 71, 72, 80, 93, 105, 172

constraints, 20

core mobile Robot, 17

Core Values, 9

dead reckoning, 108

Differential Steering, 99

Drifting, 198

effectors, 18, 19, 37, 233, 244

encoder, 82, 200

end effector, 16

End Effectors, 17

engage, 12

EV3, 16

FLU, 34, 176, 183, 184

fulcrum, 57

Fundamental Lego Units, 176, 183

Gear Slack, 114

gear train, 38, 40, 47, 49, 50

geographical proximity, 253

Going Straight Ratio, 98

Gyroscope Sensor, 84, 195

idler gear, 40, 45

Infra-Red distance measurement, 20

kinematics, 20

Large Motors, 89

launching, 12

Lego Design ID, 92, 272, 277

Line squaring, 186, 187

load distribution, 26, 27

Mindstorms, 9, 20, 75, 76, 84, 160

mission, 12

Mobile robots, 15

Motor Rotation Sensor, 83, 84, 142, 200, 204

multifunctional machines, 15

myblock, 79, 85, 132, 133, 134, 135, 136, 137, 138, 194, 200, 202, 222, 224, 225

Myblock, 135, 136, 222, 224, 225

navigational accuracy., 25

Non-mobile robots, 15

non-motorized wheel, 26

Odometery, 108

Optimal configuration, 25

parallel bar mechanism, 60, 73
PID control, 87, 153, 208
Pneumatic systems, 63
predictability, 24
proportional-integral-derivative, 208
Qualifiers, 10
Reconfigurable machines, 15
reconfigured, 12
redistribute, 12
redundancy, 20, 22
reliability, 20
repeatable performance, 3, 24
Robot Base, 10, 12
robot drift, 109
robot drift., 109
Robot Game, 14, 256, 266, 267
Robot Games, 9
Robot navigation, 20
rule of 8 out of 10, 108
self-locking, 47
Sensing unit, 20
Starting Jig, 115
synchronized, 20
tachometer, 82, 94, 200
three-point support, 25
torque, 34, 36, 37, 38, 39, 40, 45, 47, 51, 63
Turning Ratio., 101
Ultrasonic sensor, 84, 173, 174, 175, 176, 177, 178, 179, 180, 181, 208, 277
unturnable, 197
Variability, 24
wait block, 141, 142, 143, 150, 162, 177
Wall following, 120, 121, 122, 168, 177
zig zag, 198